ADVANCED NANOMATERIALS FOR BIOLOGICAL, NUTRACEUTICAL, AND MEDICINAL APPLICATIONS

ADVANCED NANOMATERIALS FOR BIOLOGICAL, NUTRACEUTICAL, AND MEDICINAL APPLICATIONS

Edited by

Sreeraj Gopi, PhD

Augustine Amalraj, PhD

T. R. Sreeraj, MSc

APPLE ACADEMIC PRESS

First edition published 2025

Apple Academic Press Inc.
1265 Goldenrod Circle, NE,
Palm Bay, FL 32905 USA

760 Laurentian Drive, Unit 19,
Burlington, ON L7N 0A4, CANADA

CRC Press
2385 NW Executive Center Drive,
Suite 320, Boca Raton FL 33431

4 Park Square, Milton Park,
Abingdon, Oxon, OX14 4RN UK

Library and Archives Canada Cataloguing in Publication

Title: Advanced nanomaterials for biological, nutraceutical, and medicinal applications / edited by Sreeraj Gopi, PhD, Augustine Amalraj, PhD, T.R. Sreeraj, MSc.
Names: Gopi, Sreeraj, editor. | Amalraj, Augustine, editor. | Sreeraj, T. R., 1992- editor
Description: First edition. | Includes bibliographical references and index.
Identifiers: Canadiana (print) 20240306996 | Canadiana (ebook) 20240307011 | ISBN 9781774913468 (hardcover) | ISBN 9781774913475 (softcover) | ISBN 9781003389163 (ebook)
Subjects: LCSH: Nanostructured materials—Therapeutic use. | LCSH: Nanobiotechnology—Therapeutic use. | LCSH: Nanomedicine.
Classification: LCC R857.N34 A38 2024 | DDC 610.28—dc23

ISBN: 978-1-77491-346-8 (hbk)
ISBN: 978-1-77491-347-5 (pbk)
ISBN: 978-1-00338-916-3 (ebk)

About the Editors

Sreeraj Gopi, PhD
Overseas Expert Professor, Shanghai University, Shanghai, China

Sreeraj Gopi, PhD, is an Industrial Scientist with a doctorate in Organic Chemistry, Nanotechnology, and Nanodrug delivery. He has been working in the area of natural products, isolation, and establishing biological activities. He has published more than 85 international articles and filed more than 75 international patents. Dr. Gopi is a fellow of the Royal Society of Chemistry.

Augustine Amalraj, PhD
Deputy Manager, Department of Research and Development, Plant Lipids Private Limited, Cochin, India

Augustine Amalraj, PhD, is currently working as a Deputy Manager in the Department of Research and Development at Plant Lipids Private Limited, Cochin, India. He obtained his doctoral degree in Chemistry from Gandhigram Rural Institute – Deemed University, Gandhigram, Tamil Nadu, India. His research interests are in applied chemistry, food chemistry, natural product chemistry, environmental chemistry, chemosensors, and polymer and nanocomposite materials. He has published more than 60 research articles in international journals as well as 15 book chapters and co-edited more than seven books.

T. R. Sreeraj, MSc
Research Associate, Research and Development Department, Plant Lipids Private Limited, Cochin, India

T. R. Sreeraj, MSc, is working as a Research Associate in the Research and Development Department at Plant Lipids Private Limited, Cochin, India. He has completed her postgraduation at Mahatma Gandhi University, India. He is working on new product development and functional foods based on natural products. He has published four research articles and two book chapters.

Contents

Contributors

Hilal Acidereli
Sen Research Group, Department of Biochemistry, Dumlupınar University, Kütahya, Turkey

Augustine Amalraj
R&D Center, Aurea Biolabs (P) Ltd., Kolenchery, Cochin – 682311, Ernakulam, Kerala, India

Hemant Borase
C. G. Bhakta Institute of Biotechnology, Uka Tarsadia University, Bardoli, Gujarat, India

Jenny Hui Foong Chau
Nanotechnology and Catalysis Research Center, University of Malaya, Kuala Lumpur, Malaysia

E. Jackcina Stobel Christy
Department of Chemistry, The Gandhigram Rural Institute – DU, Gandhigram, Dindigul, Tamil Nadu, India

N. Renuga Devi
G.T.N. Arts College (Autonomous), Dindigul – 624005, Tamil Nadu, India

S. Dharaneedharan
G.T.N. Arts College (Autonomous), Dindigul – 624005, Tamil Nadu, India

Tuerxun Duolikun
Department of Mechanical Engineering, Faculty of Engineering, University of Malaya, Kuala Lumpur, Malaysia; Nanotechnology and Catalysis Research Center (NANOCAT), Institute for Advanced Studies, University of Malaya, Kuala Lumpur, Malaysia

Asmita Duragkar
Nikalas Mahila Mahavidyalaya, Panapratap Nagar, Nagpur, Maharashtra, India

Haydar Göksu
Kaynasli Vocational College, Duzce University, Duzce – 81900, Turkey

Sreeraj Gopi
R&D Center, Aurea Biolabs (P) Ltd., Kolenchery, Cochin, Ernakulam, Kerala, India

Józef T. Haponiuk
Chemical Faculty, Gdańsk University of Technology, Gdańsk, Poland

Amal I. Hassan
Radioisotope Department, Nuclear Research Center, Egyptian Atomic Energy Authority, Cairo, Egypt

Ilyas Ilker Işler
Kaynasli Vocational College, Duzce University, Duzce, Turkey

Mohd. Rafie Bin Johan
Nanotechnology and Catalysis Research Center (NANOCAT), Institute for Advanced Studies, University of Malaya, Kuala Lumpur, Malaysia

Shintu Jude
R&D Center, Aurea Biolabs (P) Ltd., Kolenchery, Cochin, Ernakulam, Kerala, India

Chin Wei Lai
Nanotechnology and Catalysis Research Center (NANOCAT), Institute for Advanced Studies, University of Malaya, Kuala Lumpur, Malaysia

Bey Fen Leo
Nanotechnology and Catalysis Research Center (NANOCAT), Institute for Advanced Studies, University of Malaya, Kuala Lumpur, Malaysia; Faculty of Medicine, University of Malaya, Kuala Lumpur, Malaysia

Shweta Mishra
Shobhaben Pratapbhai Patel School of Pharmacy and Technology Management, SVKM's NMIMS, V.L. Mehta Road, Vile Parle (W), Mumbai, Maharashtra, India

M. Pandeeswaran
G.T.N. Arts College (Autonomous), Dindigul – 624005, Tamil Nadu, India

Prittesh Patel
C. G. Bhakta Institute of Biotechnology, Uka Tarsadia University, Bardoli, Gujarat, India

Satish Patil
School of Life Sciences, Kavayitri Bahinabai Chaudhari North Maharashtra University, Jalgaon, Maharashtra, India

Anil M. Pethe
Associate Dean, School of Pharmacy and Technology Management, SVKM's NMIMS, B-4, TSIIC, Polepally SEZ, Jadcherla, Hyderabad Campus, Telangana, India

Vijita Pillai
School of Biotechnology and Bioinformatics, D.Y. Patil Deemed-to-be University, Sector 15, CBD Belapur, Navi Mumbai, Maharashtra, India; Anant Pharmaceuticals Pvt. Ltd., W57/A, Near Police Chowki, Anand Nagar, Ambernath MIDC, Mumbai, Maharashtra, India

Anitha Pius
Department of Chemistry, The Gandhigram Rural Institute – DU, Gandhigram, Dindigul, Tamil Nadu, India

A. Rajeswari
Department of Chemistry, The Gandhigram Rural Institute – DU, Gandhigram, Dindigul, Tamil Nadu, India

Hosam M. Saleh
Radioisotope Department, Nuclear Research Center, Egyptian Atomic Energy Authority, Cairo, Egypt

Fatih Şen
Sen Research Group, Department of Biochemistry, Dumlupınar University, Kütahya, Turkey

Abhishek Sharma
C. G. Bhakta Institute of Biotechnology, Uka Tarsadia University, Bardoli, Gujarat, India

Karthik Varma
R&D Center, Aurea Biolabs (P) Ltd., Kolenchery, Cochin, Ernakulam, Kerala, India

Nagamalai Vasimalai
Department of Chemistry, B.S. Abdur Rahman Crescent Institute of Science and Technology, Vandalur, Chennai, Tamil Nadu, India

S. Vignesh
G.T.N. Arts College (Autonomous), Dindigul, Tamil Nadu, India

Abbreviations

11-MUA	11-mercaptoundecanoic corrosive
3D	three-dimensional
AFP	alpha-fetoprotein
AgNPs	silver nanoparticles
AMI	acute myocardial infarction
AuNPs	gold nanoparticles
BBB	blood-brain barrier
BMs	biomedical metals
BNC	bacterial nanocellulose
BNP	brain natriuretic peptide
CAGR	compound annual growth rate
CBER	Center for Biological Evaluation and Research
CDER	Center for Drug Evaluation and Research
CDRH	Center for Devices and Radiological Health
CDSCO	Central Drugs Standard Control Organization
CdSe	cadmium selenide
CEA	carcinoembryonic antigen
CK	creatine kinase
CK-MB	creatine kinase myocardial band
CMV	cytomegalovirus
CNMs	carbon nanomaterials
CNTs	carbon nanotubes
CPCB	central pollution control board
CPP	casein phosphopeptides
CS	chitosan
CS	chondroitin sulfate
CS-TPP	chitosan-tripolyphosphate pentasodium
CT	computed tomography
CTAB	cetyltrimethylammonium bromide
CVD	cardiovascular diseases
CVD	chemical vapor deposition
DAC	dactinomycin chemotherapy
DFDL	dual-functional drug liposomes
DMAB	didodecyldimethylammomium bromide
DMM	dimethomorph nanomaterial joint
DNA	deoxyribonucleic acid
DPPH	2,2-diphenyl-1-picrylhydrazyl

EA	ellagic acid
EC	European Commission
ECG	electrocardiograms
ECM	extracellular matrix
ECs	electrochemical capacitors
EDLCs	electric double layer capacitor
EDX	energy-dispersive X-ray spectroscopy
EGCG	epigallocatechin gallate
EGFR	epidermal growth factor receptor
ELISA	enzyme-linked immunosorbent assay
EPA	Environmental Protection Agency
EPR	enhanced permeability and retention
ESPs	electroactive smart polymers
ESR	electron spin resonance spectroscopy
EUON	EU observatory for nanomaterials
FA	ferulic acid
FAK	focal adhesion kinase
FDA	Food and Drug Administration
FIM	Foundation for Innovation in Medicine
FISH	fluorescence *in situ* hybridization
FITC	fluorescein isocyanate
FRAP	ferric reducing antioxidant power
FRET	fluorescence resonance energy transfer
FSSA	Food Safety and Standards Authority
FTIC	fluorescein isothiocyanate
FTIR	Fourier-transform infrared spectroscopy
GA	gallic acid
GI	gastrointestinal
GLN	glucosamine
GMS	glyceryl monostearate
Gr	graphene
GRAS	generally recognized as safe
HA	hyaluronic acid
HAp	hydroxyapatite
HBsAg	hepatitis B surface antigen
HIF-1α	hypoxia-inducible factor-1α
HPLC	high-performance liquid chromatography
HRP-anti-CEA	horseradish peroxidase-labeled secondary anti-CEA antibody
IDA	iminodiacetic acid
IFIC	International Food Information Council
IFN	interferon
IgG	immunoglobulin G
IND	investigational new drug

ITA	itaconic acid
ITO	indium tin oxide
JCPDS	Joint Committee on Powder Diffraction Standards
LAMEA	Latin America, the Middle East, and Africa
LC1	*Lactobacillus acidophilus*
LDH	lactate dehydrogenase
LOD	limit of detection
LSEPD	liposome-supported enzymatic peritoneal dialysis
LSPR	localized surface plasmon resonance
LZM	lysozyme
MAA	mercaptoacetic acid
MAPA	2-mthacryloyl-(L)-phenylalanine
MAPK	mitogen-activated protein kinase
MDR	multidrug-resistant
MDs	medical devices
MGCE	magnetic glassy carbon electrode
MIPs	molecularly imprinted polymers
MMP-9	matrix metalloproteinase-9
MNPs	magnetic nanoparticles
MNPs	metallic nanoparticles
MOs	metal oxides
MRI	magnetic resonance imaging
MRSA	methicillin-resistant *Staphylococcus aureus*
MSNs	mesoporous silica nanoparticles
MWCNTs	multiwalled carbon nanotubes
NAMs	nano-antimicrobials
Nano-SiO_2	silicon dioxide nanoparticles
NC	nanocellulose
NCC	nanocrystalline cellulose
NCHs	nanocomposite hydrogels
NEPA	National Environmental Protection Agency
NFC	nanofibrillated cellulose
NLC	nanostructured lipid carriers
NMR	nuclear magnetic resonance
NMs	nanomaterials
NNI	National Nanotechnology Initiative
NPs	nanoparticles
NRCG	Nanotechnology Research Coordination Group
NSE	neuron specific enolase
NSTI	Nanoscience and Technology Initiative
OECD	Organization for Economic Co-operation and Development
OmpW	outer membrane protein W
oxGCE	oxidized glassy carbon electrode

PA	phosphatidic acid
PAA-b-PBA	poly(acrylic acid)-b-poly(butylacrylate)
PBCA	poly(butyl cyanoacrylate)
PC	phosphatidylcholine
PCR	polymerase chain reaction
PDGF	platelet-derived growth factor
PDT	photodynamic therapy
PE	phosphatidylethanolamine
PEG	polyethylene glycol
PEG-PE	poly(ethylene glycol) phosphatidylethanolamine
PET	positron emission tomography
PG	phosphatidylglycerol
PGA	polyglycolic acid
PGS-co-PEG	polygycerolsebacate-co-PEG
PHDCA	poly(hexadecyl cyanoacrylate) copolymer
PHV	polymeric heart valves
PI	phosphatidylinositol
PLA	polylactic acid
PLGA	poly(lactic-co-glycolic) acid
PS	phosphatidylserine
PSA	prostate-specific antigen
PSGL-1	P-selectin super ligand
PSMA	prostate-specific membrane antigen
PtNPs	platinum nanoparticles
PUs	polyester urethane
PVA/PVOH	polyvinyl alcohol
PVP	polyvinylpyrrolidone
QDs	quantum dots
REACH	registration, evaluation, authorization, and restriction of chemicals
RF	radio frequency
ROS	reactive oxygen species
SCCS	Scientific Committee on Consumer Safety
SDS	sodium dodecyl sulfate
SEM	scanning electron microscope
SERS	surface-enhanced Raman spectroscopy
SiPy	silsesquioxane polyelectrolyte
siRNA	small interfering RNA
SLNs	solid lipid nanoparticles
SM	secondary metabolites
SMP	shape memory polymers
SNPs	semiconductor nanoparticles/single nucleotide polymorphisms
SODs	superoxide dioxide

SPDDS	smart polymeric drug delivery systems
SPF	sun protection factor
SPIO	superparamagnetic iron oxide
SPIOnps	superparamagnetic iron oxide nanoparticles
TB	tuberculosis
TE	tissue engineering
TEM	transmission electron microscope
TERM	tissue engineering and regenerative medicine
TiO_2	titanium dioxide
TMR	transparency market research
UV	ultraviolet
VBIDA	N-(4-vinylbenzyl) iminodiacetic acid
VOD	veno-occlusive disease
WHO	World Health Organization
XPS	X-ray photoelectron spectroscopy
XRD	X-ray diffraction
ZKV	zika virus
ZnONPs	zinc oxide nanoparticles
ZnS	zinc selenide

Preface

Nanotechnology offers a range of multi-interdisciplinary approaches and brings together the field of chemistry, biology, pharmaceutical, and nutraceutical sciences and clinical medicines by focusing on the design and preparation of smart materials for their biological, nutraceutical, pharmaceutical, and medicinal applications. This book discusses the advancement of nanomaterials (NMs) and their classifications, recent trends in the synthesis of nanomaterials and their therapeutic properties and biological applications in the medicinal and nutraceutical fields, the interaction of nanomaterials in biological systems and the future perspective in the applications in nutraceutical, biological, and medicinal fields. This book also addresses the clinical toxicities of nanomaterials and remediation.

The book comprises 11 chapters, each written by international experts with an exhaustive bibliography. In this book, the editors tried to collect and organize as much information as possible about advanced nanomaterials and their biological, nutraceutical, and medicinal applications, and the editors would like to place on record their gratitude to the various chapter contributors for their tremendous efforts. The book is directed to scientists, researchers, clinicians, pharmacists, nutritionists, and technical and marketing people in institutions, universities, and industries. We hope that this book will prove valuable to all those involved in nanotechnology and nanomaterials to make the world of tomorrow more extraordinary.

CHAPTER 1

Nanomaterials and Its Classification: An Introduction

JENNY HUI FOONG CHAU,[1] CHIN WEI LAI,[1] and BEY FEN LEO[2]

[1]*Nanotechnology and Catalysis Research Center, University of Malaya, Kuala Lumpur, Malaysia*

[2]*Faculty of Medicine, University of Malaya, Kuala Lumpur, Malaysia*

ABSTRACT

Nanotechnology has been expanded wide enough and into an interdisciplinary field with outstanding development to replace every traditional scientific discipline. Nanomaterials (NMs) successfully distinguished itself from other materials located within or above the microscale range and are materials with sizes ranging from 1 to 100 nm. They play a remarkable role in both the medical and biological fields because of their uncommon chemical and physical properties. High surface area to volume ratio and extraordinary physicochemical properties like high surface reaction of NMs gained more and more attention from scientists all around the world. This chapter is going to discuss the definitions, properties, toxicities, classification methods of NMs, and instruments that can be used to characterize NMs. Future trends of NMs and some potential applications of NMs in the medical field will be explored and discussed.

1.1 INTRODUCTION

Nowadays, nanotechnology plays a crucial role in the field of science and technology, where they are considered state-of-the-art. Knowledge of nanoscience and nanotechnology has been applied extensively over the past decade

Advanced Nanomaterials for Biological, Nutraceutical, and Medicinal Applications.
Sreeraj Gopi, Augustine Amalraj, and T. R. Sreeraj (Eds.)

in research, education, industry, and even the daily activities of humans. Nanotechnology generates a huge amount of interest among all scientists worldwide due to their usefulness and diversity of functionalities. Interesting features of nanomaterials (NMs) such as the large surface area to volume ratio and unique appearance make them a peculiar precursor used in different fields such as consumer products, medicine, sustainable energy, catalysis, etc. (Sajid et al., 2015). For example, nanostructured titanium dioxide (TiO_2) can be found in sunscreen and cosmetic products, nanostructured silica is used as filler for a dental filling, nanoclay composite is used to enhance the shelf-life of a tennis ball, etc. (Alagarasi, 2011). Besides large surface area and unique appearance, NMs also demonstrate other interesting properties, such as high surface energy and high reactivity (Zhou et al., 2011).

The unique characteristics of NMs create a path for the enhancement of technologies in both the medical and biological fields. For example, delivery of the drug by traditional chemotherapy induces some undesired effects for the patient, such as low efficiency for drug delivery since the drug will be destroyed by the human defence system before reaching the target organ, while we can avoid this situation from happening by using NMs in the delivery of the drug. Today, nanoparticles (NPs) have been used to deliver not only drugs but also for deliveries of proteins. Furthermore, NMs act as sensing technology for narcotics and assay for the detection of toxic substances. For biological application, NMs can be designed to imitate the structure of an extracellular matrix (ECM) for the growing of cells and transport of nutrients (Han et al., 2019; Kumar et al., 2018; Meikle et al., 2019).

The exact definition of the nanoparticles is controversial as some of the definition is size-related as well as physical and chemical properties-related (Sajid et al., 2015). Thus, the characterization process for NMs is yet a challenging practice.

1.2 DEFINITION OF NANOMATERIALS (NMS)

The prefix of "nano" in NMs refers to the material with sizes ranging from 1 to 100 nm (Dolez et al., 2015). NMs represent any material which has at least one dimension that is less than 100 nm (Alagarasi, 2011). A nanometer is one-millionth of a millimeter (Alagarasi, 2011). There is an interesting example to help us understand how small a nanometer is, as we just need to imagine the situation in which nanometer is 1,00,000 times smaller than the diameter of our hair.

The NMs are defined as a natural, engineer or artificially manufactured material containing particles present either as aggregate or in an unbound state (Filella, 2012) by the European Commission (EC) (Potocnik et al., 2011). The International Organization for Standardization defines nanostructured materials as any material which have either an external nanoscale dimension or internal nanoscale surface structure. The specific surface area to volume ratio of NMs is generally larger than 60 m^2/cm^3 (Kreyling et al., 2010).

1.3 PROPERTIES OF NANOMATERIALS (NMS)

NMs consist of different physical and chemical properties. This section will deliberate the physical, mechanical, magnetic, optical, and lastly, electrical properties of NMs (Dolez et al., 2015; Lue et al., 2007; Sajid et al., 2015). Physical properties of NMs include high surface area and quantum size effect. High surface area means a large number of atoms on the surface of NMs (Alagarasi, 2011). NMs with a size of approximately 10 nm, the amount of atom presents on the surface is account for around 20% of the total atoms contained in the NMs (Alagarasi, 2011). The band gap of NMs will tend to increase while the size of NMs is decreased due to the quantum size effect which is affected by electron confinement. This scenario is caused by the confinement of electrons and holes in the semiconductor. The energy difference between filled and empty states will increase and then broaden the band gap (Singh et al., 2018). The band gap is referred to the energy difference between the conduction band and valence band. Electrons will be excited when the adsorbed light energy is larger than the band gap energy of the semiconductor, and the excited electrons will travel from the valence band to the conduction band.

The magnetic property of NMs displays functionality in catalysis, biomedicine, biotechnology, and many more. The magnetism of NMs results from electric current through atom rotation. Nanostructured materials with one or more unpaired electrons will exhibit paramagnetism while nanostructured material with all paired electrons will exhibit diamagnetism. Material with paramagnetism behavior will be affected by an external magnetic field, whereas a material with diamagnetism behavior will not be affected by an external magnetic field (Alagarasi, 2011). Example of NMs which contained this characteristic is gold (Das et al., 2019). Gold NMs are developed as a contrast agent in multifunctional imaging such as magnetic resonance imaging (MRI) and X-ray computed tomography (CT). This material helps to solve the shortage of each imaging technique by enhancing the spatial resolution and improving the sensitivity. Hence, these imaging techniques can be used effectively in the diagnosis of cancer in the early stage (Das et al., 2019).

The mechanical property includes durability and strength (Guo et al., 2014). Durability depends on the lack of difficulty of defect formation inside a material. When the material size is reduced, the ease of defect formation will be increased. Thus, the smaller the size of NMs, the smaller the durability and the higher the strength. Chen et al. used aluminum oxide film with thickness varying from 2 to 25 nm. The results showed that when the material is small in size, the elastic modulus of the material increased or the larger the size of material, the smaller the reading of elastic modulus (Chen et al., 2015; Wheel et al., 2015).

Elastic moduli help to determine the intrinsic elasticity and stiffness of a solid or liquid. Elastic modulus relates to Hook's law which describes the elastic property of a material under tension or compression (Chen et al., 2015). Stress is divided by strain in the calculation (Chen et al., 2015; Wheel et al., 2015). In this theory, stress is directly proportional to strain (Chen et al., 2015; Wheel et al., 2015) where, the material will return to its original dimension, when the external force is removed. Stress is the applied force exerted on an analyzed sample where strain represents the changes of length of the sample after the force being exerted on it. The higher the reading of elastic modulus, the higher the strength of a material and more stress is needed to create the same amount of strain.

The optical property of NMs is mainly represented by adsorption and emission transition. These two characteristics tend to be affected by energy changes between valence and conduction band (or band gap energy) of a size-reduced nanomaterial (Dolez et al., 2015; Mayeen et al., 2018). Thus, we will be able to observe different color changes for the same type of material under different sizes. For example, gold particles will change from blue color to red color based on the thickness and size of these particles (Wheel et al., 2015). The changes in color are based on the adsorption of different wavelengths of visible light. The optical property of NMs produces applications such as optical detector, sensor, laser, solar cells, and photocatalysis. NMs such as gold, silver, graphene oxide with the optical property have been used as a biosensor for hepatitis. These materials can overcome disadvantages such as expensive, time-consuming, and complex for ordinary methods of polymerase chain reaction (PCR) and enzyme-linked immunosorbent assay (ELISA). Research showed that biosensor based on gold nanomaterial combined with fluorescence quantum dots is able to detect the presence of Hepatitis B virus. In this technique, the luminescence property of quantum dots will be triggered when there is the presence of the Hepatitis B virus (Draz et al., 2012).

The electrical property is related to the electrical conductivity of NMs. The conductivity of a large material will not depend on dimension while the conductivity of a small size material depends on dimensions such as diameter and cross-section area (Dolez et al., 2015; Sajid et al., 2015). Different sizes of NMs cause different conductivity. For example, the conductivity of a multi-walled carbon nanotube is different from the single-walled carbon nanotube. Hence, we can say that the electrical properties of NMs can be controlled by changing the nanotube geometry. NMs with high electrical conductivity can be developed to produce an effective electrochemical biosensor. This biosensor can be used to detect cancer biomarkers for lung, prostate, breast, and many more. For example, we are able to detect protein epidermal growth factor receptor (EGFR) which is produced by breast cancer tumor, by using gold nanomaterial modified with anti-EGFR. Nanobiosensor of graphite NMs coated with gold NMs has been developed for the detection of prostate cancer (Liu et al., 2012). Biosensor with an electrode made up of graphene modified with gold NMs, chitosan (CS), together with anti-carcinoembryonic antigen (CEA) and alpha-fetoprotein (AFP) biomarker can be used to detect lung cancer (Chen et al., 2013).

1.4 INSTRUMENT USED TO EXAMINE THE CHARACTERISTICS OR PROPERTIES OF NANOMATERIALS (NMS)

There are several methods that can be used to characterize NMs based on either structure or composition of the NMs. Instruments such as scanning electron microscope (SEM), transmission electron microscope (TEM), Fourier-transform infrared spectroscopy (FTIR), X-ray photoelectron spectroscopy (XPS), and X-ray diffraction (XRD) will be discussed in this section. The study of the structural and constitutions of NMs is important since these characteristics are related to its mechanical and physical properties. By understanding better of the structure and morphology of the NMs, we are able to classify the NMs into their respective categories.

Both the SEM and TEM were used to study the morphology of NMs under high magnification and resolution. TEM shoots electrons through the sample and measures how the electron beam changes as it is scattered in the sample, whereas SEM scans the sample surface with an electron beam and the electrons will tend to interact with the sample atoms (Mayeen et al., 2018; Sudha et al., 2018). These interactions will produce information such as topography, composition, and electrical conductivity of the sample.

SEM is an electron microscope that scans the sample surface with a high energy beam of electrons. This instrument requires a sample that is conductive enough to react with a scanning electron beam reaching the sample surface (Mayeen et al., 2018; Sudha et al., 2018). A coating of a conductive layer such as gold and platinum onto the non-conductive sample is required. This coating process can be carried out either by using a low vacuum sputter coating or high vacuum evaporation. The coating can help to prevent the accumulation of static electric charge on the sample surface during electron irradiation (Joshi et al., 2008).

In TEM, the analyzed material interacts with the electron beam by diffraction instead of adsorption (Joshi et al., 2008). Material prepared for TEM analysis must be thin enough for the penetration of the electron beam. The concentrated liquid sample must be diluted before dropping onto the support grid while the solid sample can be sectioned in the form of thin film by using a diamond knife (Joshi et al., 2008).

The main difference between TEM and SEM is the electron beam in TEM will pass through the sample while electrons in SEM will only scan the sample surface. Thus, TEM is normally used to examine the internal ultrastructure of a sample while SEM is generally used to determine the surface structure of a sample (Mayeen et al., 2018; Sudha et al., 2018). Besides, energy-dispersive X-ray spectroscopy (EDX) is normally linked with both the SEM and TEM in order to help us determine the unknown element(s) of a sample. It can make a conclusion for both the unknown and known elements in weight percentage or proportion.

FTIR is a remarkable and versatile tool for compositional analysis of NMs (either liquid or solid samples). FTIR determines the chemical composition and the bonding arrangement of a constituent in NMs (Mayeen et al., 2018; Sudha et al., 2018). We are able to differentiate inorganic or organic material through this testing. Motions such as stretching, vibrating, and bending of chemical bonds improve our understanding of what is going on within the molecule of nanomaterial. FTIR is used to provide functional group information in the form of peaks which belongs to a specific component of nanomaterial that will tend to adsorb the infrared source at a particular frequency. Different samples will adsorb infrared radiation at a different wavelength. The functional group of nanomaterial can be determined based on the adsorption table for infrared radiation. The unit for infrared adsorption is cm^{-1}. For example, peaks for water will be observed in the range of 1,640 cm^{-1} which represents a functional group of H and 3,300 cm^{-1} which represents a functional group of OH (Sudha et al., 2018). FTIR provides promising characteristics such as fast detection, minimal sample preparation, sensitive, non-destructive, and wide range detection of functional groups (Joshi et al., 2008; Mayeen et al., 2018; Sudha et al., 2018).

XPS can be used to detect chemical bonding involved in the nanomaterial, measure the composition providing the empirical formula of the element. Besides, XPS can be used to display both the chemical and electronic states of the elements within the analyzed sample (Mayeen et al., 2018; Sudha et al., 2018). In this technique, the X-ray beam will be irradiated onto the solid surface of the sample where emitted kinetic energy and electrons from the sample will be collected and recorded. Thus, we are able to identify and quantify any surface element with the exception of hydrogen, by analyzing the energy and intensity of the photoelectron peaks.

XRD is a remarkable and non-destructive technique. The functions of XRD are mainly on the observation crystalline phase and orientation. Moreover, functions such as determination of grain size, lattice, phase composition of NMs, measurement of thickness, and determination of atomic arrangement of nanostructures also can be obtained by using this technique (Joshi et al., 2008; Mayeen et al., 2018; Sudha et al., 2018). For example, TiO_2 can be determined to appear in either anatase, rutile, or brookite phase. This technique is based on the detection of diffraction of X-rays by the analyzed sample. Bragg's law is obeyed by this method where this law relates the wavelength of X-ray radiation to the lattice spacing and the diffraction angle of crystalline nanostructured materials (Sudha et al., 2018). We are able to identify the unknown sample by converting diffraction peaks to d-spacings since each element has its own unique d-spacing data in the standard reference provided by the Joint Committee on Powder Diffraction Standards (JCPDS) software. This well-known XRD technique composed of few advantages, such as fast and easy sample preparation, high accuracy, and its standards are available for thousands of material (Joshi et al., 2008; Sudha et al., 2018).

1.5 CLASSIFICATION OF NANOMATERIALS (NMS)

Every NMs have different physicochemical properties (e.g., dimension, shape, and size). Various NMs have been found and the number as well as the diversity of NMs are expected to increase exponentially in the future. Thus, there is a need for us to classify them based on a reliable system. NMs can be classified based on different methods such as dimension, origin (either natural or man-made), chemical composition, including composite, transition metal, metal, and carbon-based material, and potential toxicity (Miraftab et al., 2017; Nasrollahzadeh et al., 2019; Sajid et al., 2015; Saleh et al., 2016).

1.5.1 CLASSIFICATION OF NANOMATERIALS (NMS) BASED ON ORIGIN OR SOURCE

NMs can be differentiated based on their origin, namely natural origin and synthetic or engineered origin. NMs made up of natural source generally known as ultrafine particles. Examples of natural source for an ultrafine particle can be found from erupting volcanoes, soil, and forest fires (Kumar, 2009). Natural source mostly contributed by natural events occurred in our environment. Besides, some of the NMs from natural sources can also be found in a living organism. For example, iron storage protein, ferritin, as well as nanocrystalline component of bones, calcium hydroxyapatite (HAp) (Hochella et al., 2012; Miraftab et al., 2017). Furthermore, synthetic source of NMs can be led by accidental activities of humans. Daily activities such as cooking, smoking, and car driving, industrial activities such as incineration, running of machine and power generation tend to generate various types of NMs (Dijk et al., 2011; Oberdorster et al., 2005). These airborne NMs are mostly harmful. However, nowadays many useful NMs such as polymers, metal oxides (MOs), and semiconductors can also be produced through chemical means. These NMs are each designed for particular functionalities. They come in different forms: sphere, sheet, wire, rod, fiber, needle, tube, shell, ring, etc. (Dolez et al., 2015).

1.5.2 CLASSIFICATION OF NANOMATERIALS (NMS) BASED ON MATERIAL OR CHEMICAL CONSTITUENT

NMs can be classified based on materials which are organic, inorganic, carbon-based, and composite (Ashby et al., 2009; Karak et al., 2018; Nasrollahzadeh et al., 2019; Saleh et al., 2016). Nanostructured materials made up of carbon are considered carbon-based nanomaterial. Carbon black, fullerene, carbon nanotube, carbon nanofiber, carbon nanosheet, and graphene are classified as examples for this category. Graphene is made up of carbon atom sheets with a thickness of one atom per sheet (Nasrollahzadeh et al., 2019; Zhu et al., 2014). The thickness of the graphene sheet is approximately 1 nm (Nasrollahzadeh et al., 2019). Graphene contributes many useful functions by acting as an electrode for solar cells, hydrogen, and electron storage for fuel cell and supercapacitor, membrane for gas separation, etc.

Carbon nanosheet is made up of graphene sheets wound into a perfect cylinder shape connected by Van der Waals forces. On the other hand, carbon nanofiber is made up of graphene sheets without perfect arrangement compared to carbon nanosheet (Jong et al., 2001). Fullerene is formed when the graphene sheets are folded into a sphere. It can be comprised of between 28 and 1,500 carbon atoms (Georgakilas et al., 2015; Nasrollahzadeh et al., 2019). Fullerene can be used as a catalyst in water purification and conversion process of methane. Besides, fullerene also can be used as an ingredient to enhance the anti-aging property of cosmetic products (Georgakilas et al., 2015).

A carbon nanotube is made up of graphene nanofoil with a honeycomb lattice of carbon atoms (Nasrollahzadeh et al., 2019). This material consists of special physicochemical properties and hence has great usefulness in the material science field. Both ends of the nanotube can either be opened or closed (Nasrollahzadeh et al., 2019). Organic based NMs made up of organic source, thus, they are generally non-toxic and biodegradable. NMs in this group are used extensively in the field of biomedicine, especially in drug delivery system (Nasrollahzadeh et al., 2019). Liposome, dendrimer, and micelle are located within this system (Nasrollahzadeh et al., 2019). A liposome is an artificial spherical-shaped material having both the hydrophobic (non-polar) and hydrophilic (polar) characteristics. Besides, liposome possesses some promising properties such as non-toxic, biocompatible, and biodegradable, hence, the liposome is useful in delivering drug (Akbarzadeh et al., 2013).

A dendrimer is a highly branched and tree-like structure macromolecule with nanoscale range dimension. The chains on the surface of the dendrimer can be modified in order to prepare the dendrimer for performing a particular chemical function. A dendrimer is a potential drug delivery material due to some interesting properties such as high structural and chemical homogeneity and high ligand density (Vasile et al., 2018). It can be produced by using monomers through the process of polymerization (Nasrollahzadeh et al., 2019). Example of dendrimer used in the medical field is polyamidoamine-platinate (Alagarasi et al., 2011). On the other hand, a micelle is made up of linked amphiphilic block copolymers forming a nanoscopic core-shell structure (Nasrollahzadeh et al., 2019). Hence, micelle consists of both the lipophilic (fat-loving) or hydrophobic and hydrophilic properties (Nasrollahzadeh et al., 2019; Vasile et al., 2018). The core is hydrophobic while the shell is hydrophilic (Vasile et al., 2018). This unique property ensures the effectiveness of micelle in the delivery of water-insoluble or hydrophobic drug. The hydrophobic core region serves as a reservoir to store hydrophobic drugs, while the hydrophilic shell region ensures the hydrophobic core which is polar enough to allow dissolution in aqueous solution (Pepic et al., 2013).

The usefulness of nanocapsule, hollow core in the micelle, make micelle a promising material for different functions such as drug delivery in the biomedical field and food enhancement (Nasrollahzadeh et al., 2019). Ice cream and mayonnaise are the examples of food consist of high surfactant concentration which micelle will be formed. Micelle can be used to extract non-polar molecules such as cholesterol from emulsified oil and healthy food can be produced. Besides, micelle can be used to increase the concentration of non-polar molecules such as vitamin A and D, color, and flavor in food products (Dickinson et al., 1996). Micelle is biocompatible and biodegradable (Zhou et al., 2011). Example of micelle used for drug delivery is polyethyleneglygol-pluronic-doxorubicin (Zhou et al., 2011).

The inorganic NMs are made up of non-carbon and non-organic sources (Nasrollahzadeh et al., 2019). These nanostructures include metal, metal alloy, and metal oxide nanomaterial. The first category is metal and metal alloy (Nasrollahzadeh et al., 2019). Metal-based nanoparticles display extraordinary characteristics which attract scientists in both the biomedical and material science fields. Characteristics of metal-based NMs such as the large surface area to volume ratio, stability in most chemical conditions, high surface charge density, small in size, can be produced in either cylindrical or spherical shape ensure the functionality of this material. Almost all the discovered metals can be produced in the nanoscale range.

NMs such as gold, copper, cadmium, lead, silver, zinc, and iron are located within the category of metal (Dolez et al., 2015; Nasrollahzadeh et al., 2019). Metal alloy can withstand high temperature and has interesting characteristics such as high strength and light weight. Examples of metal alloy are titanium-aluminum alloy and aluminum-magnesium alloy used in aerospace (Dolez et al., 2015). The second category is MOs, including TiO_2, aluminum oxide, cerium oxide, and zinc oxide (Dolez et al., 2015; Nasrollahzadeh et al., 2019). Metal oxide can be prepared by modifying metal-based nanomaterial, for example, iron will be oxidized to iron oxide in the presence of oxygen (Nasrollahzadeh et al., 2019).

1.5.3 CLASSIFICATION OF NANOMATERIALS (NMS) BASED ON POTENTIAL TOXICITY

Despite the usefulness of NMs, they bring undesirable effects if the usage amount reaching certain permissible exposure limit as stated in Health and Safety Executives EH40 Booklet. Workplace exposure limit acts as a regulation to ensure the usage amount of certain chemicals which should not exceed and cause an undesirable effect. The function of this limit is to control the exposure of the workforce, lab user or anyone who have a chance to deal with any harmful and toxic chemical or substance. For example, painter using paints for product coating, wood dust resulted from industrial furniture production, solvents used in the printing industry, disinfectants used to clean instrument in hospital (Dolez et al., 2015; Pietrouiusti et al., 2014). The value of workplace exposure limit is expressed as a time-weighted average and divides into two types, namely long-term exposure limit and short-term exposure limit. Long-term exposure limit refers to a maximum exposure over an eight-hour period, whereas short-term exposure limit refers to a maximum exposure over a 15 minute period.

The first category refers to fiber-like nanoparticles. This category includes inflexible, biopersistent carbon nanotube, fiber-like metal oxide, and carbon nanotube without asbestos-like effect. The exposure limit for both the carbon nanotube and fiber-like metal oxide are 0.007 to 0.03 mg/m^3 or 10^4 to 10^5 fibers/m^3 whereas the exposure limit for carbon nanotubes (CNTs) without asbestos-like effect is 4×10^7 fibers/m^3 (Dolez et al., 2015; Pietrouiusti et al., 2014). The exposure limit is normally expressed in mg/m^3 for metals, salts, and compounds that are not going to produce vapor at room conditions.

The second category displays biopersistent granular nanoparticle with a density of less than 6,000 kg/m^3. The estimated exposure limit is either 0.3 to 3.5 mg/m^3 or 2 to 4×10^7 particles/m^3. Titania, aluminum oxide, zinc oxide, silica, carbon black, and polystyrene are the examples that fall within this second category. Apart from this, the second category also presents biopersistent granular nanoparticle with a density of more than 6,000 kg/m^3. Examples of this category are gold, lead, iron oxide, cerium oxide, silver, and tin oxide. The estimated exposure limit is either 0.3 mg/m^3 or 2×10^7 particles/m^3 (Dolez et al., 2015; Pietrouiusti et al., 2014).

The last category refers to CMAR nanoparticles. Each alphabet refers to different toxic characteristics for nanoparticles that fall into this category, where C stands for carcinogenic, M stands for mutagenic, A stands for asthmagenic, and R stands for reproductive toxin, respectively. This category consists of NMs such as arsenic, chromium (VI), nickel, and beryllium. The suggested exposure limit is 2 to 4×10^7 particles/m^3 or 0.3 to 3.5 mg/m^3. Last but not least, liquid, soluble nanoparticles such as hydrocarbon, fat, sodium chloride, and siloxane are classified as the last category. The proposed exposure limit is 1.5 mg/m^3 (Dolez et al., 2015; Pietrouiusti et al., 2014).

1.5.3.1 EVALUATION OF TOXICITY OF NANOMATERIALS (NMS)

The issue of toxicity of NMs is needed to be solved for the universal usage of NMs in the medical field. We need to take care of those NMs with no apparent toxic elements to prevent any inherent potential risk. This is because subcellular interaction may happen between both the human cells and NMs due to their incredible nanoscale size. NMs can enter effortlessly into our skins with only contact and create unpredictable health risks (Yang et al., 2019).

There are two main evaluation methods for NMs used in the medical field, namely *in vitro* and *in vivo* cytotoxicity testing. Cell cultures will be used in *in vitro* cytotoxicity testing while animal samples will be used in *in vivo* cytotoxicity testing. Cell such as *Escherichia coli* is normally used in *in vitro* cytotoxicity testing. Trypan blue dye can be applied to the color of both the dying and living cells. We are able to distinguish between dying and living cells since only dying cells will let the dye pass into cells due to brittle membranes and get stained. Stained cells cause a fluorescence intensity which can be observed by using fluorescence microscopy (Yang et al., 2019).

On the other hand, animals like fruit flies, mice, and zebrafish are used in *in vivo* cytotoxicity testing. This testing is having some disadvantages such as high cost, time-consuming, and involve some ethical problems (Hu et al., 2011) especially for those who are animal lovers. However, this testing is more useful if compared to *in vitro* cytotoxicity testing. This is because cells alone are not possible to mimic human characteristics exactly and hence we are not able to fully understand and observe the negative effects of toxicity of NMs applied to humans. Hence, by using animal like mice for clinical trial, which have the same organs as in human will help us understand better for the unintentional impacts of nanomedicine. Besides, the short lifetime of cells may not be able to overcome the long-term observation of toxicity of NMs since they may accumulate in the human body for a longer period before starting to affect humans.

1.5.3.2 TOXICITY OF METAL-BASED NANOMATERIALS (NMS)

A simple example will be the useful quantum dots which contain few toxic elements such as cadmium, selenium, sulfur, and tellurium. These elements are best known for their highly toxic and carcinogenic characteristics towards living organisms (Yang et al., 2019). For instance, cadmium is toxic to the kidney and causes undesirable effects like bone demineralization and lung cancer (Rafati Rahimzadeh et al., 2017). Platinum NMs is another example, which is able to destroy deoxyribonucleic acid (DNA) in human colon carcinoma cells. Researchers found that platinum NMs are present in the cells and platinum ions are released from lysosomes of the cells and later being transported to the nucleus in their experiments (Pelka et al., 2009).

1.5.3.3 TOXICITY OF GRAPHENE AND CARBON-BASED NANOMATERIALS (NMS)

Graphene and carbon-based NMs can cause brain damage. Research using largemouth bass, a type of fish, showed that these fish suffered critical headache after exposure to fullerene in the experiment. The brain cell injury was found 17 times higher compared to the control fish group without any exposure to fullerene (Oberdorster et al., 2004). The research found that both the single-wall and multiple-wall CNTs as well as graphene nanotubes have the possibility to initiate fibrosis and oxidative stress in the lungs of mice. Lung injury was mainly observed in the experiment due to the toxicity of these NMs (Duch et al., 2011). Another research of toxic effect of multiple-wall CNTs on *Stylonychia mytilus* cells showed that multiple-wall CNTs were easily ingested by these cells. The growth of cells was blocked when a certain concentration is reached and damage both the membrane and nucleus (Zhu et al., 2006).

1.5.4 CLASSIFICATION OF NANOMATERIALS (NMS) BASED ON DIMENSION

The NMs are generally classified based on their dimensions and been divided into four different unique categories, namely zero-dimension (0D), one-dimension (1D), two-dimension (2D), and three-dimension (3D) (Deepu et al., 2018; Karak et al., 2018; Nasrollahzadeh et al., 2019; Sajid et al., 2015;

Subramanium et al., 2018; Vasile et al., 2018). This classification is provided by Pokropivny and Skorokhod (Pokropivny et al., 2018).

For 0D nanomaterial, parameters of length, breadth (or width), and height (or thickness) are fixed at a single point, which can be described as a dot (Nasrollahzadeh et al., 2019; Sajid et al., 2015). All the external dimensions (x, y, and z-axis) of zero-dimensional NMs are located at a nanoscale range of 1 to 100 nm (Dolez et al., 2015; Karak et al., 2018; Saleh et al., 2016). Electrons in this nanostructure are completely constrained and being trapped in a dimensionless space which is confined in all the three directions (Nasrollahzadeh et al., 2019; Sajid et al., 2015). Nanoparticle and nanosphere are considered to have structures of 0D NMs. The spherical shape of 0D NMs is the most common shape. Metal, metal oxide, hydroxide, and carbonate are the examples that belong to this class.

Quantum dot is the best example of 0D NMs (Geogakilas et al., 2015; Sajid et al., 2015; Subramanium et al., 2018). The trapped electrons within the quantum dot cannot move freely and hence have declined the possibility for delocalization of electrons to be happened (Subramanium et al., 2018). Each electron only exists within an "infinitely deep potential well." Quantum dot is a semiconductor nanocrystal with a dimension of less than 10 nm. Examples for quantum dot are cadmium-selenium. It acts as a potential well to restrict both the electrons and holes (Smith et al., 2010) and usually being used in an electronic device such as a television. Quantum dot displays some extraordinary characteristics such as high quantum yield, broad UV excitation, high photostability, presents narrow and symmetrical emission spectra (Gerion et al., 2001; Saleh et al., 2016; Wang et al., 2014).

Zero-dimensional NMs are usually described as amorphous or crystalline (single crystalline or polycrystalline) materials. Besides, they can be a single chemical element or multi-chemical elements with various forms and shapes such as sphere TiO_2, hollow sphere carbon or fullerene, cube palladium, flowerlike zinc oxide, star-shaped vanadium oxide, tree-like magnesium oxide, and many more (Ashby et al., 2009, Dolez et al., 2015; Karak et al., 2018). Apart from this, 0D NMs can exist individually or be incorporated in another medium. Zero-dimensional NMs are able to be displayed in ceramic, metallic, or polymeric form (Ashby et al., 2009). Last but not least, all 0D NMs have an equal length to width ratio (Sajid et al., 2015).

One-dimensional NMs have two external dimensions (*x* any *y*-axis) located at the nanoscale range, 1 external dimension located at microscale range, meaning outside of nanoscale (Ashby et al., 2009; Subramanium et al., 2018). NMs with width and thickness in nm range and length in μm are hence considered 1D NMs. This nanostructure contains only one parameter which the electrons in 1D nanostructure can move along the x-axis with a distance of less than 100 nm (Nasrollahzadeh et al., 2019). Electrons are restricted within two dimensions, indicating electrons that cannot move freely in this system (Sajid et al., 2015). Common structures are nanorods, nanobelts, nanotubes, nanofibers, and nanowires. Examples of nanofiber are TiO_2, zirconium dioxide, aluminum oxide, and silicon dioxide (Dolez et al., 2015).

Nanotube demonstrates an empty cylindrical crystalline shape with all the atoms organized as either pentagons, hexagons, or heptagons (Terrones et al., 2003). Nanowire has the largest length to width ratio (normally more than 1,000) compared to others in one-dimensional nanomaterial (Hanson et al., 2007). This unique characteristic allows nanowire to confine the electron laterally. On the other hand, a nanorod has a shorter length to width ratio compared to a nanowire. The length to width ratio of nanorod usually located between the ranges of 3 to 20 (Krishnamurthy et al., 2014). Hence, we are able to differentiate both the nanowire and nanorod based on their length to width ratio.

One dimensional NMs has a longer length than width (Karak et al., 2018). They normally appear as pure material but, in some cases, they can be doped with other material like the doping in semiconductor (Karak et al., 2018; Subramanium et al., 2018). For example, zinc alone is in pure form, and at the same time, zinc can be doped with semiconductor such as TiO_2 in order to enhance the performance of semiconductor by reducing the band gap energy of TiO_2. Graphene is another famous example of this nanostructure. Similar to 0D NMs, 1D NMs can present in either the amorphous or crystalline (single crystalline or polycrystalline) form. These NMs can be either in ceramic, metallic, or polymeric state (Ashby et al., 2009).

Two-dimensional NMs consist of only one external dimension at the nanoscale range (Dolez et al., 2015). Both length and width of 2D NMs are within microscale range while the thickness is within the nanoscale range. Hence, NMs of this group is considered the thinnest material. Electron restriction in 2D NMs differentiates itself from both the 0D and 1D nanomaterial. Electrons in 2D materials are subject to both the electron confinement and delocalization. In this system, electrons are restricted within one dimension, indicating electrons cannot move freely within that particular dimension. For 2D nanomaterial, the electrons are free to move along both the x-axis and y-axis. Parameters of length and breadth exist in this nanostructure (Nasrollahzadeh et al., 2019).

The length of this 2D NMs is longer than its width (Sajid et al., 2015). Similar to both the 0D and 1D NMs, 2D NMs can be amorphous or crystalline. It can be either in the single crystalline or polycrystalline form (Sajid et al., 2015; Subramanium et al., 2018; Ashby et al., 2009). 2D NMs can present either in ceramic, metallic, or polymeric state (Subramanium et al., 2018; Ashby et al). Besides, the 2D nanostructures can be made up of different chemical compositions. It can exist in single or multiple layer forms (Ashby et al., 2009; Sajid et al., 2015). 2D NMs usually exists in the plate-like structure and hence the examples for 2D nanostructures are nanoplate, nanofilm, nanocoating, and nanosheet (Subramanium et al., 2018). Example of nanosheet is graphene.

A thin film made up of metal coating or ceramic that can be as thin as a few atomic layers (Ohrin et al., 2001). Composite such as zinc oxide-TiO_2 is used to produce nanocoating. Nanocoating can be used to improve hardness, prevent corrosion, insulate painting surface of the vehicle. The thickness of nanoplate is usually within the range of a few nanometers. Besides, the width and length of nanoplate are in the range of 70 to 600 nm (Bahinghaus et al., 2014; Uddin et al., 2008). Lastly, 2D NMs normally displays unique characteristics such as the high surface area to volume ratio and consist of an adequate number of surface functionality.

Three-dimensional NMs contains all the three parameters which are length, height, and breadth (Nasrollahzadeh et al., 2019). In this system, it only consists of the internal dimension with nanoscale range. No external dimension at nanoscale range (Dolez et al., 2015) with the three arbitrary dimensions are more than 100 nm (Karak et al., 2018; Saleh et al., 2016). All the electrons in this 3D nanostructures are completely delocalized (Karak et al., 2018; Sajid et al., 2015). The electrons are free to move along the *x*, *y*, and *z*-axis (Karak et al., 2018; Sajid et al., 2015; Subramanium et al., 2018). They are also known as bulk nanomaterial (Ashby et al., 2009). In this structure, each nanomaterial is performed as a building block, which can be combined together to form a complex 3D structure. Three-dimensional materials are hence made up of numerous nanosized crystal with diverse arrangements and orientations.

Examples of 3D nanostructures are nanocomposite, nanoflower, and nanostructured materials (Dolez et al., 2015; Subramanium et al., 2018). Each 0D, 1D, or 2D NMs can represent the nanofiller whereas matrix can be represented either by metal, ceramic, or polymer. Bone is one of the interesting examples for nanocomposite with calcium HAp nanocrystals dispersed in a collagen matrix. The main difference between nanocomposite and composite material is that nanocomposite has a larger surface-to-volume ratio compared to composite material (Dolez et al., 2015). It can be in the state of amorphous or crystalline. Besides, 3D NMs can be chemically pure or chemically impure. It can be a multi-layered composite material. Three-dimensional NMs can present in either metallic, ceramic, or polymeric states (Ashby et al., 2009). Last but not least, nanostructured materials present examples such as nanoporous structures, aerogels, and block copolymers (Dolez et al., 2015).

1.6 FUTURE RESEARCH PROSPECTS

In the 21st century, both the NMs and nanotechnology will continue to dominate compared to other industries. In 2014, the global NMs market has reached $3.4 billion, and by 2020, the NMs market is believed to reach an approximate $30 billion (Miraftab et al., 2017). There are three main unavoidable areas should be paid attention between the nanotechnology and medical fields, which include diagnosis, prevention, and treatment (Miraftab et al., 2017). Biomarker is generally used in the diagnosis stage to predict the occurrence of certain disease and illness while nanomedicine is widely used as a prescription in the treatment stage. Prevention using NMs is useful in preventing disease or undesired health effects. For an example, we can prevent growth of bacteria and prolong shelf life of food by coating of food containers with silver NMs (Bastarrachea et al., 2015).

Most pharmaceutical companies are encouraged to invest in this area in order to magnify and enhance the usefulness of NMs in both the medical and biological fields. We need to balance the actual benefits brought by the NMs used in biological and medical fields with its production cost (Fadeel et al., 2010). High production cost, long term research period, difficulty in passing the clinical trial, short period of patent, and occurrence of generic product with lower price are the problems faced these days. Long revenue time limits the attention of research in nanomedicine. Other than that, problems like NMs waste handling, accumulation of NMs in the body for a long period, long term effects of NMs, efficiency of NMs in the body, and toxicity of NMs must be addressed and settled in order to raise the popularity usage of NMs. Mechanism and physicochemical properties of NMs must be fully understood to prevent any undesirable effect.

An effective way in handling NMs waste must be developed in order to reduce contamination to both the environment and living organisms.

The research found that more than 95% of copper NMs released into the environment will enter the soil and aquatic sediments, where it could reach undesirable toxic levels with a figure of more than 50 to 500 μg/L. Besides, 97% to 99% of copper nanoparticles were found to be accumulated in the plant leaves while 1% to 3% of copper nanoparticles will be translocated to the root tissues of the plant (Zhao et al., 2016).

On the other hand, accumulation problem of NMs in the body after treatment must be solved in order to ensure the effective emission of useless NMs out of the body. This is an important issue to be resolved in order to avoid any unwanted toxic effect of NMs. For instance, high amount of accumulated gold nanoparticles (AuNPs) in the human body will tend to induce health problems like liver failure and gene modification. Research discovered that the bovine serum albumin-protected AuNPs showed extreme low performance in renal clearance with only 1% of nanostructured gold being eliminated from the body after a period of 24 hours, and an astonishing amount of 95% of this AuNPs were left in the rat body after 28 days in the experiment (Zhang et al., 2012). Detection for non-biodegradable NMs is necessary after the treatment. Ways of detection always a concern since NMs are small enough to pass through blood vessels and cell membranes, making them hard to be traced and eliminated (Sharifi et al., 2012).

Size control is an important step in enhancing the effectiveness of nanomedicine. The size of nanomedicine should be controlled in order to avoid the medicine being destroyed by the mononuclear phagocyte system of the body before implementing their functions. Besides, nanomedicine with a size of less than 250 nm can prevent the uptake by the spleen and liver while nanomedicine with a size of larger than 10 nm can prevent clearance by the kidneys. Size control is a very challenging step which we have to control the size without affecting the physicochemical properties which may affect the performance of nanomedicine in the disease treatment.

Limits of NMs used in certain applications need to be improved. For instance, gold NMs used continuously in treatment under irradiation will deform and this deformation will affect its effectiveness (Lu et al., 2010). Besides, NMs used in cancer treatment are able to pass through cancer cell membrane, leading to aggregation of NMs and hence lower down the effectiveness of treatment. Accumulated NMs located at healthy tissue may create other new problems. Future research on toxicity mechanism of NMs should be conducted at the molecular level (Yang et al., 2019). Nowadays, no standard method for toxicity determination testing of NMs. Investigation of the toxicity of NMs still lacks a proper standard. Thus, it is utmost important to create a standard method for the determination of the toxicity of NMs.

1.7 CONCLUSION

NMs contribute many advantages in the medical field to save the lives of all human beings. However, there are some limits of NMs being applied in the treatment process. Different limits of NMs in medical treatment must be addressed and resolved. Further studies on the physicochemical properties of NMs is needed in order to understand the mechanism of NMs and to prevent any undesirable effects. Careful investigation of the NMs properties is recommended for a full understanding of nanomaterial toxicity. In the future, more potential applications of NMs in the medical field are supposed to be developed and more research on the safety of NMs are supposed to be carried out.

ACKNOWLEDGMENT

This study was supported by the University Malaya Research Grant program–AET ITRC (RP045B-17AET & RP045D-17AET) and Impact-Oriented Interdisciplinary Research Grant (IIRG018A-2019).

KEYWORDS

- **classification**
- **medical**
- **nanomaterials**
- **nanotechnology**
- **toxicity**

REFERENCES

Akbarzadeh, A., Rezaeisadabady, R., & Nejatikoshki, K., (2013). Liposome: Classification, preparation, and applications. *Nanoscale Res. Lett.*, 8.

Alagarasi, A., (2011). In: *Introduction to Nanomaterials* (p. 76).
Ashby, M. F., Ferreira, P. J., & Schodek, D. L., (2009). Chapter 6 – nanomaterials: Classes and fundamentals. In: *Nanomaterials, Nanotechnologies and Design* (pp. 177–197). Butterworth-Heinemann.
Ashby, M. F., Ferreira, P. J., & Schodek, D. L., (2009). *Nanomaterials, Nanotechnologies and Design-An Introduction for Engineers and Architects*. Butterworth-Heinemann, Burlington: Elsevier, London.
Baringhaus, J., Ruan, M., Edler, F., Tejeda, A., Sicot, M., & Taleb-Ibrahimi, A., (2014). Exceptional ballistic transport in epitaxial graphene nanoribbons. *Nature, 506*, 349–354.
Bastarrachea, L. J., Denis-Rohr, A., & Goddard, J. M., (2015). Antimicrobial food equipment coatings: Applications and challenges. *Annu. Rev. Food Sci. Technol., 6*, 97–118.
Chen, X., Jia, X., Han, J., Ma, J., & Ma, Z., (2013). Electrochemical immunosensor for simultaneous detection of multiplex cancer biomarkers based on graphene nanocomposites. *Biosens. Bioelectron, 50*, 356–361.
Chen, Y., Gao, Q., Wang, Y., An, X., Liao, X., Mai, Y. W., Tan, H. H., et al., (2015). Determination of young's modulus of ultrathin nanomaterials. *Nano Lett., 15*, 5279–5283.
Das, P., Fatehbasharzad, P., Colombo, M., Fiandra, L., & Prosperi, D., (2019). Multifunctional magnetic gold nanomaterials for cancer. *Trends Biotechnol., 37*, 995–1010.
Deepu, G., Avinash, R. P., Daniel, P., Leu, S. Y., Abdul, K. H. P. S., & Sabu, T., (2018). Nanomaterials – state of art, new challenges, and opportunities. In: Sabu, T., Daniel, P., Leu, S. Y., & Deepu, G., (eds.), *Nanoscale Materials in Water Purification* (pp. 1–24). Elsevier, London.
Dickinson, E., & McClements, D. L., (1996). Surfactant micelles in food. In: *Advances in Food Colloids* (1st edn.). Springer, Blackie Academic & Professional, London, New York.
Dijk, W. D., Gopal, S., & Scheepers, P. T., (2011). Nanoparticles in cigarette smoke, real-time undiluted measurements by a scanning mobility particle sizer. *Anal. Bioanal. Chem., 399*, 3573–3578.
Dolez, P. I., (2015). Nanomaterials definitions, classifications, and applications. In: Dolez, P. I., (ed.), *Nanoengineering*. Elsevier, London.
Draz, M. S., Fang, B. A., Li, L., Chen, Z., Wang, Y., Xu, Y., Yang, J., et al., (2012). Hybrid nanocluster plasmonic resonator for immunological detection of hepatitis B virus. *ACS Nano, 6*, 7634–7643.
Duch, M. C., Budinger, G. R. S., Liang, Y. T., Soberanes, S., Urich, D., Chiarella, S. E., Campochiaro, L. A., et al., (2011). Minimizing oxidation and stable nanoscale dispersion improves the biocompatibility of graphene in the lung. *Nano Lett., 11*, 5201–5207.
Elshafey, R., Tavares, A. C., Siaj, M., & Zouro, M., (2013). Electrochemical impedance immunosensor based on gold nanoparticles-protein G for the detection of cancer marker epidermal growth factor receptor in human plasma and brain tissue. *Biosens. Bioelectron, 50*, 143–149.
Fadeel, B., & Garcia-Bennett, A. E., (2010). Better safe than sorry: Understanding the toxicological properties of inorganic nanoparticles manufactured for biomedical applications. *Adv. Drug Deliv. Rev., 62*, 362–374.
Filella, M., (2012). Nanomaterial. In: Pawliszyn, J., (ed.), *Comprehensive Sampling and Sample Preparation* (pp. 109–123). Academic Press, U.S.A.
Georgakilas, V., Perman, J. A., Tucek, J., & Zboril, R., (2015). Broad family of carbon nanoallotropes: Classification, chemistry, and applications of fullerenes, carbon dots,

nanotubes, graphene, nanodiamonds, and combined superstructures. *Chem. Rev., 115*, 4744–4822.

Gerion, D., Pinaud, F., Williams, S. C., Parak, W. J., Zanchet, D., Weiss, S., & Alivisatos, A. P., (2001). Synthesis and properties of biocompatible water-soluble silica coated CdSe/ZnS semiconductor quantum dots. *J. Phys. Chem. B., 105*, 8861–8871.

Guo, D., Xie, G., & Luo, J., (2014). Mechanical properties of nanoparticles: Basics and applications. *J. Phys. D., 47*, 013001.

Han, J., Xiong, L., Jiang, X., Yuan, X., Zhao, Y., & Yang, D., (2019). Bio-functional electrospun nanomaterials: From topology design to biological applications, *Prog. Polym. Sci., 91*, 1–28.

Hanson, G. W., (2007). *Fundamentals of Nanoelectronics*. Upper Saddle River: Prentice Hall.

Hochella, M. F., Aruguete, D. M., Kim, B., & Madden, A. S., (2012). Naturally occurring inorganic nanoparticles: General assessment and a global budget for one of earth's last unexplored geochemical components. In: Guo, H., & Barnard, A., (eds.), *Nature's Nanostructures*. Stanford Publishing, Australia, Pan.

Hu, Y., Qi, W., Han, F., Shao, J., & Gao, J., (2011). Toxicity evaluation of biodegradable chitosan nanoparticles using a zebrafish embryo model. *Int. J. Nanomed., 6*, 3351–3359.

ISO/TS 229 80004-1:2010 – Nanotechnology-Vocabulary – Part 1: Core Terms, (2010). International Organization for Standardization.

Jong, K. P., & Geus, J. W., (2001). Carbon nanofibers: Catalytic synthesis and applications. *Catal. Rev. Sci. Eng., 42*, 481–510.

Joshi, M., Bhattacharyya, A., & Ali, S. W., (2008). Characterization techniques for nanotechnology applications in textiles. *Indian J Fibre Text., 33*, 304–317.

Karak, N., (2018). Fundamentals of nanomaterials and polymer nanocomposites. In: Karak, N., (ed.), *Nanomaterials and Polymer Nanocomposites* (pp. 1–45). Elsevier, London.

Kreyling, G. W., Semmler-Behnke, M., & Chaudhry, Q., (2010). A complementary definition of nanomaterial. *Nano Today, 5*, 165–168.

Krishnamurthy, N., Vallinayagam, P., & Madhavan, D., (2014). *Engineering Chemistry*. New Delhi: Raj Press.

Kumar, C. S. S. R., (2009). Nanostructured oxides. In: *Nanomaterials for the Life Sciences*. Wiley, Weinheim.

Kumar, V., Kumar, P., Pournara, A., Vellingiri, K., & Kim, K. H., (2018). Nanomaterials for the sensing of narcotics: Challenges and opportunities. *Trends Analyt. Chem., 106*, 84–115.

Liu, B., Lu, L., Hua, E., Jiang, S., & Xie, G., (2012). Detection of the human prostate-specific antigen using an aptasensor with gold nanoparticles encapsulated by graphitized mesoporous carbon. *Microchimica Acta, 178*, 163–170.

Lu, W., Singh, A. K., Khan, S. A., Senapati, D., Yu, H., & Ray, P. C., (2010). Gold nano-popcorn based targeted diagonosis, nanotherapy treatment and in-situ monitoring of photothermal therapy response of prostate cancer cells using surface enhanced Raman spectroscopy. *J. Am. Chem. Soc., 132*, 18103–18114.

Lue, J. T., (2007). Physical properties of nanomaterials. In: *Encyclopedia of Nanoscience and Nanotechnology* (Vol. X, pp. 1–46). American Scientific Publishers, U.S.A.

Mayeen, A., Shaji, L. K., Nair, A. K., & Kalarikkal, N., (2018). Morphological characterization of nanomaterials. In: Bhagyaraj, S. M., Oluwafemi, O. S., Kalarikkal, N., & Thomas, S., (eds.), *Characterization of Nanomaterials* (pp. 335–363). Woodhead Publishing, United Kingdom.

Meikle, T. G., Strachan, J. B., & Conn, C. E., (2019). Lipid nanomaterials for the delivery of antimicrobial peptides. In: Gurtler, V., Ball, A. S., & Soni, S., (eds.), *Methods in Microbiology* (pp. 2–293). Elsevier, London.
Miraftab, M., (2017). Summary and future of nanomaterials in medicine/biomaterials. In: *Nanobiomaterials Science, Development, and Evaluation* (pp. 301–313). Elsevier, London.
Nasrollahzadeh, M., Issaabadi, Z., Sajjadi, M., Sajadi, S. M., & Atarod, M., (2019). Types of nanostructures. In: *Interface Science and Technology* (pp. 29–80). Elsevier, London.
Oberdorster, E., (2004). Manufactured nanomaterials (fullerenes, C60) induce oxidative stress in the brain of juvenile largemouth bass. *Environ. Health Perspect., 112*, 1058–1062.
Oberdorster, G., Orster, E. O., & Orster, J. O., (2005). Nanotoxicology: An emerging discipline evolving from studies of ultrafine particles. *Environ. Health Perspect., 113*, 823–839.
Ohrin, M., (2001). *Materials Science of Thin Films*. Boston: Academic Press.
Pelka, J., Gehrke, H., Esselen, M., Türk, M., Crone, M., Bräse, S., Muller, T., et al., (2009). Cellular uptake of platinum nanoparticles in human colon carcinoma cells and their impact on cellular redox systems and DNA integrity. *Chem. Res. Toxicol., 22*, 649–659.
Pepic, I., Lovric, J., & Filipovic-Grcic, J., (2013). How do polymeric micelles cross epithelial barriers? *Eur. J. Pham. Sci., 50*, 42–55.
Pietroiusti, A., & Magrini, A., (2014). Engineered nanoparticles at the workplace: Current knowledge about workers' risk. *Occup. Med. (Lond), 64*, 319–330.
Pokropivny, V. V., & Skorokhod, V. V., (2007). Classification of nanostructures by dimensionality and concept of surface forms engineering in nanomaterial science. *Mater. Sci. Eng. C., 27*, 5–8.
Potocnik, J., (2011). Commission recommendation of 18 October 2011 on the definition of nanomaterial. *Official Journal of the European Union*, 38–40.
Rafati Rahimzadeh, M., Rafati Rahimzadeh, M., Kazemi, S., & Moghadamnia, A. A., (2017). Cadmium toxicity and treatment: An update. *Caspian Journal of Internal Medicine*, *8*(3), 135–145.
Sajid, B., & Liu, J. B., (2015). Nanomaterials and their application. In: *Advanced Nanomaterials and Their Applications in Renewable Energy*. Elsevier Science, London.
Saleh, T. A., & Gupta, V. K., (2016). Synthesis, classification, and properties of nanomaterials. In: *Nanomaterial and Polymer Membranes* (pp. 83–133). Elsevier, London.
Sharifi, S., Behzadi, S., Laurent, S., Forrest, M. L., Pieter, S. P., & Morteza, M. M., (2012). Toxicity of nanomaterials. *Chem. Soc. Rev., 41*, 2323–2343.
Singh, M., Goyal, M., & Devlal, K., (2018). Size and shape effects on the band gap of semiconductor compound nanomaterials. *J. Taibah Univ. Sci*., 470–475.
Smith, A. M., & Nie, S., (2010). Semiconductor nanocrystals: Structure, properties, and band gap engineering. *Acc. Chem. Res., 43*, 190–200.
Subramanian, A., Sambandam, A., & Muthupandian, A., (2018). Introduction to advanced nanomaterials. In: Bharat, A. B., Vijay, P., Sanjay, J. D., Shirish, H. S., & Muthupandian, A., (eds.), *Nanomaterials for Green Energy* (pp. 1–53). Elsevier.
Sudha, P. N., Sangeetha, K., Vijayalakshmi, K., & Barhoum, A., (2018). Nanomaterials history, classification, unique properties, production and market. In: Makhlouf, A. S. H., & Barhoum, A., (eds.), *Emerging Applications of Nanoparticles and Architectural Nanostructures* (pp. 341–384). Elsevier.
Terrones, H., & Terrones, M., (2003). Curved nanostructured materials. *New J. Phys., 5*, 126.1–126.37.

Uddin, F., (2008). Clays, nanoclays, and montmorillonite minerals. *Metall. Mater. Trans. A Phys. Metall. Mater. Sci.*, *39*, 2804–2814.

Vasile, C., (2018). Polymeric nanomaterials: Recent developments, properties and medical applications. In: Vasile, C., (ed.), *Polymeric Nanomaterials in Nanotherapeutics*. Elsevier, London.

Wang, Y., & Hu, A., (2014). Carbon quantum dots: Synthesis, properties and applications. *J. Mater. Chem.*, *2*, 6921–6939.

Wheel, M. A., Frame, J. C., & Riches, P. E., (2015). Is smaller size always stiffer? On size effects in supposedly generalized continua. *Int. J. Solids Struct.*, *67*, 84–92.

Yang, L. X., Luo, X. B., & Luo, S. L., (2019). Assessment on toxicity of nanomaterials. In: Luo, X., & Deng, F., (eds.), *Nanomaterials for the Removal of Pollutants and Resource Reutilization* (pp. 273–289). Elsevier.

Zhang, X. D., Wu, D., Shen, X., Liu, P. X., Fan, S. Y., & Fan, S. J., (2012). *In vitro* renal clearance, biodistribution, toxicity of gold nanoclusters. *Biomaterials*, *33*, 4628–4638.

Zhao, L., Ortiz, C., Adeleye, A., Hu, Q., Zhou, H., Huang, Y., & Keller, A., (2016). Metabolomics to detect response of lettuce (*Lactuca sativa*) to $Cu(OH)_2$ nanopesticides oxidative stress response and detoxification mechanisms. *Environ. Sci. and Technol.*, *50*, 9697–9707.

Zhou, Z. Y., Tian, N., Li, J. T., Broadwella, I., & Sun, S. G., (2011). Nanomaterials of high surface energy with exceptional properties in catalysis and energy storage. *Chem. Soc. Rev.*, *40*, 4167–4185.

Zhu, Y., James, D. K., & Tour, J. M., (2012). New routes to graphene, graphene oxide and their related applications. *Adv. Mat.*, *24*, 4924–4955.

Zhu, Y., Zhao, Q., Li, Y., Cai, X., & Li, W., (2006). The interaction and toxicity of multi-wall carbon nanotubes with *Stylonychia Mytilus*. *J. Nanosci. Nanotechnol.*, *6*, 1357–1364.

CHAPTER 2

Advanced Therapeutic Properties of Nanomaterials

VIJITA PILLAI[1,2]

[1]*School of Biotechnology and Bioinformatics, D.Y. Patil Deemed-to-be University, Navi Mumbai, Maharashtra, India*

[2]*Anant Pharmaceuticals Pvt. Ltd., Ambernath MIDC, Mumbai, Maharashtra, India*

ABSTRACT

Since the past four decades, when the World Health Organization (WHO) approached a group of pharmaceuticals and therapeutics experts, till now, the field of pharmacology has only bettered. Over the years, a lot has changed in terms of our demands from the type of treatments, drugs, and their delivery mechanisms, especially post the advent of molecular biology and nanotechnology. Humans were introduced to some unprecedented insights into cellular behavior, individual health conditions, genomic profile, biomedical tools, and approaches to fight diseases or conditions such as cancer, to name a few. Overall, an impactful change in global health care was presented, and research was shifted from "one drug for most" to now be customized for personalized health care. Nanomaterials, which are really small in size, gained a lot of attention with their ability to interact with cells at a molecular level and enabling research possibilities that would change the face of medical science and pharmacology. Nanotechnology could play a pivotal role in the shift to modern treatment using drug delivery systems because of an array of advantages that it offers including reduced toxicity, improved efficiency, low drug dosages and decreased treatment expenses. These benefits guaranteed the evolutionary leap that we believe science

Advanced Nanomaterials for Biological, Nutraceutical, and Medicinal Applications.
Sreeraj Gopi, Augustine Amalraj, and T. R. Sreeraj (Eds.)

will have with the assistance of nanotechnology. To further ascertain this advancement, constant efforts are being made to improve the properties of nanomaterials (NMs). In this chapter, we shall discuss in detail these properties that make NMs an extremely suitable tool and drug delivery system in modern therapeutics and theranostics.

2.1 INTRODUCTION

Nanotechnology is a field of developing technologies that enables the use of very small particles to explore materials in the macroworld at a smaller scale, with modifications that present us with useful characteristics, different from its macroworld counterparts. Nano-(dwarf, in Greek) refers to the measurement in the range of as miniscule as 10^{-9} units (Trotta & Mele, 2019). Materials that are 1 nm to 100 nm in size are referred to as nanomaterials (NMs), which are scaled down to over 10,000 times lesser than the diameter of human hair! (Figure 2.1).

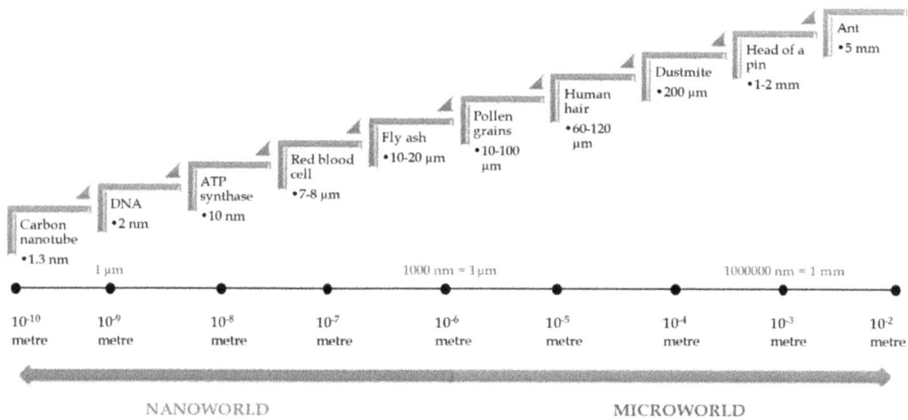

FIGURE 2.1 Size comparison of different-sized objects.

A material with one, two, or three external dimensions in nanoscale is referred to as a nano-object, while a nano-object with all three external dimensions in the nanoscale is called a nanoparticle (Trotta & Mele, 2019). These possess desirable properties like high mechanical strength, high current density and high thermal conductivity. At the nanoscale, properties such as color of the particles, optical dispersion and size-tuned light emissions, magnetic permeability, high quantum yield and even chemical reactivity may differ from that of its relative macromolecule (Grandhi & Rege, 2014; Matea et al., 2017; Raghav et al., 2017). It is believed that the drastic change in characteristic properties of the NMs from its macro-material varieties is owing to mainly three features, viz., the increased surface area to volume ratio, quantum confinement of electrons at the minute scale and the tendency of certain molecules to self-assemble into various shapes and structures with at the expense of least or no external forces (Grandhi & Rege, 2014). The properties of nanoparticles can be further changed by coating the surface. The changes in physiochemical properties of the nanoparticles can in turn affect its cytotoxic and pharmacokinetic properties (Gatoo et al., 2014).

The advent of nanotechnology has opened up a wide range of possibilities for novel and specialized applications, to boost performance parameters that may possibly give rise to highly efficient systems. Its impact has been realized on numerous applications in the field of medicine, pharmaceuticals, agriculture, cosmetics, food, paint, and dye, and even extended for molecular profiling studies and multiplexed biological assays (Rathod et al., 2011). In the field of genetics, it has been possible to replace normal fluorescent dye with AuNPs for better results in DNA sequencing (Nikalje, 2015). It has numerous practical applications in medicine and related fields. Figure 2.2 summarizes the utilization of nanotechnology in sectors such as pharmacy, medicine, and diagnostics. Reports suggest that over 100 patents were granted by the USPTO in a five-year span between 2001 and 2005 (Rathod et al., 2011), which suggests not just fast growth, but also competitiveness and newer techniques or nanoproducts that are being developed.

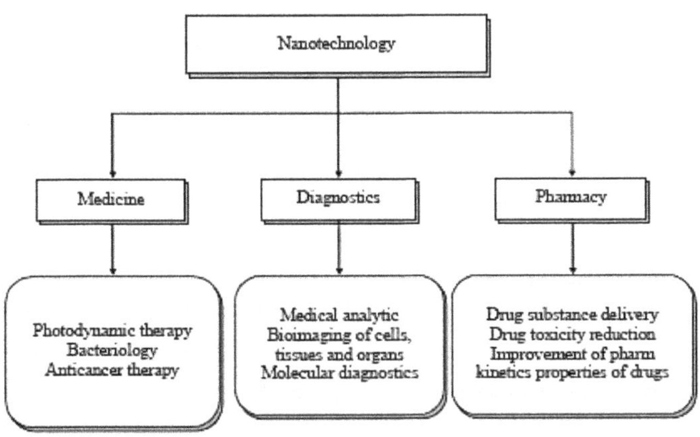

FIGURE 2.2 Nanotechnology in medicine and associated fields (Raghav et al., 2017).

Source: Reprinted from Raghav et al., 2017. https://creativecommons.org/licenses/by/4.0/

The growth of this field and advancements has been so tremendous and unfathomable in case of few fields some years back, like the use of "DNA origami" technique to target cancerous cells using nanorobots that function within the body (Li et al., 2018). Newer methods to ensure safety and secured usability of nanotechnology on living subjects are also being developed.

One of the fast-growing effects of nanotechnology has been witnessed in the field of therapeutics. Few core areas in medicine and biology where NMs have been used are enlisted by Salata (2004). These include fluorescent biological labels, bio-detection of pathogens, detection of protein, probing of DNA structure, drug delivery, tissue engineering (TE), tumor destruction, separation, and purification of biological molecules, imaging of cells and biomolecules, and phagokinetic studies. If we draw a comparison between this report and recent trends in therapeutics, it can be seen that for the same functions that have been enlisted, we have come very far in terms of advanced design constructs as well as modified properties of the NMs used. It is interesting to see how nanoparticles have aided in drug delivery and diagnostics, or even replaced the traditional technologies in medicinal and therapeutic areas. To have medicines or drugs directed at the absolute diseased site, without causing any side-effects, or affecting the specificity, sensitivity, efficiency or stability of the drug is a challenging task. These together determine the success of translation of novel therapeutic drugs from laboratory to clinics (Zhang et al., 2008; Grandhi & Rege, 2014). Nanotechnology-based therapeutic and diagnosis has been possible for a range of health issues like cancer, diabetes, asthma, allergy, infections, skeletal injuries, pains, etc. (Zhang et al., 2008). Rizvi & Saleh (2018) have listed antiretroviral drugs that have also been entrapped in nanoparticles and used against severe conditions like that of HIV.

We have introduced our readers to the wide range of applications of NMs, especially in therapeutics. Now the following sections will explain the various properties or characteristics of the different NMs or nanoproducts that have been attempted so far. The nine key properties that will be discussed are:

- Surface area: volume ratio;
- Optoelectronic properties;
- Electrical conductivity;
- Chemical properties;
- Water solubility;
- Magnetic properties;
- Mechanical strength;
- Antimicrobial properties;
- Biocompatibility, biodistribution, and bioavailability.

2.2 SURFACE AREA: VOLUME RATIO

In nanoparticles, as mentioned earlier, the surface area to volume ratio is significantly high. This allows a relatively higher binding capacity and leads to excellent dispersibility of these particles in solutions. Surface area: volume is a relative parameter and at times a term used to define NMs too. A particle that has the ratio of 60 m^2/cm^3 is considered as a nanoparticle, nanomaterial or a nanoproduct (Soarse et al., 2018).

Size tuning in nanoparticles has been achieved by various methods. The small sizes of the materials or carriers offer them greater mobility, thus allowing more interaction with the cell membranes (Seleci et al., 2016). Further, internalization of nanoparticles by the cell is dependent on the bending energies of the constructs during endocytosis. The rod, cube, and disk-shaped nanoparticles may need to overcome a greater deal of the membrane bending energy. Since spherical particles overcome this easily, they are internalized faster (Seleci et al., 2016). For engineered NMs, that might play a crucial role in cellular responses by penetration and entering circulation, size is a critical factor. Since different NMs may behave differently as in case of their oxidative mechanisms, etc., under *in-vivo* conditions, each of these must be studied separately. Shape of the nanomaterial affects cellular uptake, biocompatibility, and its retention at the target site (Jeevanandam et al., 2018).

A recent development based on thermal decomposition of iron carboxylate has helped in preparing better quality iron oxide particles with tuned sizes (particles that contain maghemite and magnetite, ranging in diameter from 1–100 nm) and crystalline structures. The large surface area of gold nanoparticles (AuNPs) has been useful in modifying the molecule by conjugating different biomolecules or coating it with kinds of varying drugs or enhancers, etc. The very small size also allows it to interact with the DNA in the nucleus, while if tuned to be a larger molecule, it loses the cytotoxicity effect, only penetration ability remains (Khan et al., 2019).

2.3 OPTOELECTRONIC PROPERTIES

The light that is incident on a nanoparticle is either absorbed or scattered, based on the particle diameter and shape. A particle having less than 20 nm diameter while absorbs the light, the particles with diameter over 100 nm scatters it. Gold particles, in particular have been found to scatter light with much higher intensity than chemical fluorophores (Fadeel & Garcia-Bennett, 2010). Gold nanorods are both good photo absorbers, as well as scattering contrast agents. While the former characteristic has been attributed to nanorods with smaller volume are higher aspect ratio, the latter has been showcased by nanorods that have high volumes as well as high aspect ratios (Kim et al., 2019). It has also been noted that aggregation of nanoparticles, which often leads to increased size of the particle, also causes light extinction by scattering. Quantum confinement effect is another way in which the light absorption or emission property can be changed in nanoparticles of semiconductors, that form quantum dots. The shift in this case can be made from UV to visible light, and even into IR region. Quantum dots of CdSe have been targeted for emission in the whole EMR spectrum ranging from 2 nm particles emitting blue radiation, while 8 nm or even higher emitting red light.

Selective reflection of a part of the spectrum is also possible using mono disperse, magnetically responsive nanoparticles. The best example is the use of silica particles in opal gemstones that display multi-colored iridescence type optical response, as displayed in Figure 2.3.

FIGURE 2.3 Film formation with optical iridescence using nanoparticles.

Source: Picture courtesy: nanocomposix.com.

The changes in color and properties of colloidal nanoparticles of some metals have been used and researched on for various purposes. One such metal nanoparticle is of gold, the interaction of which, with light is governed by the physical dimension of the molecule and the environment in which it is interacting. AuNPs are also not susceptible to photobleaching (Fadeel & Garcia-Bennett, 2010). Since ages the use of colored gold solution has been in use for numerous works like staining glass or in photography, etc. It wasn't however, until Michael Faraday's work that it gained importance in the scientific sector. In the most recent times, the AuNPs have been employed in the hi-tech applications such as sensory probes, as therapeutic agents or drug delivery system and for some medicinal applications too. It is the surface modification that these particles undergo that make them a suitable candidate for the wide applications. The modification is instead based on the principle of surface plasmon resonance (Fadeel & Garcia-Bennett, 2010). It is the oscillation of the conduction electrons that are in resonance with the visible light frequency, when a light ray passes into the colloid of nanoparticle. An interaction is caused between the electric field of the light ray, which is in oscillation, and the nanoparticles. The resonance causes absorption of a particular wavelength of light, while reflecting the opposite end of the spectrum. For instance, when a blue-green light is absorbed, the red light is reflected, thus imparting a red color to the solution. Likewise, when red light is absorbed and blue light is reflected, the solution appears to be blue in color. Surface plasmon resonance has been observed in gold, silver as well as aluminum nanoparticles.

The color of the colloid, in terms of the shades of red or blue, can further be varied by changing the extent of resonance which is indirectly governed by the parameters mentioned earlier. Property of aggregation is another parameter that might affect the resonance. In gold particles, for instance addition of other molecules may also result in a change of color, like excess of salt when added to the gold solution turns it to blue from red (Figure 2.4). This is because the particles aggregate as the surface charge starts to become neutral on the addition of salt. Aggregation of particles can further be minimized by coating the gold surface with small molecules or typically, polymers, and other biologically important molecules, as the application demands. Such additives and coatings have modified nanoparticles to help find larger avenue in medical and therapeutic applications and are still being researched on.

FIGURE 2.4 Colors of various sized monodispersed gold particles.

Source: Reproduced with permission from Merck KGaA, Darmastat, Germany.

AuNPs especially have been used as probes in transmission electron microscopes (TEMs), coated with therapeutic elements, such as peptides, proteins or antibodies to act as a potent targeted drug delivery system, or simply to function as appropriate cellular targets. These nanoparticles have been used as substrates in SERS for detection of proteins and pollutants, as biomarkers to detect infectious agents and in the diagnosis of cardiac illnesses or even cancers (Fadeel & Garcia-Bennett, 2010). Their properties have been utilized even in simple routine immunoassays, like that of home pregnancy test kit. In short, the optical "bar coding" feature by nanoparticles such as quantum dots and others, to a great extent replaced organic dyes and other labeling material in various bio-tagging assays and high throughput screening technologies involving sequence data (Salata, 2004; Matea et al., 2017).

Recently, quantum dots free of heavy metal and cadmium have also been used due to their biocompatibility and good photoluminescence quantum yield in lymph node mapping, in *ex-vivo* mouse imaging models (Matea et al., 2017). Hybrid silica nano-capsules, ZnO, Mn:ZnS, GdNS, Graphene quantum dots are all other recently used NMs in the treatment of cancer, all of which have been used for their properties of fluorescence coupled with theranostic usage (Matea et al., 2017; Raghav et al., 2017). Another promising theranostic agent is the carbon nanotube, which can incorporate inorganic materials, and can also be used in selective uptake by the cells (Menon et al., 2013). A multifunctional theranostic agent has been depicted in Figure 2.5.

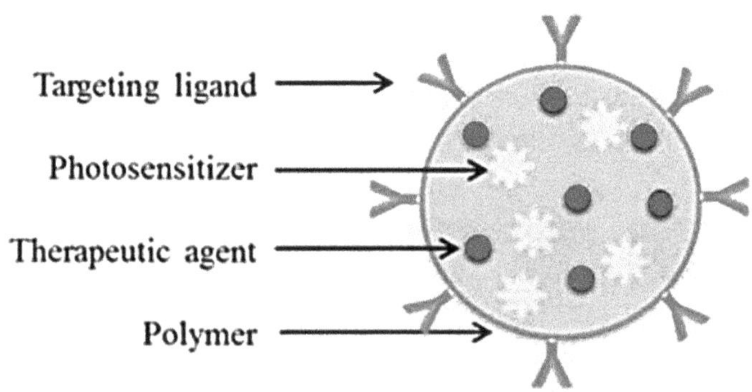

FIGURE 2.5 Theranostic nanoparticle that has multifunctional activities (Menon et al., 2013). In the image, targeting ligand could be folate or hyaluronic acid derivative, photosensitizer could be gold or phthalocyanines, therapeutic agent could be paclitaxel or doxorubicin, and polymer could be PEG, chitosan, or dextran.

Source: Reprinted from Menon et al., 2013. Copyright © Ivyspring International Publisher. http://creativecommons.org/licenses/by-nc-nd/3.0/

In a nutshell, photo-based nanoparticles ranging from metal-based to polymer-based, have been used in the delivering fluorescent dyes and photosensitizers for photoimaging and therapeutic applications. It is their ability to deliver high resolution images and the minimal-invasive, non-contact model of functioning, that nanoparticles have gained significance in photo-based diagnosis and treatment methods (Menon et al., 2013).

2.4 ELECTRIC CONDUCTIVITY

Usually, electrical properties of NMs depend upon the diameter of the materials and may render features between metallic to semiconductors. The tight packing of materials often in NMs and the scarce defects in crystals offer it extremely high electrical conductivity as well as enhanced thermal conductivity due to the heavy vibration of covalent bonds. The thermal conductivity of such extremely solid NMs with extreme strain is believed to be at least 10 times higher than metals.

Carbon nanotubes (CNTs) are NMs that conduct electricity with very little resistance and thus have been widely used in biomedical applications of building devices and sensors for medical imaging. CNTs are the allotropic crystalline form of carbon that are available either as single-walled nanotubes or multi-walled nanotubes, with electrical characteristic ranging from conducting to semi-conducting to insulating. For its electric conductivity and electrochemical stability, CNTs have been utilized in the preparation of electrochemical glucose. The method was proven to have fast response and less detection limit with enhanced stability and sensitivity (Raghav et al., 2017). Nanomix is a technology developed on the basis of, and that relies on the electrical properties of a random network of CNTs. It has been used to detect target molecules in gas or liquid (Harris & Graffagnini, 2007).

2.5 CHEMICAL PROPERTIES

An important job of the researchers working in nanobiotechnology is to understand clearly the physicochemical properties of NMs synthesized. This is crucial in evaluating the toxicological and ecological hazards. The challenge sets in when the nanoparticles are to be especially used a carrier of biomolecules and needs to be functional *in-vivo*. Under such circumstances, these particles often physically alter their conformation to suit the environment that they are in. The changes in physical arrangements are in turn affect the stability and chemical nature of the particles too (Kumar & Dixit, 2017). Chemical properties that are important in characterizing materials are composition and structure of the molecule, molecular weight, boiling, and melting points, vapor pressure, octanol-water partition coefficient, water-solubility, reactivity, and stability. The significance of NMs arises from their specific chemical properties, such as decreased melting points, increased heat capacity, decreased Debye temperature, decreased Curie temperature, increased conductivity, and improved catalytic activity (Trotta & Mele, 2019).

In his prediction of the thermodynamic model, Pawlow had proven that the surface area: volume ratio differences in the NMs to its bulk counterparts is what affects the thermodynamic and thermal properties, like melting temperature (Figure 2.6). Besides, the larger surface area and improved reactivity also aid in creating better catalysts of these NMs (https://www.nano.gov). In case of metal nanoparticles, the catalytic activity is dependent on the structure, shape, size distribution, as well as the environment (El-Nour et al., 2010). In addition, when considering a formulation used for say, pharmaceutical or therapeutic purpose, it is important to understand its product variability, purity, performance, and extent of application.

FIGURE 2.6 Effect on surface and melting properties with change in dimensions.

Source: Reprinted from https://www.nano.gov/about-nanotechnology/what-is-so-special-about-nano. https://creativecommons.org/licenses/by/3.0/

The surface of nanoparticles is charged and enables them to attract a thin layer (Stern layer) of counter ions to these particles. The double layer of ion passes with the molecules and disperses in the solution. The magnitude of zeta potential created at the double layer is what determines the colloidal stability of the solution. Nanoparticles that have > +30 mV and < –30 mV have high degrees of stability (Seleci et al., 2016; Kumar & Dixit, 2017). The ones that have zeta potential values of > +25 mV and < –25 mV forms agglomerate (Kumar & Dixit, 2017).

Nanoparticles can be functionalized by the addition of some organic functional groups, which confers reactivity to the nanostructure. When such groups are added, it is important to check their efficiency in performing the function well within the modified limits of the nano-molecule. Also, the amounts of functional groups added and their binding strength to the nanoparticle should also be quantified (Fadeel & Garcia-Bennett, 2010). Although a costly affair, the AuNPs perform well as probes in biosensing and imaging and are sensitive, giving reliable results, due to their ability to bond with alkenes (Khan et al., 2019).

Organic NMs are governed by hydrophobic effect, electrostatic attraction and reduction in interfacial tension in surfactants stabilized water-in-oil system (Grandhi & Rege, 2014). Hybrid organic-non organic NMs have been devised like for example functionalizing AuNPs with attachment of probe molecules like antibodies, enzymes or nucleotides. These are best suited for active functioning in biosensor assays, drug, and gene delivery systems, advanced microscopic tools and other biomaterial-based imaging systems (Fadeel & Garcia-Bennett, 2010).

Nanocomposites are such molecules that are made of bulk materials incorporated with NMs and having dimensions larger than 100 nm. The presence of NMs can regulate the properties of the bulk, like nanoparticles of reducing agents dispersed in hydrogel enhances dye reduction capacity or organic dyes when compared to other molecular systems. Polymers like PVDF can also be used to disperse NMs. Titanium dioxide (TiO_2) particles when embedded in the PVDF polymer improves its surface hydrophobicity and permeability, along with decreased absorption of proteins. This is advantageous as low protein binding to the membrane would nearly indicate the anti-fouling nature of such membranes and can be used in various detection and diagnosis purposes (Trotta & Mele, 2019).

TiO_2 has been much researched as a material of interest in the creation of nanoparticles due to its photocatalytic and dielectric properties, off course, optical features too. These may be either of the three forms: rutile, anatase, and brookite, each differing in the crystal pattern and related photoelectrochemical behavior and applications. Anatase, for example, have lower surface energy compared to the other two forms, which makes it the most toxic among the three. While the dimension of the particle is one of the major parameters governing toxicity of the nanoparticles generally, in Titania nanoparticles, cytotoxicity depends on its phase composition. In case of anatase, a wider absorption gap and smaller electron effective mass could possibly lead to increased mobility of charged carriers which in order enhances generation of reactive oxygen species (ROS). Thus, Titania nanoparticles can lead to killing of the cells using photocatalytic chemistry (Kwatra et al., 2013). Titania nanoparticles have a high photooxidation efficiency and are photoexcited by UV radiations, for which they have been considered in photodynamic therapy (PDT) (Fadeel & Garcia-Bennett, 2010; Kwatra et al., 2013). Moreover, the rutile crystal structure of TiO_2 have been found to cause lipid peroxidation and micronuclei formation in the absence of light, which anatase of same size and composition cannot bring about (Gatoo et al., 2014). Other NMs which have the crystal structure are iron oxide, silicon dioxide, and carbon.

Similar to anatase, AuNPs also display the property of high production of ROS when comes in contact with organisms, say bacteria and can kill them. This is a marked difference from gold counterpart because gold is known to be chemically inactive. The reactivity offers AuNPs antimicrobial property, which is by virtue of its ultra-small size and shape, but not the chemical composition (Trotta & Mele, 2019).

Graphene oxide, a cationic polymer can be electrostatically bonded to positively charged polyethyleneimine, the construct which can be used to transfer gene or plasmid DNA as a part of gene therapy. The charges play immensely in this case and generate such attraction that the DNA easily condenses onto the surface of the graphene oxide nanoparticle (Raghav et al., 2017).

Rare earth metals and high-Z elements like hafnium oxide, which has the photo-luminescent property has also been used in designing nano-carrier vehicles for the various molecules. These are good radiosensitizing nanoparticles which can damage the cellular components, for instance when tumors are to be destroyed, by thermal induced stress (Kwatra et al., 2013).

Mesoporous silica nanoparticles (MSNs) can be described as a structure of organic micellar template supported by an amorphous silica wall. The porous nature of mesoporous silica with two- and three-dimensional sharply distributed pores and cages provide high surface area. Further, the interior of these structures are alkoxysilanes, which provide the functionality to the nanoparticle by allowing to manipulate the surface properties (Fadeel & Garcia-Bennett, 2010). In porous NMs, it has been observed that the smaller the pore size and with the decrease in pore diameter, melting temperature decreases (Trotta & Mele, 2019).

One of the recent interests of researchers are also the modifications of nanoparticles to be able to use it as best means of medication for diseases such as Alzheimer's and Parkinson's, where antioxidants can be delivered to control the oxidative stress as well as nitrative stress accompanying these conditions (Sadowska-Bartosz & Bartosz, 2018). Redox-active polymeric nanoparticles have been persuasive from a therapeutic perspective even for delivery for tumor-targeted drugs (Cho et al., 2012; Kumar et al., 2014). Polyethylene glycol (PEG)-modified polylactic acid (PLA) diblock copolymer, PEG-Polypropylene glycol-PEG-modified PLA tetrablock copolymer, polygycerolsebacate-co-PEG (PGS-co-PEG) self-assembly are all good examples of modified polymeric nanoparticles that have proven effect as intracellular delivery systems (Kumar et al., 2014; Desai et al., 2018). Increasing chain length of PEG have shown to enhance peptide (anti-cancer NuBCP-9, which converts protector cell to killer cell and carries out apoptosis) encapsulation by the copolymers (Kumar et al., 2014). Attempts of successful release of paclitaxel at the tumor site have been conducted with the use of redox-sensitive polymeric nanoparticles (Cho et al., 2012). PGS-co-PEG nanoparticles are biocompatible molecules and find use in immune modulation, cancer therapy, stem cell therapy and regenerative medicine (Desai et al., 2018). Chitosan (CS) is another biodegradable and compatible molecule with positive surface charges. The nanoparticles using CS to deliver Paclitaxel have proven more efficient and shown improvement in the intestinal pharmacokinetic profile, when the hydrophobic long acyl chains were grafted to the original CS molecule (Lomabado et al., 2019). The enhancement of their chemical properties has aided all of the above discussed nanoparticles to be significant in tumor targeting, imaging, and theranostics.

2.6 WATER SOLUBILITY

Since bioavailability is a critical parameter in drug delivery mechanisms, it is necessary to avoid issues arising due to particle's low solubility. This is mainly where nanoparticles are useful. The nanodrugs often have better or high dissolution capacity and demands only smaller drug doses (Shubhika, 2013). For efficient bio-accessibility of nanoparticles, it is in the best interest that the water solubility of these be further improved. One of the strategies involving monolayer polymer coating has worked well on organic ligand-coated iron oxide nanoparticles for this purpose. High stability of these particles impart tolerance to conditions unusual such as high pH and temperatures, thus making them available to be used in conjugation with any other biomolecules too. In case of iron oxide nanoparticles, they have been coated with a variety of biomolecules, for different activities including, but not limited to use as drug carriers for target specific drug delivery, gene carriers for gene therapy, therapeutic agents for hyperthermia-based cancer treatments, nano-adjuvant for vaccine and antibody production.

Forms of titanium oxides have been discussed in the section above. One of the forms, the rutile is known to be poorly soluble in water. So, when used with an aqueous medium, these particles may form agglomerates and sediment down. This will leave the cells or whatever sample is being used, to be in contact with the nanoparticle only for a very short time (www.nanopartikel.info).

In most cases, agglomeration is not a desirous property. To counteract this, mechanisms like electrostatic repulsion and steric stabilization are often attempted by addition of stabilizers into the medium or colloidal suspensions, along with nanoparticles (El-Nour et al., 2010; Kumar & Dixit, 2017). Commonly used stabilizers include ionic stabilizers such as sodium dodecyl sulfate (SDS), sodium lauryl sulfate, lecithin, and docusate sodium; as well as, non-ionic stabilizers like Tween 80, PEG, polyvinyl alcohol (PVA), polyvinyl pyrrolidone, hydroxypropyl cellulose and hydroxypropyl methyl cellulose (Kumar & Dixit, 2017). Another method to tackle the sedimentation issue due to agglomerate formation is by freeze drying the nanosuspension or by transforming it to dry and solid form by some apt means (Kumar & Dixit, 2017).

In the case of nutraceuticals that are usually water-insoluble but are good health supplements, the bioavailability and efficacy are augmented by the use of nanoparticles. Liposomes and dendrimers have been used in better solubility of compounds that are fat containing or lipid-based, such as polyunsaturated fatty acids, vitamins A, D, E, and K, as well as phytochemicals. Examples are curcumin and Resveratrol (Rizvi & Saleh, 2018), which has been explained under liposomes in a later section.

2.7 MAGNETIC PROPERTIES

The very minute nanoparticles, as small as 2–20 nm may display super paramagnetism, meaning their stability is enhanced in the presence of an external magnetic source, which allows it to get magnetized. The best use of paramagnetization has been noticed in iron oxide nanoparticles. These have been increasingly used in biomedical application based on its super-paramagnetism thereby stability; biocompatibility and non-toxicity; and are referred to as SPIONs, meaning super-paramagnetic iron oxide nanoparticles (Hajipour et al., 2012; Raghav et al., 2017). They have been relied on for imaging in hi-tech analytical techniques like magnetic resonance imaging (MRI), for their ability to deliver specific and sensitive results. To increase the tumor detection capacity by better penetration and retention brought about by changes in charge or other surface properties, SPIONs are often modified by coating with biologically active biomolecules (Raghav et al., 2017). The modification of SPIONs allows them to bind to large group of molecules such as proteins, peptides, antibodies, and anticancer drugs (Lombardo et al., 2019).

Magnetoreception is a phenomenon that utilizes magnet nanoparticles labeled or coated with therapeutic genes or other active biomolecules, to be specifically targeted towards an affected cell, tissue or organ, under the influence of a gradient external magnetic field (Raghav et al., 2017). Virus like adenovirus, has also been explored as a component biomolecule in the composite construct. The release of biomolecules at the target site occurs either by hydrolysis or enzymatic degradation. Even in TE, nanogels have been applied that are magnet stimulus responsive. These have been put to use for controlled release of osteogenic growth factors. Magneto-responsiveness is achieved in the nanogels by encapsulating it with iron oxide particles that are super-paramagnetic in nature. These nanogels have the advantage of high loading efficiency and protection of growth factors or other biomolecules to be delivered from proteolytic degradation (Casanellas et al., 2018). Nanogels can withstand high temperatures too, like in radiation and imaging, and can be used in photo-therapy and chemotherapy (Menon et al., 2013). Magnetic nanoparticles (MNPs) have also been tried in the processes of isolation and grouping stem cells (Nikalje, 2015). More on TE, stem cells and controlled release of drugs will be explained in later sections.

Other metal particles have also been tried to bring about magnetoreception. Addition of porphyrins and thiols to gold and nickel particles have been attempted for a very long time, to make nanowires of high residual magnetization, and that are labeled. Weak magnetic field could be used further to control the self-assembly of nanowires too, when in solution (Salata, 2004).

MNPs that are put into application as anti-tumor drug carrying particles or any other therapeutic function, are usually kept below a certain threshold. The effect of these particles are defined by their specific absorption rate, also known as specific loss power, and are dependent on the heating efficiency of the particle owing to its magnetic, physical, and hydrodynamic properties (Bogart et al., 2020).

2.8 MECHANICAL STRENGTH

One of the major medical aspect where the use of nanotechnology is most felt, especially recently, is in the fabrication of tissue scaffolds. With the many limitations of organ and tissue transplants, the modern therapeutic route has been in discovering materials that mimic the natural extracellular matrix (ECM) composition of a tissue. The construction of scaffold for these have been possible with the use of nanoparticles only due to their appropriate mechanical strength (Fathi-Achachelouei et al., 2019). Nano-materials play an important role as additives to the materials that are generally used for these purposes (check the section on Biocompatibility), to make nanocomposites. This is done so as to improve the mechanical stability of the material being used in the scaffold and also enhance biological activity (van Rijt & Habibovic, 2017).

Mechanically strong transparent or translucent nanocomposites have been designed by embedding silicon and/or zirconium oxide particles in a photopolymerizable resin matrix. These have found use in dental restoration. Other application of nanoparticles in dentistry involves employing hydroxyapatite (HAp) nanoparticles and calcium phosphate in bone and tooth filling (Scaletti, 2016).

2.9 ANTIMICROBIAL PROPERTY

Metal nanoparticles such as silver colloids have anti-bacterial properties, which are dependent on the size. The smaller the nanoparticle size, the higher is the anti-bacterial activity (El-Nour et al., 2010). Silver nanoparticles (AgNPs) have been used for the disinfection of medical devices (MDs) due to its anti-bacterial, antiviral, and antifungal properties (El-Nour et al., 2010; Roy, 2017). Silver and zinc nanoparticles can produce free radicals, which could also induce oxidative stress by the formation of ROS, leading to the destruction of the bacterial cell, as shown in Figure 2.7 (Hajipour et al., 2012). Commonly used silver forms or colloids are silver nitrates, silver oxides and silver sulfadiazine (Lara et al., 2010; Roy, 2017).

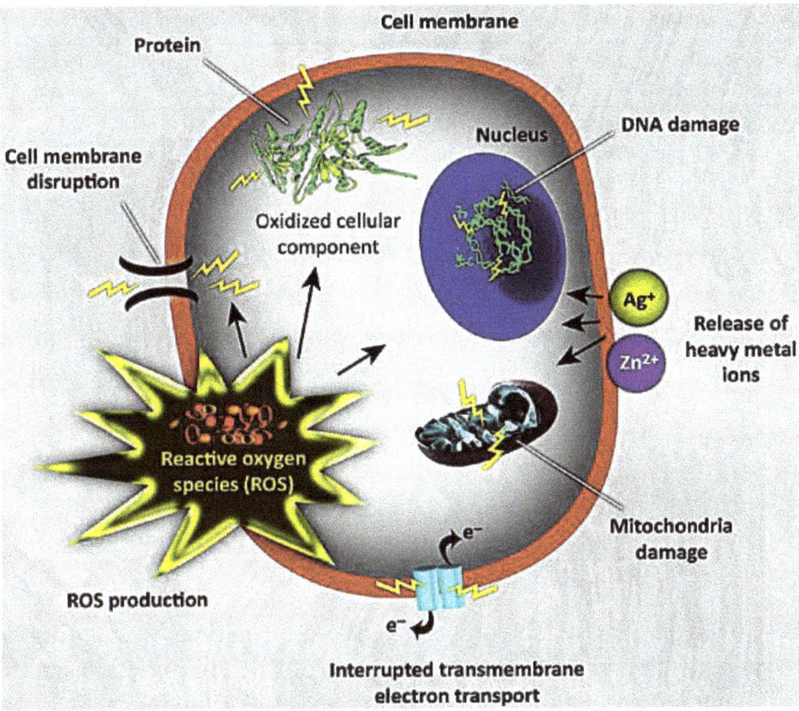

FIGURE 2.7 Mechanism of toxicity of metal nanoparticles against bacteria (Hajipour et al., 2012).

Source: Reprinted with permission from Hajipour et al. 2012. Copyright © 2012 Elsevier Ltd.

In a review by Roy (2017), the various micro-organisms against which AgNP is observed to be effective, has been mentioned. Table 2.1 shows a compilation of the activity of AgNP against these organisms. Silver ions work by affecting the membrane permeability of the organism, like bacteria, blocking the microbial respiratory chain system, and regulating organism to precipitate out its cellular proteins (Fathi-Achachelouei et al., 2019).

The mode of action of AgNP against HIV has not been fully understood, but a few strategies that could possibly result in its efficient functioning could be inhibiting the binding of the virus to the cell. This can be brought about by the binding of positively charged silver ions in the nanoparticles to the sulfhydryl groups of the virus with which it attacks the cells. This interaction should lead to reduction of disulfide bonds, which causes the viral protein denaturation. The AgNP is also capable of suppressing the expression of the cytokine TNF-α, which plays a vital role in HIV pathogenesis (Lara et al., 2010).

TABLE 2.1 List of Micro-Organisms Against which Silver Nanoparticles have been Used

Nature of Anti-Microbial Activity	Effective Against	References
Antibacterial	*Escherichia coli, Staphylococcus aureus, Staphylococcus epidermis, Streptococcus pyrogens, Salmonella typhi, Klebsiella pneumoniae, Bacillus megaterium, Proteus vulgaris, Shigella sonnei*	Roy (2017)
Antifungal	*Candida albicans, Trichophyton mentagrophytes, Phoma herbarum, Phoma glomerate, Fusarium semi-tectum, Trichoderma species, Alternaria alternata, Macrophomina phaseolina, Sclerotinia sclerotiorum, Botrytis cinerea, Rhizoctonia solani, Curvularia lunata, Aspergillus fumigates, Penicillium brevicompactum, Chaemotomium globosum, Cladosporium cladospooides, Stachybotrys chartarum, Mortierella alpine*	Roy (2017)
Antiviral	HIV, Hepatitis B virus, Respiratory syncytial virus, Herplex simplex virus type I, Monkeypox virus	Lara et al. (2010)

AuNPs have been used in composite colloid formations with antibiotics in order to enhance their action against micro-organisms, like bacteria and fungi, most of them that are pathogenic. A greater amount of literature is also available on the effect of nanoparticle colloids on biofilm former *Pseudomonas*, enteropathogenic strains of *Enterococcus* and *Escherichia*, as well as fungal strains of *Aspergillus niger* and *Aspergillus fumigates*. Vancomycin is one of the most effective antibiotics against entero-pathogens. Antibiotics against bacteria have especially shown tremendous improvement in their action against *Escherichia coli* and *Staphylococcus aureus* (Raghav et al., 2017).

Graphene and its various oxide derivatives are NMs that have been used for numerous purposes. They have also proven to be useful in the preparation of antibacterial sheets, which are nanowalls that can be coated on stainless steel surfaces, to be functioning as antibacterial agents (Raghav et al., 2017).

2.10 BIOCOMPATIBILITY, BIODISTRIBUTION, AND BIOAVAILABILITY

Let's first learn these terms for better understanding:

- Biocompatibility can be defined as the property of an external material introduced into a host system to not have a toxic or injurious effect on the body by eliciting appropriate biological host responses.

- Biodistribution can be defined as an *in-vivo* method of testing, tracking or detecting the position of a compound of interest inserted or injected into a subject's body to travel within so as to meet the therapeutic or diagnostic function.
- Bioavailability can be defined as the active effect demonstrated by a drug or other biomolecule that had been introduced into the subject's body, and has been absorbed into systemic circulation.

Although it is the properties discussed above that offer the nanoparticles its biocompatible nature, in this section, we will be considering the achievements so far in nanotechnology, due to its compatibility. Few therapeutic benefits that gives nanoparticles an upper hand over its macro-counterparts is its enhanced molecular stability, solubility, diffusivity, cell infiltration and permeability, retention effect, multifunctional capability, biodistribution, nutrient, and waste exchange when required, release characteristics, reduced toxicity and immunogenicity (Valavanidis & Vlachogianni, 2016, van Rijt & Habibovic, 2017; Casanellas et al., 2018). Stability of the nanoparticles and reduction in its toxicity can also be brought about by employing hydrophilic and flexible PEG, or other surfactant copolymers like polyethylene and poloxamers, to surface coat the nanoparticles (Gatoo et al., 2014).

Delivery by NMs promise specificity to the target and precision in terms of administering doses, where, small doses also prove to have far reaching effects than when using non-nano-delivery drugs moieties (Rathod et al., 2011). The targeted- or stimuli-response delivery potential, along with low toxicity and flexible characteristics has also facilitated the use of nanoparticles in tissue engineering and regenerative medicine (TERM). During TE, it is important to monitor and control properties of the scaffold. Also, since healing and repair takes time, equally significant is the delivery of a bioactive agent like genes, cytokines, chemokines, growth factors or inhibitors, etc. (Fathi-Achachelouei et al., 2019; van Rijt & Habibovic, 2017). A schematic of such a process is represented in Figure 2.8 highlighting the use of NMs in bone regeneration. Recent innovations also include the use of bio-nanoceramics that are re-absorbable in nature and, if made of glass, can be used in colloid formation that has dual appearance of injectable in molten state and solidifies like gel when cooled. This has been applied in healing wounds lately (Fathi-Achachelouei et al., 2019).

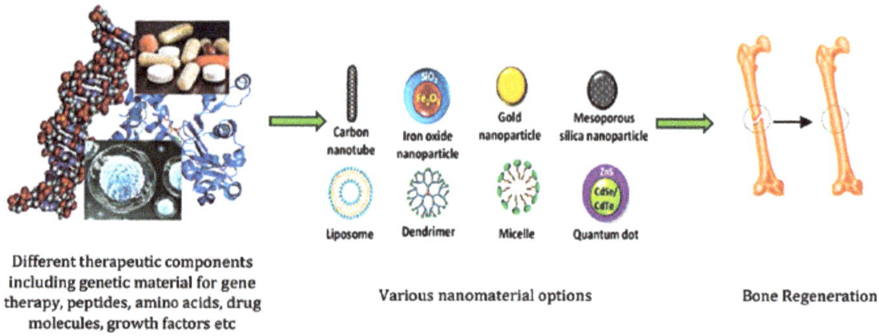

FIGURE 2.8 Nanoparticles as carriers in bone regeneration (in the image are the different small molecules and therapeutic components that can be loaded on to the various nanomaterials available for its release into the site of medical improvement or correction, such as bone regeneration in this case. The release mode can be sustained, controlled or dual).

Recently, long fibers of NMs have been designed by the method of electrospinning, which results in high porosity mesh and improved mechanical strain for regeneration of bones. The application of electrospinning for this purpose has been successful in cases of gelatin/polylactic acid, polycaprolactone/gelatin solutions as well as coral microparticles (Casanellas et al., 2018). The use of scaffolds (Figure 2.9) and NMs either natural or synthetically designed that can replicate the behavior of a normal body part or organ is referred to as Biomimicry.

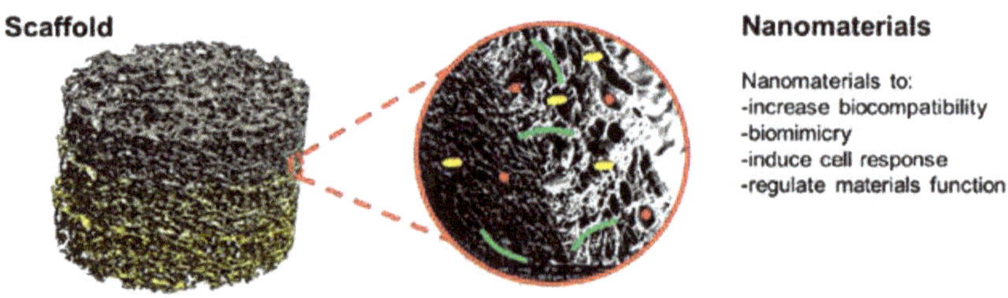

FIGURE 2.9 Scaffolds used in musculoskeletal regeneration (Casanellas et al., 2018).

Source: Reprinted from Casanellas et al., 2018. https://creativecommons.org/licenses/by/4.0/

Contrast agents for imaging purposes during TE and further to monitor the progress, is another requisite in TERM. Tissue regeneration is usually associated with musculoskeletal injuries and usually materials needed for this have to be such that they meet the chemical and mechanical requirements of the native tissue environment. This is the only way to ensure firm implantation with high reabsorption rates (Casanellas et al., 2018). All of these can be met with nanoparticles made out of polymers, metals, and/or ceramics, and their different composites (Rathod et al., 2010; Fathi-Achachelouei et al., 2019). Nanoparticles when coupled with these or other biomolecules aid in their ability to adhere to the target surface, say the cell surface; and also bring about differentiation and integration, and tracking of stem cells (Figure 2.10) into surrounding environment (Nikalje, 2015; van Rijt & Habibovic, 2017). Other naturally available nanostructure that can be utilized in TERM is bacterial nanocellulose (BNC). BNC has been worked on since the past few years for the preparation of artificial body parts and essential internal components such as skin, blood vessels, cornea, bone, cartilage, heart valve, etc. They have been alternatively used successfully in delivering proteins and drugs (Mohanty et al., 2016).

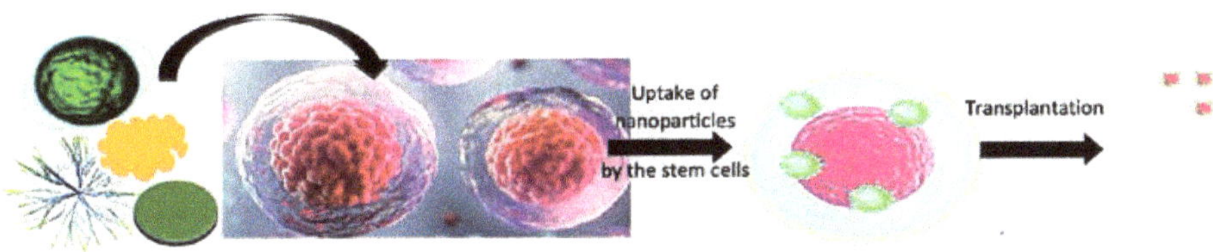

FIGURE 2.10 Nanoparticles being used in stem cell tracking and therapy (labeled cells after internalization of the therapeutic nanoparticles are transplanted into the host body, where the stem cell and particle movement can be tracked using various analytical techniques).

Once it enters the body, the nanoparticles are capable of movement in and out of the blood vessels. They also have the ability to interact with biomolecules within the cell and on the surface. They also display deep tissue penetration abilities. Such molecules also help improve the pharmacokinetic and pharmacodynamic properties of the drug they carry (Shubhika, 2013). With advantages like improved transport efficiency and the capacity to offer pharmacokinetics profiles, NMs may be explored in the healthcare and pharmaceutical sector to not just detect diseases, but also to provide treatment and offer preventive care. Some of the NMs that are being used in the healthcare sector have been listed in Table 2.2.

The responsive devices and drug delivery system that has been utilized in detection, capture, isolation, and treatment of undesirable biologicals by the nanoparticles by specific attraction and release of therapeutics is termed as "intelligent therapeutics" (Rathod et al., 2011). The therapeutic effect is created by either degradation of the particle or by induction of heat and/or

TABLE 2.2 Nanoparticles Used in Therapeutics and Healthcare Sector

Type of Nanomaterial	Applications in Healthcare
Metallic particles (iron oxide, gold, or silver)	Hyperthermia cancer treatment, selective magnetic bioseparation, membrane transport studies, drug delivery, magnetic resonance imaging contrast agent.
Silver nanoparticles	Anti-microbial agent, Incorporated into a wide range of medical devices, including bone cement, surgical instruments, surgical masks.
Gold shell nanoparticles	Improve solubility of drugs, can be modified further with conjugation with other molecules.
Carbon nanomaterials [fullerenes and carbon nanotubes (CNTs)]	Buckyballs are used in drug delivery systems to support the optimal transport and release of medicines to the right target inside the body, coatings for prosthetics and surgical implants, functionalized CNTs can be used for drug delivery, gene therapy, and biomedical applications.
Quantum dots	Tag multiple biomolecules to monitor complex cellular changes and events associated with diseases, optics technology, disease diagnosis and screening technologies.
Dendrimers	Polymerized macromolecules that are used as carriers for anti-cancer, anti-viral, anti-bacterial drugs with capacity to improve solubility and bioavailability.
Lipid-based nanoparticles	Can fuse with the cell membrane and deliver molecules directly inside the cells.
Ceramic nanoparticles	Inorganic systems used as drug vehicles (if porous and biocompatible); used in cosmetic applications (zinc oxide, titanium dioxide).
Nanotubes, nanowires, magnetic nanoparticles	Disease diagnosis and screening technologies.

Source: Adapted from Lauterwasser (2005); Jadoun et al. (2020); and Mody et al. (2010).

light (Shubhika, 2013). An ideal nanoparticle should be able to reach, recognize, bind, break through the larger epithelial junctions of the damaged or inflamed cell/tissue and finally, deliver the load to the target site. The release of drug from the nanoparticle is mostly based on the type of nanoparticle used, among other parameters such as physical conditions of the nanoparticle solution/formulation, binding capacity of the drug and drug diffusion through the nanoparticle matrix (Rizvi & Saleh, 2018).

2.10.1 LIPOSOMES AND NIOSOMES

These are 50–200 nm phospholipid vesicles that offer good entrapment efficiency of drugs and are biocompatible (Nikalje, 2015). Liposomes are non-toxic, biodegradable, and non-immunogenic, have self-assembling capacity, can carry multiple active agents, and can be used to deliver both hydrophilic and hydrophobic biomolecules due to the aqueous interior and lipid exterior (Zhang et al., 2008; Hua & Wu, 2013). These vesicles like structures are, therefore, often referred to as polymerosomes (Lombardo et al., 2019). Due to the amphiphilic property, they have been used in active as well as passive delivery of gene, protein, peptides, and other biomolecules (Rathod et al., 2011). The liposome delivery system is also such that once the function is met, it is degraded and cleared from circulation (Hua & Wu, 2013; Robson et al., 2018). The best functioning of liposomes *in-vitro* and *in-vivo* is however determined and guaranteed by homogeneity across a set of characteristics, including its drug encapsulation efficiency, ligand-conjugation, lipid composition, surface charge, and other structural properties (Robson et al., 2018). Usually, high drug/lipid ratio in the liposomes can be brought about by maintaining a lower pH inside the nanoparticle as compared to outside. This will allow easy passage of the drugs into the liposome but will prevent its movement back into the solution (Grandhi & Rege, 2014). It is also the responsibility of the carrier liposome to take the therapeutic biomolecule at the target site and not be destructed in the way by the extremely low pH of lysosome. Direct delivery of the biomolecule or drug is often possible by fusogenic liposomes that can bind to the lipid bilayer of cells (Grandhi & Rege, 2014).

Liposomes can be used even for controlled or sustained release of drug. The half-life of the drug in circulation can be increased and systemic toxicity can be alleviated, all while having reduced volume of distribution. Moreover, till the time that drug is within the liposome, it is not degraded, inactivated or diluted by the external contents, say blood or other body fluids (Zhang et al., 2008; Hua & Wu, 2013). Commercially available liposome formulations are Doxil® and Myocet®, both anticancer therapeutics, as well as AmBisome®, which is an antifungal agent (Hua & Wu, 2013). The combination of Doxil, which was doxorubicin HCl with liposome injection, and Abraxane which was paclitaxel protein bound, was the first-generation anti-cancer commercialized drug approved by FDA in 1995, for the treatment of AIDS associated with Kaposi's sarcoma (Zhang et al., 2008; Nikalje, 2015; Valavanidis & Vlachogianni, 2016). It is the success of this drug that actually paved the way for nanotechnology into medicine to improve the clinical upshots in cancer therapy as well as cure for other diseased conditions. Recently, cationic liposomes have also been designed to target the surface receptors that are overexpressed in some solid tumors like that in lung cancer. These contained small interfering RNA (siRNA) developed to target the surface receptor by conjugation of a MAL-thiolated antibody from the preformed liposome chain (Lombardo et al., 2019).

Liposomes have also been used to deliver pain relieving drugs or agents, including treatments for rheumatoid and adjuvant-induced arthritis, anti-inflammatory drugs and anesthetics, few of which have been listed in Table 2.3.

Liposomes have been utilized in the transport of a plant-derived component called Resveratrol, which is anti-cancer, anti-oxidant, anti-inflammatory in nature. Its formulation can be used even for cardio-protection. However, its use as a therapeutic agent was for long limited due to the fast elimination from the body. With liposomes and certain other nanoparticles, Resveratrol could be attached and its life span increased in body circulation (Rizvi & Saleh, 2018). Liposome delivery of other chemotherapeutic drugs have also demonstrated reduced systemic toxicity, while having longer life in circulation.

Liposome is the most employed and preferred choice of nanodrug delivery for brain. The tightness of the blood-brain barrier (BBB) by tight junctions guards the passage of molecules, by reduced permeation of solutes, thus preventing many attempts of pharmacological therapies. Natural polymer or protein-based polymers have been used in the construction of nanostructures with liposomes which can effortlessly cross this barrier (Garbayo et al., 2014). Most used biodegradable synthetic polymers include that of poly-lactic acid (PLA), polyglycolic acid (PGA), polylactic-co-glycolide (PLGA) copolymers (Lombardo et al., 2019).

While liposomes are the most used nano-constructs in therapeutics, it has two major limitations:

- Fast clearance from the blood by phagocytic cells; and
- Cost.

These issues can however be overcome by adopting some strategies. Long-circulating liposomes can be designed by coating its surface with inert polymers like PEG, which are also biocompatible (Grandhi & Rege, 2014). Other alternatives that are recently being considered for their antifouling properties and ability to enhance biocompatibility are carboxy- and sulfo-betaines. Both PEG and betaines are soluble in water and function well as surface ligands by resisting absorption by adventitious proteins (Bogart et al., 2020). In this manner the liposome is protected from opsonin recognition and attack. Alternately, liposome attached with targeted ligands can be allowed to accumulate in the cells/tissues or organs (Zhang et al., 2008).

TABLE 2.3 Liposomes as Carriers for Pain Management and Their Benefits

Biomolecule or Drug Type	Glucocorticoids	NSAIDs	Opioids (CNS)	Opioids (PNS)	Local Anesthetics	References
Liposomes used	PEGylated nanoliposomes	Hyaluronate gel	Polysorbate 80-coated nanoparticles	ICAM-1 targeted nanoliposomes	Uni/multilamellar liposomes	Dong et al. (2013); Avnir et al. (2008); Tosi et al. (2007); Hua & Wu (2013); de Paula et al. (2012)
Advantages of using nanoparticles	Decrease required dosage	Decrease systemic toxicity	Enhance penetration across blood-brain barrier	Sustained analgesic effect at inflammatory site	Increase duration of action	Ding et al. (2006); Hua et al. (2010); de Paula et al. (2012); Hua & Wu (2013)
	Improve therapeutic benefit	Improve therapeutic benefit	Improve therapeutic benefit	Decrease systemic toxicity	Decrease required dosage	

Liposome treatment might be costly, although extremely significant and competent. As a result, these are of late being replaced by non-ionic surfactant vesicles called niosomes (Figure 2.11), which too are useful targeted and controlled delivery systems (Shubhika, 2013; Ag Seleci et al., 2016; Obeid et al., 2019).

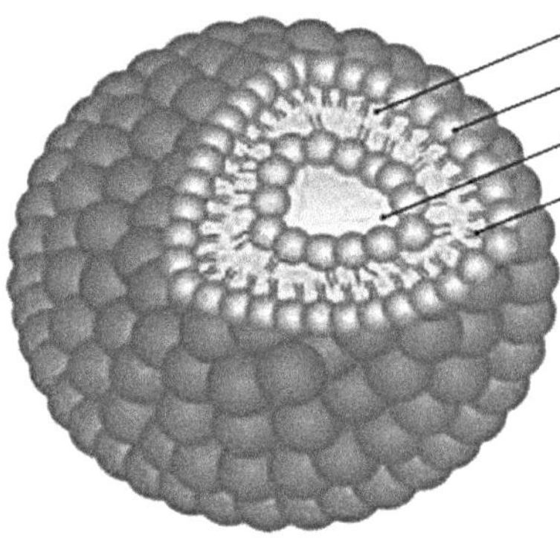

FIGURE 2.11 Structure of niosome (Ag Seleci et al., 2016).

Source: Reprinted from Ag Seleci et al., 2016. Copyright © 2016 Didem Ag Seleci et al. https://creativecommons.org/licenses/by/4.0/

Niosomes are lipid-based nanoparticles that contain cholesterol in the membrane bilayer structure, and requires a hydration medium as its main component (Ag Seleci et al., 2016; Obeid et al., 2019). The addition of cholesterol or any other lipid, however, is dependent on the physicochemical characteristics of the surfactants and the drug that needs to be loaded. Like liposomes, these too are non-immunogenic, biodegradable, and biocompatible, having long shelf-life and high stability (Ag Seleci et al., 2016). Niosomes behave like the liposomes, by allowing various types of molecules to be conjugated or embedded with/in it. This is due to the arrangement of the bilayer, leaving an aqueous core, as shown in Figure 2.11. The multifunctional ways of noisome usability can be seen in Figure 2.12.

An example of the use of niosome is as a carrier for a biologically active, chemical compound found in *Curcuma longa*. Curcumin has limited water solubility and gets fast degraded inside the body, but has benefits like being an anti-tumor and anti-inflammatory molecule. It also has antioxidative, antiangiogenic, and antiapoptotic properties (Rizvi & Saleh, 2018). In an attempt to explore curcumin for its therapeutic benefits, niosomes have played a role in successfully carrying curcumin and in establishing slow release of the molecule for the set purposes (Obeid et al., 2019). However, the use of

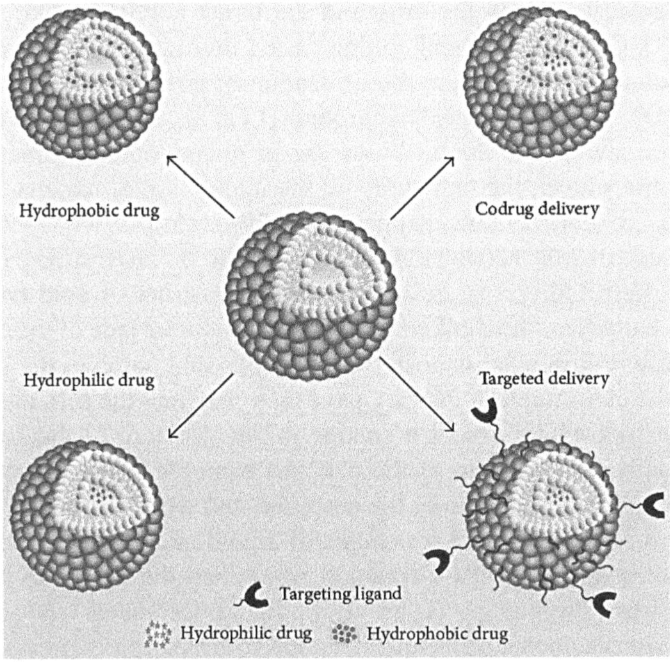

FIGURE 2.12 Niosome as drug carrying molecule (Ag Seleci et al., 2016).

Source: Reprinted from Ag Seleci et al., 2016. Copyright © 2016 Didem Ag Seleci et al. https://creativecommons.org/licenses/by/4.0/

niosomes is not just limited to anticancer therapeutics, it has also demonstrated efficiency as a carrier for anti-viral, anti-inflammatory drugs and antibiotics, singly, and in multidrug combinations (Ag Seleci et al., 2016).

2.10.2 GOLD NANOPARTICLES (AUNPS)

In the light of the most-deadliest disease that this century is witnessing, let's just say that it wouldn't be astonishing if the drug to cure COVID-19 was a nanoparticle formulation! AuNPs have the potential to be used as an efficient anti-viral drug. A research group from Switzerland had developed AuNPs capable of destroying viral particles. Its *modus operandi* is to behave like human cells once injected into the body and get the virus to bind to them, and subsequently degrade them. These broad-spectrum antiviral drugs are proposed to be instrumental in treating infections caused by viruses, including HIV and Ebola.

Gold is an inert metal and is seen to have low systemic clearance, which allows the bound drug to be slowly absorbed into the target site. AuNPs also offer a variety of shapes and sizes of particles to choose from, which can be selected based on the drug and the target cell/tissue/tumor (Kwatra et al., 2013). Of the recent technologies used involving nanoparticles in the treatment of some diseases are the utilization of gold nanostars (Dam et al., 2015) and DNA-origami-based nanorobots (Li et al., 2018). While nano-stars had been in news since 2010, for its use in magnetomotive imaging or as plasmonic transducer, the first study on the interaction of nanoparticles with the nucleus of cancer cells happened in 2012 (https://www.sciencedaily.com/releases/2012/04/120405131540.htm). That is when a research group at Northwestern University designed a nano-construct of gold that worked (what they called) as "hitchhikers." A protein on the surface of the cancer cell attracts these drug-loaded nanostars to the cells and helps in the passage of these into the cell's nucleus. As they pave their way into the cell, the nanostars release the drug which causes a change in the structure of the cell nucleus and eventually, in the death of the cell. An example is the use of a highly anisotropic gold nano-construct for detection and treatment of breast cancer tumors. In this, the construct was made of a gold nanostar core and a ligand shell of a G-quadruplex DNA aptamer, which had the ability to both target and to facilitate the therapy (Dam et al., 2015). Extended work on AuNPs based on their plasmonic transduction property have been carried out by Kim et al. (2019), however, for use in biomedical sensing and imaging activities.

Nanorobots have also been a promising avenue in cancer treatment. These are constructed using DNA origami technique where the drug is specifically delivered to the tumors. Like the nanostars, these are meant to act as both the targeting domain and the molecular trigger for opening and drug delivery. An example is the use of nanorobot that has a DNA aptamer on the outside and thrombin on the inside. The mode of action is such that the nanorobot targets nuleolin which is a protein expressed only on tumor-associated endothelial cells, and bind to it by DNA aptamer, which is multifunctional and does both the attracting and also destructing activities. So, the DNA nanorobot is mechanically opened after aptamer binds nucleolin, following which the thrombin in the inner cavity is exposed and starts coagulation at the tumor site. This induces intravascular thrombosis which then leads to tumor necrosis and inhibition of tumor growth (Li et al., 2018).

Another example of gold nanoparticle biocompatibility is its use in the treatment of damaged skin. In this, AuNPs are used to reconstruct collagen soft tissues. It is believed that the resistance of collagen to degradation improves due to the antioxidant and antimicrobial properties of the nanoparticles. For the very same reason that the collagen tissues gain much more stability, the technique has been employed as an anti-aging treatment.

AuNPs are also antiangiogenic and have been used in the cure of rheumatoid arthritis (Raghav et al., 2017).

AuNPs have also been used in the making of artificial saliva. The benefit of this would be in enabling people with salivary secretion disorder to swallow, eat, and speak. The nature of the synthetic saliva includes greasing the mouth, acting as an anti-microbial in the prevention of dental plaque, thus maintaining hygiene. Earlier, AuNPs were to be made with the help of cetyltrimethylammonium bromide (CTAB), a chemical that is toxic (Menon et al., 2013). However, Sona Nanotech has successfully devised nanoparticles without adding this toxic chemical. This adds to the hope that a greater number of non-invasive medical techniques can now be probably carried out, targeting a cell, organ or may be a tumor, etc. (www.gold.org, 2018).

2.10.3 IRON OXIDE NANOPARTICLES

It is a metallic nanoparticle majorly used for its super-para magnetism property, as discussed in the section above. Like other hybrids or composites that we have seen earlier, iron oxides can also be conjugated with a variety of compounds. Its stability can be improved when coated with dextran, phospholipids or surfactants; and is significant both as passive and targeting agent (Zhang et al., 2008). The addition of polymers, surfactants, or inorganic shells to iron oxide nanoparticles has aided in enhancing their hydrophilicity, biocompatibility as well as binding of bioactive functional groups (Raghav et al., 2017). A schematic of such a conjugated nanocomposite material can be seen in Figure 2.13.

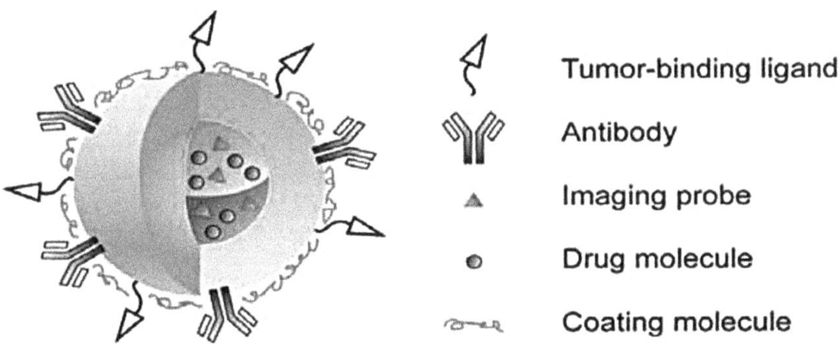

FIGURE 2.13 Depiction of a multifunctional nanoparticle having imaging probes and drugs encapsulated within and tumor-specific antibodies being expressed/presented on the surface (Rathod et al., 2011).

Source: Reprinted with permission from Rathod et al., 2011.

Iron oxide nanoparticles are well suited for studying the bacterial resistance to antibiotics by identifying the minimum inhibitory concentration of the drug within a couple of hours. It is also used to assess levels of telomerase activity in tumors. In fact, there's a great interest in utilizing iron oxide nanoparticles in conjugation with soft polymers or polymer-coated fluorophore, for *in-vitro* as well as *in-vivo* imaging of cancer cells (Rathod et al., 2011).

2.11 CONCERNS IN HEALTH CARE SECTOR DUE TO NANOTECHNOLOGY

With advancement in the construction and application of various nano-products, concern started rising with regard to its use in humans. The toxicity levels of not all the nano-molecules that are used to date, is known. This raises immense anxiety among toxicologists and health agencies, for not just the receivers of the treatment, but also people involved in the construction of these NMs, the medicos who are involved in the process, etc. Researchers involved in nanotechnology are still in search of concrete answers for the definition of each of these products, their characterization, efficiency, toxicity, and regulations (Patra et al., 2018). NMs are still treated as conventional chemicals. The WHO experts meeting held in 2012 had also concluded that there's uncertainty working with nano-molecules and that the risk assessment is a challenging task. Such conclusions have led to a lack of clear specific guidelines for having policies and regulations, as well as appropriate standard working and precautionary protocols (Valavanidis & Vlachogianni, 2016; Soarse et al., 2018). The extent of such uncertainties in defining toxicity limits often is the failure of many potential nano-drugs at late-phase of clinical trials (Patra et al., 2018).

Based on their properties, interaction with other moieties and the target site, and on their toxicological effects, the NMs have been broadly classified as:

- Soluble;
- Biopersistent – with high aspect ratio;
- Passive – with obvious biological effects;
- Active – with specific toxic effects.

A grouping of this sort facilitates targeted testing of NMs, so that enough data is available for risk assessment. The concern lies mainly with non-biodegradable type nanoparticles. The once that are biodegradable are eliminated from the body after their function is met with. Even then, should it cause any infection or inflammation in the body, then post its elimination, the source would come back to its normal untreated state (Soarse et al., 2018).

Engineered nanoproducts are expected to be harmful in many ways (Buzea et al., 2007; Valavanidis & Vlachogianni, 2016):

- Respiratory toxicity when small nanoparticles particles like that of CNTs sustain in air and deposit in the respiratory tract, or even enter the lungs of the user and the consumer. For instance, multi-walled or grounded CNTs that are not eliminated out of the body easily can increase the risk of lung cancer by its prolonged presence in alveoli.
- Application on skin when its penetration and toxicological impact are still unclear. However, most of the penetration occurs at the hair follicles, and through mechanical deformation of skin, such as flexed and broken skin. However, it has been found that application f sunscreen containing up to 8% of nanoparticles do not penetrate into the skin and are completely safe for use.
- Organ penetration, cytotoxicity, accumulation, and metabolism, all of which are unanswered yet, although signs of oxidative stress and cellular toxicity have been observed in cell lines, due to the formation of free radicals, depletion of antioxidants, ending with loss of cell viability. Example, a C_{60} molecules that enter the cells can be found in the nucleus and along the nuclear membrane. Similarly, a nanoparticle that has not been phagocytosed can remain in the cells and be deposited in various organelles, leading to localization in the endoplasmic reticulum, mitochondria, cytoplasm or even nucleus, which could lead to cell damage.

In Table 2.4, are listed some of the potential health hazards that could result from the various NMs that are commonly used for therapeutics and other health care related purposes.

As of now, all the regulations concerning compatibility, safety, and toxicity of the nanomedicines and nanodrugs are being performed as that for any conventional drugs, laid down by the FDA. Only a molecule that gains the investigational new drug (IND) status, is allowed to move on to clinical trials (Patra et al., 2018).

TABLE 2.4 Potential Health Hazards Caused by Use of Some Nanomaterials

Example of Nanomaterials	Therapeutic Application	Potential Health Hazards and Risks	References
Carbon nanomaterials	Used in gene therapy. Widely used theranostic agent.	If inhaled, it can lead to lung disorders, including asbestos-like effects.	Madani et al. (2013)
Dendrimers	Used extensively for delivery of anti-cancer drug.	The use of dendrimers in the human body is restricted because of their inherent toxicity. Exposure to dendrimers may lead to erythema-multiforme-like contact dermatitis.	Ashrafizadeh et al. (2020); Di Giampaolo et al. (2019)
Silver nanoparticles	Known for its antimicrobial nature.	At high doses, can cause adverse health effects, such as pulmonary edemas and skin stains.	Panyala et al. (2008); Mohamed et al. (2021); European Commission (2012)
	Commonly used anti-bacterial agent in wound dressings to protect patients with severe burns against infections.	Prolonged nanosilver exposure causes irreversible and incurable condition of argyria or argyrosis (i.e., gray or gray–blue discoloration or black pigmentation of the skin, nails, eyes, mucous membranes, or internal organs by silver deposits). Silver nanoparticles show the capability to reach the brain by the upper respiratory tract.	
Titanium dioxide (TiO_2)	Used against pathogens and tumors.	It is carcinogenic when inhaled.	Behnam et al. (2018); International Agency for Research on Cancer (2006)
Gold nanoparticles	For multiple uses ranging from drug delivery to biosensing and bioimaging.	Could accumulate in lungs and kidneys.	Dykman & Khlebtsov (2011); Sung et al. (2011)

2.12 CONCLUSION

This chapter has focused on all the important properties of nanoparticles that make them an important entity in modern therapeutics. We have also discussed many different types of nanomolecules, for various biological and medical applications. Although we have made huge progress in developing better and more advanced mechanisms to tackle health problems and go deep down to the nanoscales to find solutions, there is yet no drug for all! In terms of pharmaceutical development, we are yet to find means for scalability and reproducibility of the nanoproducts that have been tried and tested for the numerous health conditions (Soarse et al., 2018). A lot of these problems and other queries related to nanodrugs, nanomedicines, and all other nano-constructs, like their delivery efficiency, pharmacokinetics, and pharmacodynamics, etc., can be answered once the clarity is gained on its toxicity and correct usage doses, which has to be set by a common authority, for usage, globally.

KEYWORDS

- **biocompatibility**
- **gene therapy**
- **gold nanoparticles**
- **intelligent therapeutics**
- **liposome**
- **niosomes**

REFERENCES

Ag Seleci, D., Seleci, M., Walter, J. G., Stahl, F., & Scheper, T., (2016). Niosomes as nanoparticular drug carriers: Fundamentals and recent applications. *Journal of Nanomaterials, 2016*.

Ashrafizadeh, M., Zarrabi, A., Hushmandi, K., Hashemi, F., Rahmani, M. E., Raei, M., Kalantari, M., et al., (2020). Progress in natural compounds/siRNA co-delivery employing nanovehicles for cancer therapy. *ACS Combinatorial Science, 22*(12), 669–700.

Avnir, Y., Ulmansky, R., Wasserman, V., Even-Chen, S., Broyer, M., Barenholz, Y., & Naparstek, Y., (2008). Amphipathic weak acid glucocorticoid prodrugs remote-loaded into sterically stabilized nanoliposomes evaluated in arthritic rats and in a beagle dog: A novel approach to treating autoimmune arthritis. *Arthritis & Rheumatism, 58*(1), 119–129.

Behnam, M. A., Emami, F., Sobhani, Z., & Dehghanian, A. R., (2018). The application of titanium dioxide (TiO_2) nanoparticles in the photo-thermal therapy of melanoma cancer model. *Iranian Journal of Basic Medical Sciences, 21*(11), 1133.

Bogart, L. K., Pourroy, G., Murphy, C. J., Puntes, V., Pellegrino, T., Rosenblum, D., Peer, D., & Lévy, R., (2014). Nanoparticles for Imaging, Sensing, and Therapeutic Intervention.. *ACS nano, 8*(4), 3107–3122.

Buzea, C., Pacheco, I. I., & Robbie, K., (2007). Nanomaterials and nanoparticles: Sources and toxicity. *Biointerphases, 2*(4), MR17–MR71.

Casanellas, I., García-Lizarribar, A., Lagunas, A., & Samitier, J., (2018). Producing 3D biomimetic nanomaterials for musculoskeletal system regeneration. *Frontiers in Bioengineering and Biotechnology, 6*, 128.

Cho, H., Bae, J., Garripelli, V. K., Anderson, J. M., Jun, H. W., & Jo, S., (2012). Redox-sensitive polymeric nanoparticles for drug delivery. *Chemical Communications, 48*(48), 6043–6045.

Dam, D. H. M., Culver, K. S., Kandela, I., Lee, R. C., Chandra, K., Lee, H., Mantis, C., et al., (2015). Biodistribution and *in vivo* toxicity of aptamer-loaded gold nanostars. *Nanomedicine: Nanotechnology, Biology and Medicine, 11*(3), 671–679.

De Paula, E., Cereda, C. M., Fraceto, L. F., De Araujo, D. R., Franz-Montan, M., Tofoli, G. R., Ranali, J., et al., (2012). Micro and nanosystems for delivering local anesthetics. *Expert Opinion on Drug Delivery, 9*(12), pp.1505–1524.

Desai, P., Venkataramanan, A., Schneider, R., Jaiswal, M. K., Carrow, J. K., Purwada, A., Singh, A., & Gaharwar, A. K., (2018). Self-assembled, ellipsoidal polymeric nanoparticles for intracellular delivery of therapeutics. *Journal of Biomedical Materials Research Part A, 106*(7), 2048–2058.

Di Giampaolo, L., Di Gioacchino, M., Mangifesta, R., Gatta, A., Tinari, N., Grassadonia, A., Niu, Q., et al., (2019). Occupational allergy: Is there a role for nanoparticles? *Journal of Biological Regulators and Homeostatic Agents, 33*(3), 661–668.

Ding, B. S., Dziubla, T., Shuvaev, V. V., Muro, S., & Muzykantov, V. R., (2006). Advanced drug delivery systems that target the vascular endothelium. *Molecular Interventions, 6*(2), 98.

Dong, J., Jiang, D., Wang, Z., Wu, G., Miao, L., & Huang, L., (2013). Intra-articular delivery of liposomal celecoxib–hyaluronate combination for the treatment of osteoarthritis in rabbit model. *International Journal of Pharmaceutics, 441*(1, 2), 285–290.

Dykman, L. A., & Khlebtsov, N. G., (2011). Gold nanoparticles in biology and medicine: Recent advances and prospects. *Acta Naturae (англоязычная версия), 3*(2 (9)).

El-Nour, K. M. A., Eftaiha, A. A., Al-Warthan, A., & Ammar, R. A., (2010). Synthesis and applications of silver nanoparticles. *Arabian Journal of Chemistry, 3*(3), 135–140.

European Commission, (2012). *Types and Uses of Nanomaterials*, Including Safety Aspects.

Fadeel, B., & Garcia-Bennett, A. E., (2010). Better safe than sorry: Understanding the toxicological properties of inorganic nanoparticles manufactured for biomedical applications. *Advanced Drug Delivery Reviews, 62*(3), 362–374.

Fathi-Achachelouei, M., Knopf-Marques, H., Riberio De, S. C. E., Barthès, J. G. D., Bat, E., Tezcaner, A., & Vrana, N. E., (2019). Use of nanoparticles in tissue engineering and regenerative medicine. *Frontiers in Bioengineering and Biotechnology, 7*, 113.

Garbayo, E., Estella-Hermoso De, M. A., & J Blanco-Prieto, M., (2014). Diagnostic and therapeutic uses of nanomaterials in the brain. *Current Medicinal Chemistry, 21*(36), 4100–4131.

Gatoo, M. A., Naseem, S., Arfat, M. Y., Mahmood, D. A., Qasim, K., & Zubair, S., (2014). Physicochemical properties of nanomaterials: Implication in associated toxic manifestations. *BioMed Research International, 2014*.

Grandhi, T. S. P., & Rege, K., (2014). Design, synthesis, and functionalization of nanomaterials for therapeutic drug delivery. In: *Nanomaterial* (pp. 157–182). Springer, Dordrecht.

Hajipour, M. J., Fromm, K. M., Ashkarran, A. A., De Aberasturi, D. J., De Larramendi, I. R., Rojo, T., Serpooshan, V., et al., (2012). Antibacterial properties of nanoparticles. *Trends in Biotechnology, 30*(10), 499–511.

Harris, D. L., & Graffagnini, M. J., (2007). Nanomaterials in medical devices: A snapshot of markets, technologies and companies. *Nanotech. L. & Bus., 4*, 415.

Hua, S., & Cabot, P. J., (2010). Mechanisms of peripheral immune-cell-mediated analgesia in inflammation: Clinical and therapeutic implications. *Trends in Pharmacological Sciences, 31*(9), 427–433.

Hua, S., & Wu, S. Y., (2013). The use of lipid-based nanocarriers for targeted pain therapies. *Frontiers in Pharmacology, 4*, 143.

International Agency for Research on Cancer, (2006). Monographs on the evaluation of carcinogenic risks to humans. http://monographs.iarc.fr/ENG/Classification/index.php (accessed on 13 February 2022).

Jadoun, S., Verma, A., & Arif, R., (2020). Modification of textiles via nanomaterials and their applications. *Frontiers of Textile Materials: Polymers, Nanomaterials, Enzymes, and Advanced Modification Techniques*, 135–152.

Jeevanandam, J., Barhoum, A., Chan, Y. S., Dufresne, A., & Danquah, M. K., (2018). Review on nanoparticles and nanostructured materials: History, sources, toxicity and regulations. *Beilstein Journal of Nanotechnology, 9*(1), 1050–1074.

Khan, I., Saeed, K., & Khan, I., (2019). Nanoparticles: Properties, applications and toxicities. *Arabian Journal of Chemistry, 12*(7), 908–931.

Kim, M., Lee, J. H., & Nam, J. M., (2019). Plasmonic photothermal nanoparticles for biomedical applications. *Advanced Science, 6*(17), 1900471.

Kumar, A., & Dixit, C. K., (2017). Methods for characterization of nanoparticles. In: *Advances in Nanomedicine for the Delivery of Therapeutic Nucleic Acids* (pp. 43–58). Woodhead Publishing.

Kumar, M., Gupta, D., Singh, G., Sharma, S., Bhat, M., Prashant, C. K., Dinda, A. K., et al., (2014). Novel polymeric nanoparticles for intracellular delivery of peptide cargos: Antitumor efficacy of the BCL-2 conversion peptide NuBCP-9. *Cancer Research, 74*(12), 3271–3281.

Kwatra, D., Venugopal, A., & Anant, S., (2013). Nanoparticles in radiation therapy: A summary of various approaches to enhance radio sensitization in cancer. *Transl. Cancer Res., 2*(4), 330–342.

Lara, H. H., Ayala-Nuñez, N. V., Ixtepan-Turrent, L., & Rodriguez-Padilla, C., (2010). Mode of antiviral action of silver nanoparticles against HIV-1. *Journal of Nanobiotechnology, 8*(1), 1.

Lauterwasser, C., (2005). *Small Sizes That Matter: Opportunities and Risks of Nanotechnologies*. Report in Co-operation with the OECD International Futures Program.

Li, S., Jiang, Q., Liu, S., Zhang, Y., Tian, Y., Song, C., Wang, J., et al., (2018). A DNA nanorobot functions as a cancer therapeutic in response to a molecular trigger *in vivo*. *Nature Biotechnology, 36*(3), 258.

Lombardo, D., Kiselev, M. A., & Caccamo, M. T., (2019). Smart nanoparticles for drug delivery application: Development of versatile nanocarrier platforms in biotechnology and nanomedicine. *Journal of Nanomaterials, 2019.*

Madani, S. Y., Mandel, A., & Seifalian, A. M., (2013). A concise review of carbon nanotube's toxicology. *Nano Reviews, 4*(1), 21521.

Matea, C. T., Mocan, T., Tabaran, F., Pop, T., Mosteanu, O., Puia, C., Iancu, C., & Mocan, L., (2017). Quantum dots in imaging, drug delivery and sensor applications. *International Journal of Nanomedicine, 12*, 5421.

Menon, J. U., Jadeja, P., Tambe, P., Vu, K., Yuan, B., & Nguyen, K. T., (2013). Nanomaterials for photo-based diagnostic and therapeutic applications. *Theranostics, 3*(3), 152.

Mody, V. V., Siwale, R., Singh, A., & Mody, H. R., (2010). Introduction to metallic nanoparticles. *Journal of Pharmacy and Bioallied Sciences, 2*(4), 282.

Mohamed, E. M., Kattaia, A. A., Abdul-Maksoud, R. S., El-Baset, A., & Samia, A., (2021). Cellular, molecular and biochemical impacts of silver nanoparticles on rat cerebellar cortex. *Cells, 10*(1), 7.

Mohanty, S., Nayak, S. K., Kaith, B. S., & Kalia, S., (2015). *Polymer Nanocomposites Based on Inorganic and Organic Nanomaterials*. John Wiley & Sons.

Nikalje, A. P., (2015). Nanotechnology and its applications in medicine. *Med. Chem., 5*(2), 081–089.

Obeid, M. A., Khadra, I., Albaloushi, A., Mullin, M., Alyamani, H., & Ferro, V. A., (2019). Microfluidic manufacturing of different niosomes nanoparticles for curcumin encapsulation: Physical characteristics, encapsulation efficacy, and drug release. *Beilstein Journal of Nanotechnology, 10*(1), 1826–1832.

Optical properties of nanoparticles from: https://nanocomposix.com/pages/introduction-to-nanoparticle-optical-properties#target (accessed on 13 February 2022).

Panyala, N. R., Peña-Méndez, E. M., & Havel, J., (2008). Silver or silver nanoparticles: A hazardous threat to the environment and human health? *Journal of Applied Biomedicine, 6*(3).

Patra, J. K., Das, G., Fraceto, L. F., Campos, E. V. R., Del Pilar Rodriguez-Torres, M., Acosta-Torres, L. S., Diaz-Torres, L. A., et al., (2018). Nano based drug delivery systems: Recent developments and future prospects. *Journal of Nanobiotechnology, 16*(1), 71.

Raghav, R., Painuli, R., & Kumar, D., (2017). Multifunctional nanomaterials for multifaceted applications in the biomedical arena. *Int. J. Pharmacol., 13*, 890–906.

Rathod, K. B., Patel, M. B., Parmar, P. K., Kharadi, S. R., Patel, P. V., & Patel, K. S., (2011). Glimpses of current advances of nanotechnology in therapeutics. *Int. J. Pharm. Pharm. Sci., 3*(1), 8–12.

Realizing the Healthcare Potential of Gold Nanoparticles, (2018). https://www.gold.org/goldhub/research/gold-investor/gold-investor-july-2018 (accessed on 13 February 2022).

Rizvi, S. A., & Saleh, A. M., (2018). Applications of nanoparticle systems in drug delivery technology. *Saudi Pharmaceutical Journal, 26*(1), 64–70.

Robson, A. L., Dastoor, P. C., Flynn, J., Palmer, W., Martin, A., Smith, D. W., Woldu, A., & Hua, S., (2018). Advantages and limitations of current imaging techniques for characterizing liposome morphology. *Frontiers in Pharmacology, 9*, 80.

Roy, A., (2017). Synthesis of silver nanoparticles from medicinal plants and its biological application: A review. *Res. Rev. Biosci., 12*(4), 138.
Sadowska-Bartosz, I., & Bartosz, G., (2018). Redox nanoparticles: Synthesis, properties and perspectives of use for treatment of neurodegenerative diseases. *Journal of Nanobiotechnology, 16*(1), 87.
Salata, O. V., (2004). Applications of nanoparticles in biology and medicine. *Journal of Nanobiotechnology, 2*(1), 3.
Scaletti, F., (2016). When nano becomes smarter: Nanomaterial-containing medical devices. *Medical Devices and Diagnosis Engineering*. Editorial, ISSN: 2399-6854.
Seleci, M., Seleci, D. A., Joncyzk, R., Stahl, F., Blume, C., & Scheper, T., (2016). Smart multifunctional nanoparticles in nanomedicine. *BioNano Materials, 17*(1, 2), 33–41.
Shubhika, K., (2012). Nanotechnology and medicine-The upside and the downside. *Int. J. Drug Dev. Res., 5*, 1–10.
Soares, S., Sousa, J., Pais, A., & Vitorino, C., (2018). Nanomedicine: Principles, properties, and regulatory issues. *Frontiers in Chemistry, 6*, 360.
Sung, J. H., Ji, J. H., Park, J. D., Song, M. Y., Song, K. S., Ryu, H. R., Yoon, J. U., et al., (2011). Subchronic inhalation toxicity of gold nanoparticles. *Particle and Fiber Toxicology, 8*(1), 1–18.
Tosi, G., Costantino, L., Rivasi, F., Ruozi, B., Leo, E., Vergoni, A. V., Tacchi, R., et al., (2007). Targeting the central nervous system: *In vivo* experiments with peptide-derivatized nanoparticles loaded with loperamide and rhodamine-123. *Journal of Controlled Release, 122*(1), 1–9.
Trotta, F., & Mele, A., (2019). Nanomaterials: Classification and properties. *Nanosponges: Synthesis and Applications*.
U.S. Government Research and Development Initiative for Nanotechnology. What Is So Special about "Nano"? https://www.nano.gov/about-nanotechnology/what-is-so-special-about-nano (accessed on 13 February 2022).
Valavanidis, A., & Vlachogianni, T., (2016). Engineered nanomaterials for pharmaceutical and biomedical products new trends, benefits and opportunities. *Pharmaceutical Bioprocessing, 4*(1), 13–24.
Van, R. S., & Habibovic, P., (2017). Enhancing regenerative approaches with nanoparticles. *Journal of the Royal Society Interface, 14*(129), 20170093.
Zhang, L., Gu, F. X., Chan, J. M., Wang, A. Z., Langer, R. S., & Farokhzad, O. C., (2008). Nanoparticles in medicine: Therapeutic applications and developments. *Clinical Pharmacology & Therapeutics, 83*(5), 761–769.

CHAPTER 3

Biosensors: A Breakthrough in the Detection of Biological Threat Agents through Nanomaterials

ILYAS ILKER IŞLER,[1] HILAL ACIDERELI,[2] HAYDAR GÖKSU,[1] and FATIH ŞEN[2]

[1]*Kaynasli Vocational College, Duzce University, Duzce, Turkey*

[2]*Sen Research Group, Department of Biochemistry, Dumlupınar University, Kütahya, Turkey*

ABSTRACT

Nanomaterial-based biosensors have become one of the major topics in the field of biological threat agents. With increasing threats and costs, the demand for biosensors has increased in the process of detecting nanomaterial threats. There are four categories of biosensor research on nanomaterials (NMs): gold nanoparticles (AuNPs), carbon nanotubes (CNTs), photonic crystals, and graphene. In recent years, new research and new technologies have been introduced along with new developments in the design of biosensors on NMs. Also, the biosensors are combined with new technologies such as molecular biology, NMs, agricultural production, food, and environmental monitoring activities, animals, plants, foods, soil, the air in water, pesticides, antibiotics, pathogens, toxins, proteins, malodors, germs, and more. Also, ongoing research has brought new designs of biosensors, which has a low cost.

3.1 INTRODUCTION

Significant developments have been observed in the use of biosensors in the detection of chemical agents after World War II, in the determination

Advanced Nanomaterials for Biological, Nutraceutical, and Medicinal Applications.
Sreeraj Gopi, Augustine Amalraj, and T. R. Sreeraj (Eds.)

of agricultural production, food processing, drug testing, and bioterrorism threats.

Biosensors are devices that combine a sensitive biological recognition component and physical transducer to detect analytes of interest (Giraldo et al., 2014; Iverson et al., 2013; Bozkurt et al., 2017; Ulissi et al., 2014; Zhang et al., 2013; Baskaya et al., 2017; Sen et al., 2012, 2013; Savk et al., 2019; Ayranci et al., 2019; Kaya et al., 2019). The results are analyzed by directing biological reactions to quantifiable and quantitative determinations and directing them to a measurable signal. There are also nucleic acids, enzymes, antibodies, receptors, microorganisms, their tissues, biosensors, biological recognition sensitivity.

Living things separate the biological materials that provide the perception of the warnings in the life process. Then, these parts and the relations between them are examined as a whole. Current technological developments, thanks to the biosensors for the very fast development. The inclusion of sensor technology in the fields of chemistry, biology, materials science, engineering, and updating with new technologies in these disciplines has significantly contributed to the development of biosensors.

Biosensors such as amperometric, potentiometric, piezoelectric, thermal, acoustic, and optical sensors. Different immunological agents have emerged as a result of bringing together the existing structures such as chemical receptors and biological systems such as enzymes, cells, microorganisms. Biosensors are defined as the conversion of a biological event in the living body into an electrical signal. By combining many sciences such as physics, chemistry, biology, biochemistry, and engineering, advanced electronic processing capabilities have developed with the selectivity of biological molecules or systems.

Biosensors are bioanalytical devices developed for use in many different industrial areas besides medicine and pharmacy. Over time, electronic circuits have resulted in small-sized biosensors structures. Biosensors are mounted on the chips using nano-dimensional work. However, the availability of different biological structures and different response capabilities have been developed in the electronic field. Biosensors are affected by nanotechnology. Thanks to nanotechnologies, biosensors can be made smaller, cheaper, sensitive, and long-lasting.

3.2 BIOSENSOR

In general, the biosensor is a device or instrument comprising a biological detection material associated with a chemical and physical transducer that converts the biological, chemical, and biochemical signal to a measurable and processable electrical signal.

Biosensors used in bio-analysis materials include enzymes, antibodies, nucleic acids, whole cells, receptors, tissues, organelles, and more (Turner et al., 1987; Datta et al., 1990; Wagner et al., 1994; Coulet et al., 2010).

The transducer used in biosensors can be electrochemical (voltammetry, amperometric, potentiometric, conductive, capacitive, impedance), optical (absorption, surface plasmon resonance, chemical luminescence), biological luminescence (bioluminescence), fluorescence, and piezoelectric (quartz, crystal microbalance, surface sound, calorimetric, magnetic, and others). A processing unit generally requires data acquisition and control, database, and use of modeling, and wireless communication with a network connection. In order to bind the biosensing materials to the transducers, an effective fixing technique should be used. The sampling process is critically important in the sense of the concentration and concentration of target analytes in obtaining specific and detectable signals. As the need for smaller and automatic biosensors increases, the need for advanced software systems and mix production facilities will increase.

A biosensor can be a system containing sampling detection dialing and calculation. Biosensor technology is based on a multidiscipline approach between biology, chemistry, and engineering (Li et al., 2006). Typical structure and components of a biosensor are depicted in Figure 3.1.

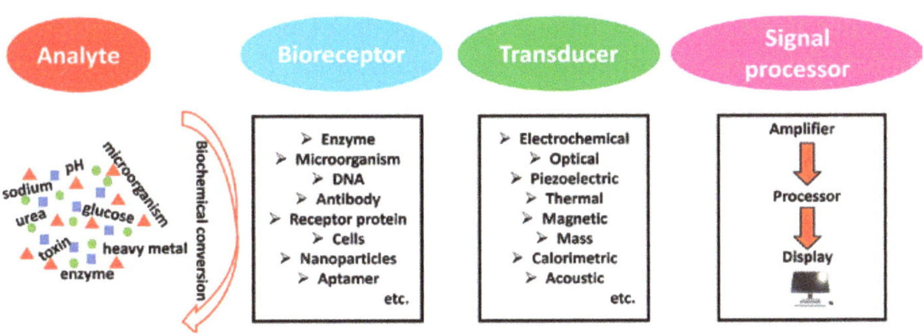

FIGURE 3.1 Typical structure and components of a biosensor.

3.2.1 THE HISTORY OF BIOSENSORS

The first biosensor was identified by Clark and Lyons in their work on an enzyme electrode. In this study, an oxide-reducing enzyme was retained in a membrane sandwich next to a platinum electrode. The purpose is to react the enzymes with the polarized platinum cathode. In this system, which is the main target of glaucoma, the glucose oxidation reaction is catalyzed by glucose oxidase (Clark et al., 1962).

The voltage applied between the platinum and silver electrodes is sufficient to reduce the oxygen in the device. In this way, the electric current can be measured in proportion to the oxygen concentration. An enzyme layer is placed on the membrane in the device coated with a permeable oxygen membrane such as electrode, polythene, and cellophane, and the acetate is then attached to a second membrane (Li et al., 2006).

Another important discovery in the development of biosensors is the potentiometric, biologically defined molecule, released by Guibalt & Montalvo (1969). The above electrodes are classified as first-generation biosensors (Scheller et al., 1991). As second-generation biosensors, the immobilization of the biosensing material was obtained by using cross-linked reagents or biodegradable materials by incorporating into a polymer matrix on the modified transducer interfaces or the transducer surface thereof (Li et al., 2006).

For the third-generation biosensors, the biomolecule becomes an internal part of the biosensing material, as in the surface plasmon resonance biosensors. The fourth-generation biosensors are expected to have more features with MEMS/NEMS/BioNEMS (micro, nano, or bio nanoelectromechanical systems), nanotechnology, and biotechnology (Li et al., 2006).

After the increase of biotechnological developments, the structure of biomolecules and identification of nanomaterials (NMs) became easier. The biosensor technologies provided molecular definitions and increased signal processing capacity of the electronic and optical technology of most chemical reactions. After the harmonization of multi-disciplinary structures, large-scale creative biosensors for development have emerged with the support of disciplines such as engineering, biochemistry, physics, biology, computer, and engineering (Li et al., 2006).

Biosensors are devices that are used to control target analyte materials in biological reactions and are capable of detecting the targets. It consists of two key structures with intertwined biochemical and electrochemical properties. The biochemical part interacts with the substance to be analyzed and ensures its recognition. The biochemical product may be produced during this process. The electrochemical part of the biosensors, provide a readable (measurable) numerical value (Mavrogiannis et al., 2016). Biosensors are called modern structures, which are developed by combining the selectivity of biological molecules or the mechanisms used and the processing capabilities of advanced electronic methods used. Biosensors, which are developed by today's technology and have a wide usage area, find many applications (Turner et al., 1989). The basic working principle of a biochemical sensor is represented in Figure 3.2. Biosensors perform their functions as a result of their selective, rapid, and continuous reaction to a substance with a biological or chemical effect. It has to have bioactive and biosensing material when performing its function. These materials should be able to recognize analytes and be in close contact with the transducer. Otherwise, there may be problems during the process on the quantities obtained in the biosensor applications (Griffiths et al., 1993; Nakamura et al., 2003).

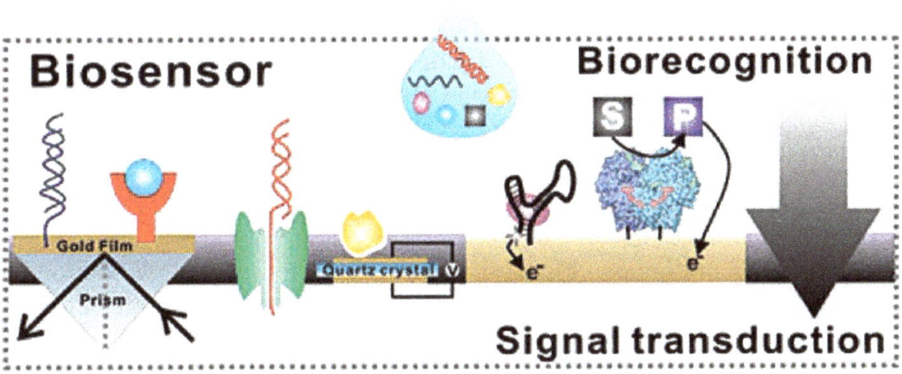

FIGURE 3.2 Schematic representation of the working principle of a biochemical sensor.

Source: Reprinted with permission from Soleymani et al., 2017. Copyright American Chemical Society.

Biosensors have emerged as a result of the use and experience of many scientific or technical sciences. Due to the long warehouse stability, short analysis time, originality of the substances processed in reactions and high repeatability possessed with biosensors, it can be easily applied in all laboratory environments. This method does not require much time to implement. The number of process steps is less during the measurement. Hence, high-priced devices are made requiring fewer chemical materials (Hobson e al., 1990; Ivnitski et al., 1999).

Biosensors can make very important determinations in clinical diagnoses. Especially glucose sensors are the most successful and widely used biosensors today. The nanospheres biosensors are of interest in the detection and identification of NMs with their selectivity, sensitivities, and rapid yielding potentials. Biosensors allow for the analysis of complex sample matrices. Biosensors can be used to analyze environmental samples in a short time to protect against pathogens (Li et al., 2006).

Factors determining the use potential of immunosensors; in general, the antigen-antibody reaction is specific and selective and depends on the high sensitivity of the method (Alocilja et al., 2008). Biosensor technology was used in the analysis of blood samples, the diagnosis of infectious diseases, and drug screening. Beyond this, biosensors can be used for agricultural production, food analysis, and environmental monitoring. It has been adopted in the determination of biological and chemical active substances in the fields of mining, biotechnology, biosphere, and country security. Over the last decade, research on biosensors has been intensively involved in many disciplines such as scientific and engineering continued. After 40 years of development, biosensors are biospecific; it has become more powerful devices and instruments for monitoring interactions and identifying biological and chemical agents in all areas (Li et al., 2006).

3.2.2 STRUCTURE AND FUNCTIONS OF BIOSENSORS

Biosensing technologies, biosensing materials, transducer devices, and immobilization methods are being developed as a result of multidisciplinary research in the fields of biology, chemistry, and engineering. Therefore, biosensors can be classified in a variety of ways based on both biosensing materials and applied transformation tools (Zhou et al., 2006).

Biosensors consist of biocomponents (receptors) and physical components (transducers). Receptors are called bioresceptors because they have a biomolecular structure. Bioreceptors have biomolecules which are capable of detecting the analyte. Bioreceptors, which have a key role in the development of biosensors, are highly sensitive biological structures that can interact selectively with the substance to be analyzed. In bioreceptor structures, biological molecules (such as antibodies, enzymes, proteins, nucleic acids) or living biological systems (such as cells, tissues, and microorganisms) can be used today, biosensors contain many different substances and systems as biosensors and transducers. The most important of these are given in Table 3.1 (Scott et al., 2012).

TABLE 3.1 Biosensor Components

Analyte	Receptor	Signal Connected System
Metals	–	Electrochemical-based
Hormones	Enzymes	Electrochemical-based
Enzymes	Antibody	Semiconductor-based
Coenzymes	Cell	Optical
Substrate	Tissue sections	Photometry
Activator	Receptors	Fluorometer
Inhibitor	Microorganisms	Fluorometer
Antibody-antigen	Nucleic acid	Piezoelectric
Nucleic acid	Lipitor	Quartz crystal microbalance
Microorganisms	Cell organelles	Microclimate ever
L-amino acids	Horseradish peroxidase	Piezoelectric quartz crystal
Herbist	Antibody	Potentiometric
Glucose	Glucose oxidase	Amperometric

The fabrication and detection procedure of QCM biosensor is presented in Figure 3.3 (Xi et al., 2021).

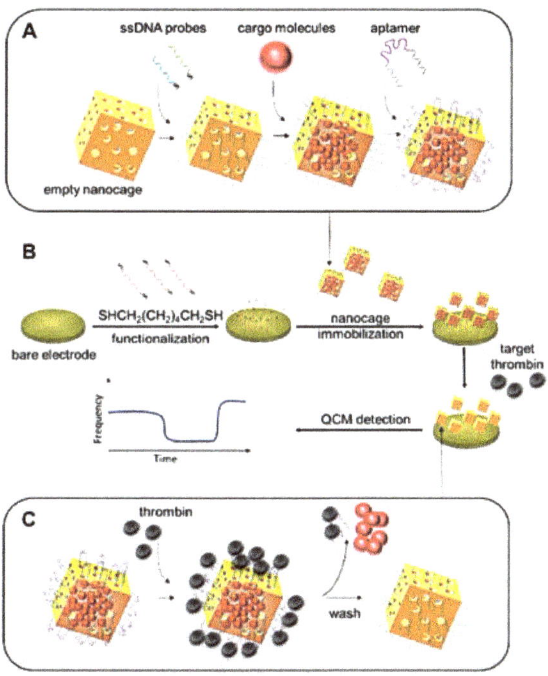

FIGURE 3.3 Illustration of the QCM biosensor fabrication and detection procedure.

Source: Reprinted with permission from Xi et al., 2021. Copyright American Chemical Society.

Three different formats of fluorescent biosensors are illustrated in Figure 3.4.

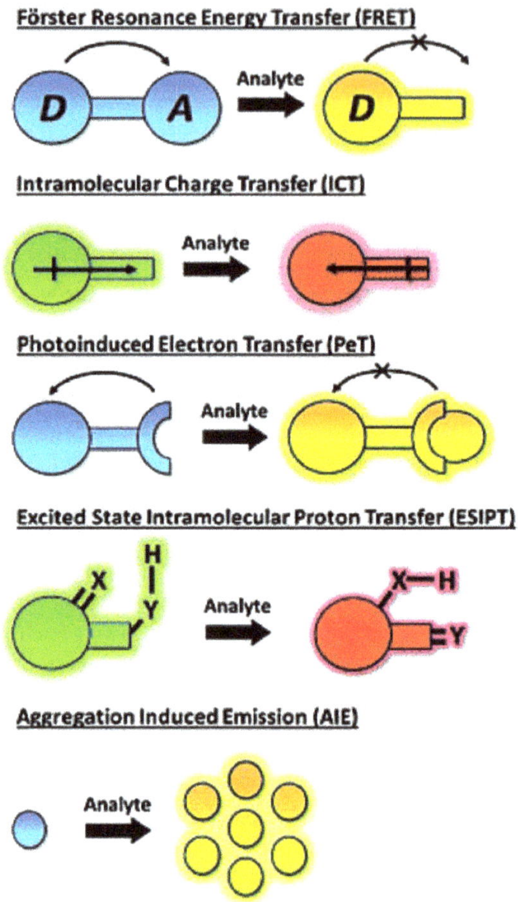

FIGURE 3.4 Three formats of fluorescent biosensors with use of the evanescent wave: (a) excitation through the waveguide; (b) excitation and emission through the waveguide; and (c) collection of the emission through the waveguide.

Source: Reprinted with permission from Tian et al., 2021. Copyright The Royal Society of Chemistry. https://creativecommons.org/licenses/by/3.0/

Biosensors convert a biological event into an electrical signal. Generally, it is formed by combining the signal to be analyzed with a transmitter which transmits the signal resulting from the interaction of the bioactive component which interacts with the optional substance and using it together with the measurement devices. Depending on the system features, such as an amplifier, microprocessor, the digital imager can be included in the system when needed. These physical components, which are subsequently incorporated into the system, convert the biomarkers into physical signals depending on their ability to measure their biological function. Appropriate physical size is selected by considering the biochemical reaction. Electrodes used in the system as a physical size (transducer) are used in measurement operations such as amperometric and potentiometric (Zhou et al., 2006). In addition, electronic components such as transistors and thermistors can be used as physical components in the system. In biosensors, electronic devices such as electrochemical, optical, calorimetric, and piezoelectrics are used to measure and transmit the signal resulting from the interaction of the agent that determines the bioactive components. These devices are used in the so long-lasting assays system finalized in shorter time (Zülfü et al., 2017).

In biosensor applications, the substances and structures to be analyzed are defined as analytes. The components forming the compound or mixtures in the biosensor structure are called bioreceptors. Bioreceptors convert the analyte substances into a different format. These operations commonly use analytes, enzymes, and antibodies. The first step in the relationships between the enzyme-substrate and the antibody-antigen is the binding of analytes to protein molecules. The catalytic antibodies developed in recent years not only adopt antigens but also allow the reaction to take place or change the rate of a reaction without modifying a substance which reacts chemically with them. Large molecules are formed as a result of polymerization of small materials with protein structure (Kindschy et al., 2004). Also, nucleic acids and carbohydrates are used in areas such as gene-specific gene ring analyses and cell surface characterization. Thus, biosensor application areas are developed. Many classifications are made for biosensors, according to the type of bioreceptors (antibody/antigen interaction, nucleic acid interaction, enzymatic interactions, cellular interactions, biomimetric material interaction) or based on the type of transducer (optical measurements, electrochemical, mass-based measurements) (Gutés et al., 2005).

Transducers are structures that measure biological reactions generated by receptors and transform them into physical signals. After recognition of the analyte, a physical and chemical change is detected, measured, and converted into digital signals. Transducer selection is made by considering biochemical reactions. Enzyme electrodes are used in amperometric and potentiometric measurement processes. Biosensor devices that convert an energy form into another form of energy (transducers), electrochemical (amperometric, voltammetry, potentiometric, capacitive, conductive, impedance, including those), optical (surface plasmon resonance, absorption, chemical luminescence, biological luminescence, fluorescence, optical fiber), piezoelectric (quartz crystal microbalance, surface sound wave) exhibits calorimetric and magnetic properties (Updike et al., 1967).

Biosensors are currently used mainly for diagnostic purposes in biomedical areas. Of course, biomedical device technology is one of the best application areas for biosensors. The first sensor in the biomedical field is enzyme biosensor. The first biosensor produced for commercial purposes is the glucose oxidase electrode developed for the diagnosis of diabetes, glucose determination in blood, and urine. These enzyme sensors are frequently used to monitor biological products such as glucose, urea, and sugar in the body, to monitor cancers and to detect cases caused by microbes. When we look at the historical development of biosensor technology, it is seen that the first study in this field was started with enzyme sensors. As a result of the studies conducted in this field, new analytical systems created, biological system (enzyme) and many types of biosensors are developed by performing applications for many biological parameters such as selectivity and sensitivity of the determination of the physical system (sensor) (Updike et al., 1967).

3.3 NANOMATERIALS (NMS) AND IMMUNOSENSORS

The use of nano-scale materials in chemical detection, biomedical, and biological analysis is seen as efficient (Shimizu et al., 2008; Schultes et al., 2009). In recent years, many studies have widely used NMs; has been successfully used in the production of different auxiliary structures, composites, and especially various sensors. Advances in material synthesis allow nano-sizes of 1 to 100 nanometers, which provides controlled production and characterization of the structures (Marquis et al., 2009). With this controlled production, very rare, size-dependent features can be obtained, which can contribute to the development of many new technologies. The inorganic compounds in the nanometer size are different from the other materials. These compounds are called art nanoparticles (Portakal et al., 2008). Since the 1970s, scientists have been studying the nanoparticles and trying to determine the properties of nanoparticles. Fine tuning of the physical and chemical properties of NMs allows the production of nanostructured sensors capable of detecting very low analyte concentrations they can be used for non-degenerative biological analysis, chemical decomposition and many other promising applications (Asefa et al., 2009).

Nanoparticles are compounds with properties less than 100 nm and specific in size (electron-retaining effect, transient magnetization property, surface plasmon resonance). They can easily integrate into biological systems due to their size and their properties (Marquis et al., 2009). Based on the functions, we can classify them into nanoparticles, nanorods, nanowires, nanoporous structures, and core-shell models. Selectable sensors for specific analytes can be produced using several NMs and various ligands or bioactive groups. With the combination of two or more core-shell models and other composite nanostructures, it is possible to obtain sensor features that increase accuracy and reduce response time (Asefa et al., 2009). For the diagnosis of bacterial pathogens in nutrients in recent years, it has a high degree of sensitivity. The use of microarray-based sensors significantly shorten this time. Multiple advantages of biosensors are another advantage. Immunosensors are based on the interaction between the antigen and the antibody where this interaction has been detected or measured by the sensors. There are a wide variety of immunoassay methods, and these techniques are widely used in areas including toxicology and pathogen detection (Patel et al., 2006).

3.4 BIOLOGIC WEAPONS AND AGENTS

Biological Weapons can be defined as the effects on the target they are directed to, the microorganisms that cause disease and the weapons that depend on the transmission and disease-making abilities of the self-replicating biological structures such as viruses, infectious nucleic acids and prions (Bourgeois et al., 2005).

Biological weapons used to damage the health integrity of other living organisms used in bacteria, viruses, and so on infectious agents. This definition usually includes toxins and poisons obtained by biological experiments. While some authors, mostly consider toxins chemically, biological threat vehicles include living microorganisms (bacteria, protozoa, rickettsia, viruses, and fungi) as well as toxins produced by plants and animals (chemicals).

3.4.1 BIOLOGIC WEAPONS AND AGENTS IN HISTORY

The biological weapons convention conducted in 1972 has significantly dealt with the norms of development, production, and stockpiling of biological and toxin weapons as well as their destruction (Rissanen, 2001).

There are many biological agents mentioned in the literature (Table 3.2). The danger dimensions of these agents are often fatal. Some agents are used as capacity reducing or dose reducing agents. Genetically modified methods used in the literature for the assimilation of certain ethnic groups agents are mentioned. There are conditions of infectivity, virulence, toxicity, pathogenicity, and stability, which increase the potential of biological agents as weapons. Biological agents may remain in a particularly favorable (temperate) environment for a long time and increase their effects. The effects of chemical weapons on fatal consequences are well known by the community (Erkekoğlu et al., 2018). However, even the smallest component of biological weapons can produce worse results than chemical weapons. For example; Bacillus anthrax caused by the bacilli in the case of anthrax, airborne symptoms show signs of influenza infection in the initial stage and then lead to fatal consequences. When anthrax passing through the inhalation phase gets into the human body, the risk of death reaches 100%. The effects of botulinum toxin, a type of biological agent, are less known. However, this toxin has 3 million times more harmful effects than the chemical toxin that society faces. People who are exposed to botulinum toxin usually suffer from paralysis due to ruptures in their muscles, and to death by up to 60% in the course of progress (Erkekoğlu et al., 2018).

TABLE 3.2 Biological Warfare Agents

Anthracosis	*Botulinum toksinleri*
Brucella species	*Vibrio cholera*
	Clostridium perfirenges
	Salmonella typhi
	Pseudomonas psoudomallei
	Pseudomonas mallei
	Yersinia pestis
	Francisella tularensis
	Coxiella burnetti
Smallpox virus	
Congo-Crimean Hemorrhagic	
Ebola	
Staphylococcal enterotoxin type B	
Rift Valley	
Trichothecene mycotoxins	
Venezuelan Equine Encephalitis	
Cryptococcus	
Coccidiomycosis	
Plasmodium vivax	
Ricin	
Saxitoxin	

Throughout history, naturally occurring biological diseases have been used by some people as an effective weapon that will make a difference in wars. Biological agents were used in World War I and to a limited degree in the Second World War. Their first use only goes back to ancient times. In the sixth century B.C., societies have poisoned the enemies during the war.

In the same way, poisons obtained from various plants have been used in. The immersion of the arrows into the dead, the fogging of the combat devices and the deportation of them to the rotting flesh are commonly used in ancient societies for killing. Human bodies have also been used as biological agents throughout history. In particular, the bodies of people who died due to diseases such as plague, and the combat vehicles used to dispose them are widely used as biological agents (Frischknecht et al., 2008). It is an example of these agents that the Europeans spread the measles microbe after the discovery of the American and caused mass deaths (Barras et al., 2014).

In particular, with the developing technologies from the beginning of the industrial revolution to today, countries have produced biological weapons in a laboratory environment to increase their power. For example, during the First World War, the dependence of the Allied countries on horses and mules in the transportation of military materials was determined by the Germans, and the biological war campaign was initiated by using the glanders virus against these animals. As the part of biological warfare program, even a private laboratory was established in the US state of Maryland. There were facilities for the research and production of the anthrax glanders organism also. During the First World War, there were observations by the British in Çanakkale that biological weapons had been used by the Germans and Japanese during the Second World War. Only these have not been officially proved (Barras et al., 2014).

Looking at the second world war, the Japanese army emitted 16 different biological viruses in different parts of the far east, using different methods, and even experimented with infectious diseases such as anthrax, plague, tularemia, cholera, scarlet, and tetanus on Chinese prisoners in the Manchuria region of China (Daniel et al., 2004).

During the Second World War, Russian forces used a tularemia virus to stop the German army in the battle of Stalingrad, but because the wind changed its direction suddenly, this attack was a historical fact in which Russian soldiers were damaged.

According to a study by the World Health Organization (WHO) in 1970, 29 different microorganisms (15 viruses, 11 bacteria, 2 parasites, and 1 fungus) have the property of being used as biological weapons. The most important threat among these microbes is the microorganisms such as anthrax, smallpox (bloom), plague, and botulinum. In this context, the US defense ministry declared anthrax as the number one biological weapon threatening the army in 1996 and initiated the necessary activities for an effective struggle. The seriousness of the danger posed by the biological weapons against humanity was reached during the first world war, and after this period, international studies on the prohibition of the production and use of biological microorganisms were initiated (Erkekoğlu et al., 2018).

The first successful international initiative in this field is the Geneva protocol signed in 1925. The protocol mentioned above aims to prevent the use of poisonous and other gases by respiration and biological weapons in wars. This protocol has been approved by 132 countries as of 2001 (Fidler et al., 1999).

The second important step was the Convention on Biological Weapons, signed in 1972. This contract, which aims to prevent the development, production, and storage of biological weapons, has been ratified by 143 states, including Turkey (Hamburg et al., 2002).

Several microorganisms or biomasses have been used as a weapon in some terrorist incidents or individual crimes. However, none of these initiatives have had a significant impact on target populations, and individual offenses have been largely due to non-political reasons. It can be said that the anthrax terror, which started with the attacks of September 11 in the USA, which has been brought to its goals by 'letters,' is the most effective biological terror event to date.

3.4.2 THE MOST DANGEROUS BIOLOGICAL WEAPONS

1. **Chimera Viruses:** In modern genetic science, the chimera virus is called organisms that have genetically different tissues. Genetic scientists, by combining genes, theoretically created two viruses that trigger two diseases at a time. In the late 1980s, within the framework of the Soviet Union's Chimera Project, the researchers conducted a study of the combination of smallpox and Ebola into a single super virus.

2. **Nipah Viruses:** This is a class C biological weapon. In 1999, the epidemic, a disease that attracted the attention of the world health institutions, started in the Nipah region of Malaysia and infected 265 people and killed 105 people. The disease appears with fever,

muscle pain, and progressive stage of brain inflammation, in a duration of 6 to 10 days. In more serious cases, the patients are relieved, the muscles contract, and the patients eventually go into a coma and even death. So far, the standard treatment and vaccine of the disease could not be found.

3. **Beef Plaque:** The disease is caused by a virus associated with measles and affects cattle and other horned, ruminant animals such as goats, giraffes, and bison. Fever, anorexia, dysentery, and mucous membranes inflammation are other symptoms. The duration of the animal's death from dehydration varies between 6 and 10 days.

4. **Rice Blast:** Especially America and Russia showed great interest in research against insects and diseases targeting agricultural products. This virus causes disease by a fungus known as Magnaporthe grisea. In the leaves of the affected plant, grayish lesions consisting of thousands of fungal spores are developing.

5. **Botulinum Toxin:** Completely colorless and odorless, this lethal bacterium can kill people between 24 hours and 72 hours. A class A biological weapon is spread through bacteria, food-borne, wound, and newborn.

6. **Tularemia:** This, claimed to have a 5% mortality rate, is one of the most infectious bacteria in the world. This bacterium, which was seen in 10,000 people in Russia in 1941, reached up to 1,00,000 people in the battle of Stalingrad.

   Symptoms of the disease are sudden fever, severe headache, malaise, nausea, vomiting, and diarrhea. Untreated, respiratory failure, shock, and death occur. The disease can be transmitted by direct contact with animals, and can also be carried by flies and ticks.

7. **Plague:** This, which can be treated with modern antibiotics, is known as black death and caused the death of half of the European population in the 14th century. The disease, which manifests itself as bubonic, septicemic, pneumonic, and gastro-intestinal, is spread by flea bites. The disease can be transmitted from person to person by body fluids. The disease manifests itself in two to eight days. There is no vaccine.

8. **Anthrax:** Due to its high mortality and environmental stability, anthrax bacteria, a class A biological weapon, live in the soil.

Animals are contacting bacterial spores when they dig to find food. People are caught in anthrax by touching sports, breathing, or swallowing them. It is not easy to get anthrax, and it can be transmitted from soil or by inhalation. There is no effective protection vaccine available in the world.

9. **Smallpox:** The disease is shown as class A biological weapon due to high mortality and airborne contamination. Symptoms of the disease include high fever, body pain, and hives with water-filled swells, while the wounds formed become permanent pits. The most common form of the disease is associated with the mortality rate of 30% (Erkekoğlu et al., 2018).

3.5 CONCLUSION

Nowadays, it is possible to use many microbiological agents and toxins of these agents as biological warfare agents. In particular, volatilized bacteria, viruses, and toxins can be released into the environment by different methods and may cause serious fatal effects. Although there are many studies on biological warfare agents, the studies related early detection of these agents and volatile determination in nature are relatively new.

Bioterrorism is more dangerous than chemical and physical terror attacks. In the past, state resources have been needed for the use and dissemination of biological threat agents, today biological threat agents can be produced without the need for a state. A large number of resistant microorganisms can be produced by developing techniques. With varying climate conditions and ecological changes, the biological agent tends to be increased in threatened diseases. İn order to prevent these threats, biosecurity measures should be implemented at international, national, and environmental levels.

In recent years, the microorganisms and biochemical compounds used in the detection of nanoscale materials have been proven to have adjustable and improved properties. Therefore, the use and development of biosensors in the process of biosensing have been one of the important research subjects.

Fast and reliable detection of biosensors and various biological threat agents is possible. These sensors can interact selectively with the target agent, and they can be used for early detection identification. In addition, with advanced sensor models, the potential for preventing threats in the health, environment, and food area has increased.

In the development of biosensors, it is thought that the engineering studies of biosensing materials will play an important role. The ultra-small size and ultra-precision of the materials that prevent the development of biosensors alone is an important problem in this case, increasing the costs of using biosensors.

The importance of antibodies in the biosensing process is great. Scientific studies on the use of sensors in the early diagnosis of high quality and selective antibodies are carried out. These studies are aimed at producing cheap and manpower-free results that minimize errors. The biosensor technology developed as a result of these studies will contribute to many industries rather than blocking biointensive agents.

KEYWORDS

- **biologically active substances**
- **biosensing supplies**
- **biosensor**
- **carbon nanotube**
- **chemical active ingredients**
- **graphene**
- **nanomaterial**

REFERENCES

Alocilja, E., & Muhammad-Tahir, Z., (2008). Label-free microbial biosensors using molecular nanowire transducers. In: Zourob, M., Elwary, S., & Turner, A. P. F., (eds.), *Principles of Bacterial Detection: Biosensors, Recognition Receptors and Microsystems* (pp. 377–413). Springer, New York.

Asefa, T., Duncan, C. T., & Sharma, K. K., (2009). Recent advances in nanostructured chemosensors and biosensors. *Analyst, 134*, 1980–1990.

Ayranci, R., Demirkan, B., Sen, B., Şavk, A., Ak, M., & Şen, F., (2019). Use of the monodisperse Pt/Ni@rGO nanocomposite synthesized by ultrasonic hydroxide assisted reduction method in electrochemical nonenzymatic glucose detection. *Mater. Sci. Eng. C., 99*, 951–956.

Barras, V., & Greub, G., (2014). History of biological warfare and bioterrorism. *Clin. Microbiol. Infect., 20*, 497–502.

Bourgeois, S. L., & Doherty, M. J., (2005). Bioterrorism and biologic warfare. *Oral Maxillofac. Surg. Clin. North Am., 17*, 299–330.

Bozkurt, S., Tosun, B., Sen, B., Akocak, S., Savk, A., Ebeoğlugil, M. F., & Sen, F., (2017). A hydrogen peroxide sensor based on TNM functionalized reduced graphene oxide grafted with highly monodisperse Pd nanoparticles. *Anal. Chim. Acta, 989,* 88–94.

Clark, L. C., & Lyons, C., (1962). *Electrode Systems for Continuous Monitoring in Cardiovascular Surgery (*Vol. 102, p. 29). Annals of the New York Academy of Sciences, Ala.

Coulet, P. R., (2010). Enzyme biosensors. In: Scott, A. O., (ed.), *Biosensors for Food Analysis* (p. 28). Woodhead Publishing.

Coyle, S., Curto, V. F., Benito-Lopez, F., Florea, L., & Diamond, D., (2014). Wearable bio and chemical sensors. In: Edward Sazonov, (eds.), *Wearable Sensors* (pp. 65–83). IOP Publishing Ltd.

Daniel, B., (2004). *A Plague Upon Humanity: The Secret Genocide of AXIS JAPAN'S GERM WARFARE Operation.* Souvenir.

Datta, A. K., (1990). Novel chemical and biological sensors for monitoring and control of food processing operations. *J. Food Eng., 12*, 223–238.

Erkekoğlu, P., & Koçer-Gümüşel, B., (2018). Biyolojik savaş ajanlari: Tarihçeleri, patofizyolojileri, tanilari, tedavileri ve önlemler. *FABAD J. Pharm. Sci., 43*, 171–201.

Fidler, D. P., (1999). Facing the global challenges posed by biological weapons. *Microbes Infect., 1*, 1059–1066.

Frischknecht, F., (2008). The history of biological warfare. In: Richardt, A., & Blum, M., (eds.), *Decontamination of Warfare Agents* (pp. 1–10). Wiley-VCH Verlag GmbH & Co. KGaA, Weinheim, Germany.

Giraldo, J. P., Landry, M. P., Faltermeier, S. M., McNicholas, T. P., Iverson, N. M., Boghossian, A. A., Reuel, N. F., et al., (2014). Plant nanobionics approach to augment photosynthesis and biochemical sensing. *Nat. Mater., 13*, 400–408.

Griffiths, D., & Hall, G., (1993). Biosensors – what real progress is being made? *Trends Biotechnol., 11*, 122–130.

Guilbault, G. G., & Montalvo, J. G., (1969). A urea-specific enzyme electrode. *J. Am. Chem. Soc., 91*, 2164–2165.

Gutés, A., Céspedes, F., Alegret, S., & Del, V. M., (2005). Determination of phenolic compounds by a polyphenol oxidase amperometric biosensor and artificial neural network analysis. *Biosens. Bioelectron., 20*, 1668–1673.

Hamburg, M. A., (2002). Bioterrorism: Responding to an emerging threat. *Trends Biotechnol., 20*, 296–298.

Hobson, N. S., Tothill, I., & Turner, A. P. F., (1990). Microbial detection. *Biosens. Bioelectron., 11*, 455–477.

Iverson, N. M., Barone, P. W., Shandell, M., Trudel, L. J., Sen, S., Sen, F., Ivanov, V., et al., (2013). *In vivo* biosensing via tissue-localizable near-infrared-fluorescent single-walled carbon nanotubes. *Nat. Nanotechnol., 8,* 873–880.

Ivnitski, D., Abdel-Hamid, I., Atanasov, P., & Wilkins, E., (1999). Biosensors for detection of pathogenic bacteria. *Biosens. Bioelectron., 14*, 599–624.

Kaya, S. I., Demirkan, B., Bakirhan, N. K., Kuyuldar, E., Kurbanoglu, S., Ozkan, S. A., & Sen, F., (2019). Highly sensitive carbon-based nanohybrid sensor platform for determination of 5-hydroxytryptamine receptor agonist (eletriptan). *J. Pharm. Biomed. Anal., 174*, 206–213.

Kindschy, L. M., & Alocilja, E. C., (2004). A review of molecularly imprinted polymers for biosensor development for food and agricultural applications. *Trans. ASAE., 47*, 1375.
Li, Y., (2006). Section 2.3 biosensors. In: Munack, A., (ed.), *CIGR Handbook of Agricultural Engineering* (Vol. VI, p. 52). St. Joseph: Michigan.
Marquis, B. J., Love, S. A., Braun, K. L., & Haynes, C. L., (2009). Analytical methods to assess nanoparticle toxicity. *Analyst, 134*, 425–439.
Mavrogiannis, N., Crivellari, F., & Gagnon, Z. R., (2016). Label-free biomolecular detection at electrically displaced liquid interfaces using interfacial electrokinetic transduction (IET). *Biosens. Bioelectron., 77*, 790–798.
Nakamura, H., & Karube, I., (2003). Current research activity in biosensors. *Anal. Bioanal. Chem., 377*, 446–468.
Patel, P. D., (2006). Overview of affinity biosensors in food analysis. *J. AOAC Int., 89*, 805–813.
Portakal, O., (2008). Bioassays and nanoparticles. *Turkish J. Biochem.*, 35–38.
Rissanen, J., (2003). *Biological Weapons Convention.* Nuclear Threat Initiative.
Savk, A., Özdil, B., Demirkan, B., Nas, M. S., Calimli, M. H., Alma, M. H., Asiri, A. M., & Şen, F., (2019). Multiwalled carbon nanotube-based nanosensor for ultrasensitive detection of uric acid, dopamine, and ascorbic acid. *Mater. Sci. Eng. C., 99,* 248–254.
Scheller, F. W., Hintsche, R., Pfeiffer, D., Schubert, F., Riedel, K., & Kindervater, R., (1991). Biosensors: Fundamentals, applications and trends. *Sens. Actuators B Chem., 4,* 197–206.
Schultes, G., Koppert, R., Goettel, D., Freitag-Weber, O., Probst, A. C., & Werner, U., (2009). New perspectives for pressure and force sensors thin films combining high gauge factor and low TCR. In: Schmid, U., (ed.), *Smart Sensors, Actuators and MEMS IV.*
Scott, A. M., Wolchok, J. D., & Old, L. J., (2012). Antibody therapy of cancer. *Nat. Rev. Cancer., 12*, 278–287.
Sen, F., Boghossian, A. A., Sen, S., Ulissi, Z. W., Zhang, J., & Strano, M. S., (2012). Observation of oscillatory surface reactions of riboflavin, trolox, and singlet oxygen using single carbon nanotube fluorescence spectroscopy. *ACS Nano, 6*, 10632–10645.
Sen, S., Sen, F., Boghossian, A. A., Zhang, J., & Strano, M. S., (2013). Effect of reductive dithiothreitol and Trolox on nitric oxide quenching of single-walled carbon nanotubes. *J. Phys. Chem. C., 117*, 593–602.
Shimizu, K. I., Chinzei, I., Nishiyama, H., Kakimoto, S., Sugaya, S., Yokoi, H., & Satsuma, A., (2008). Hydrogen sensor based on WO_3 subnano-clusters and Pt co-loaded on ZrO_2. *Sens. Actuators B Chem., 134*, 618–624.
Turner, A. P. F., (1989). Current trends in biosensor research and development. *Sens. Actuators, 17*, 433–450.
Turner, A., Karube, I., & Wilson, G. S., (1987). Biosensors: Fundamentals and applications. *Anal. Chim. Acta.*
Ulissi, Z. W., Sen, F., Gong, X., Sen, S., Iverson, N., Boghossian, A. A., Godoy, L. C., et al., (2014). Spatiotemporal intracellular nitric oxide signaling captured using internalized, near-infrared fluorescent carbon nanotube nanosensors. *Nano Lett., 14*, 4887–4894.
Updike, S. J., & Hicks, G. P., (1967). The enzyme electrode. *Nature, 214*, 986–988.
Wagner, G., & Guilbault, G. G., (1994). *Food Biosensor Analysis*. M. Dekker, New York.
Wang, L., Wei, Q., Wu, C., Ji, J., Liu, Q., Yang, M., & Wang, P., (2007). Detection of *E. coli* O157:H7 DNA by a novel QCM biosensor coupled with gold nanoparticles amplification. In: 7th *IEEE Conference on Nanotechnology*, U.S.

Zhang, J., Landry, M. P., Barone, P. W., Kim, J. H., Lin, S., Ulissi, Z. W., Lin, D., et al., (2013). Molecular recognition using corona phase complexes made of synthetic polymers adsorbed on carbon nanotubes. *Nat. Nanotechnol., 8,* 959–968.

Zhou, Y. L., & Zhi, J. F., (2006). Development of an amperometric biosensor based on covalent immobilization of tyrosinase on a boron-doped diamond electrode. *Electrochem. Commun., 8,* 1811–1816.

Zülfü, T., (2017). Biosensors and nanotechnological interaction, *Bitlis Eren University Journal of Science and Technology, 6*, 71–80.

CHAPTER 4

Correlations Between Nanoscience and Nanotechnology

NAGAMALAI VASIMALAI

Department of Chemistry, B.S. Abdur Rahman Crescent Institute of Science and Technology, Vandalur, Chennai, Tamil Nadu, India

ABSTRACT

Nanotechnology is one of the fastest emerging fields and has a wide range of applications in various aspects. The materials which range from 10 nm to 100 nm in size are known as nanomaterials (NMs). The use of nanoscale materials provides greater surface coverage, better efficiency, and selectivity. Moreover, the surface of NMs can be tuned to attain desired and physical, chemical, and mechanical properties. Therefore, NMs are extensively used in several fields, including agriculture, food science, biomedical sector, energy, and sensor applications. The outline of this chapter is: (i) introduction of nanoscience and nanotechnology; (ii) difference between the nanoscience and nanotechnology and the classification of NMs based on their structural morphology; (iii) nanotechnology for agriculture and biomedical applications; (iv) nanotechnology for energy and sensor applications; and (v) conclusions and opportunities in this exciting and emerging research field.

4.1 INTRODUCTION

Nanotechnology is one of the most exciting and rapidly evolving techniques in the field of physics, chemistry, engineering, medicine, and biological sciences (Figure 4.1). Nanotechnology is the control and understanding of matter at the nanoscale dimensions (Porter et al., 2007; Neethirajan & Jayas, 2010). As nanomaterials (NMs) provide unique physical, chemical,

Advanced Nanomaterials for Biological, Nutraceutical, and Medicinal Applications.
Sreeraj Gopi, Augustine Amalraj, and T. R. Sreeraj (Eds.)

and biological properties, it provides various opportunities to apply nanotechnology in different fields (Gordon et al., 2009). However, the fabrication and characterization of NMs remain as a challenging task. Therefore, efforts have been made to discover the synthetic methods for nano-architectures by many researchers globally (Suehiro, 2010). Applications of NMs have become a part of our day-to-day life.

4.1.1 WHAT IS NANOSCIENCE?

Nanoscience is an emerging field of science which deals with nanoscale range materials. Nanoscience is the study of physical and chemical properties of NMs and understanding of molecular level interactions (Whitesides, 2005). Unique properties have been observed in nanoscale materials, which are differing from their bulk metals. Nanoscience is an atomic physics, which deals with the complex systems of physics and chemistry.

In 1959, Feynman's talk about nanoscience but the article was published in 1992 (Ozin et al., 2005). The remarkable technological implications laid out by Feynman's. Hence, he is considered as the father of nanotechnology and nanoscience.

4.1.2 WHAT IS NANOTECHNOLOGY?

Nanotechnology is not a single discipline or technology it encloses various technologies that intersect sectors such as medicine, fabrication, electronics, devices, communications, NMs, and energy which has the capacity to measure and rule matter at the nanometer scale (Huang et al., 2018; Heath, 1998). The prefix word of nanotechnology, 'nano' was acquired from the Greek word dwarf, where the term nanotechnology was worn by the Japanese researcher Norio Taguchi in 1974 (Rai & Ingle, 1998). The Hypothesis of nanotechnology was perceived by the well-known physicist Richard Feynman in 1959, where he brought up the possibility of manipulating material at the level of single molecules and atoms (Toumey, 2009).

The significant drive for nanotechnology came from the industry of electronics for the progress of miniaturized electronic devices based on silicon chips. The remarkable growth of nanotechnology in the past two decades (Yang et al., 2004; Bohr, 2002), has emerged in various applications in diagnostics (Johnson et al., 2008), drug delivery (Subramani, 2006; Saeedi et al., 2019; McNeil, 2010; Lobenberg, 2003), biosensors (Chao et al., 2016; Vashist et al., 2012), healthcare (Raffa et al., 2010; Horton et al.,

2006; Gardner, 2015; Munir & Yasin, 2007; Prasad et al., 2018; Venugopal & Ramakrishna, 2016), medicine (Pustovalov & Babenko, 2004), food packaging and food safety (Duncan, 2011), therapeutics (Bamrungsap et al., 2012; Sonvico et al., 2005), environmental monitoring (Andreescu et al., 2009), textiles (Sawhney et al., 2008) and information and communication technologies.

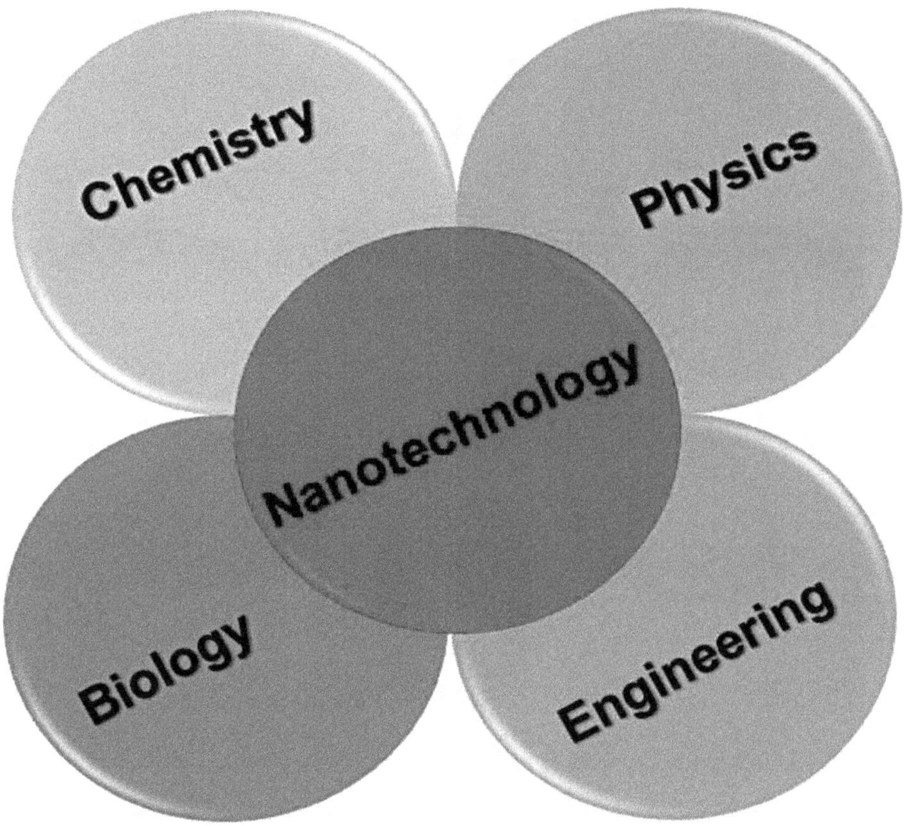

FIGURE 4.1 Nanotechnology involves different fields of science and engineering.

The outbreak of nanotechnology explains how small our interpretation is on the detailed process by which molecules arrange and gather among themselves, which ways to the exciting field of operation of complex nanostructured systems and interesting development of quantum devices (Fu, 2014; Sun et al., 2005). It is the change in properties upon the material size that leads to the investigation of a wide new formation of different possibilities. Nanotechnology is extensively used in several fields (Figure 4.2), includes optical engineering, textile industry, medical field, fertilizer industry, bioengineering, energy sector, devices, and food industry, etc.

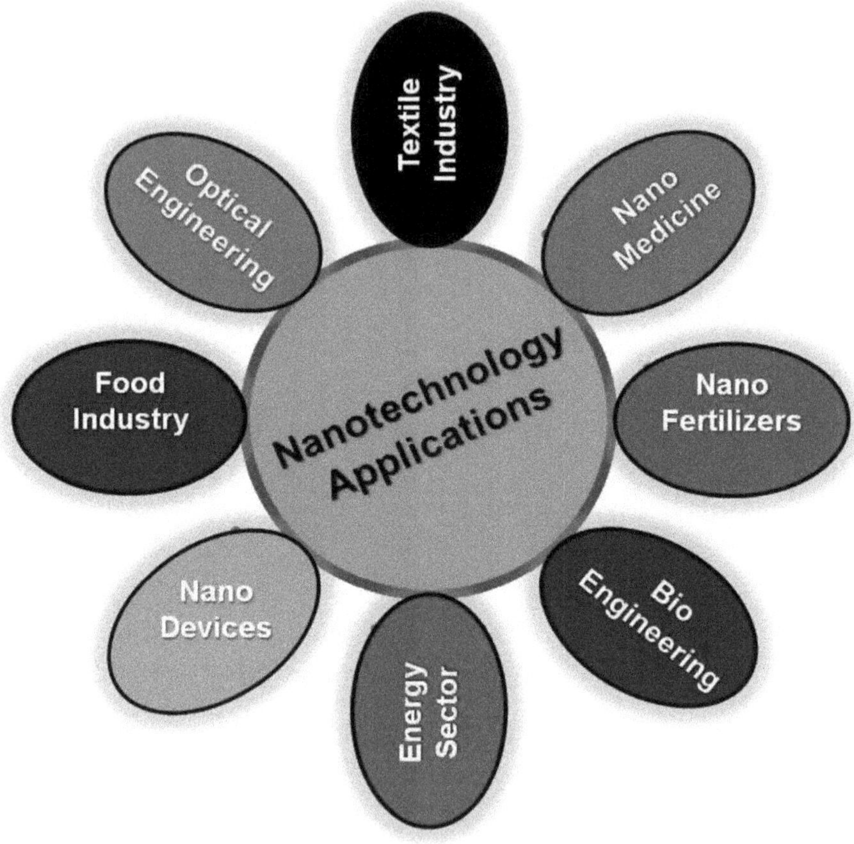

FIGURE 4.2 Variety of applications to nanotechnology.

4.1.3 DIFFERENCE BETWEEN NANOTECHNOLOGY AND NANOSCIENCE

For the past two decades, there has been a lot of stories surrounding the world of nanotechnology and how this arrival of innovative technology is making the changes around the world with numerous applications (Sparrow, 2007). This technology has blended into our everyday lives in different aspects such as Nanoparticles carrying dirt-resistant glass paint with nano-coating, sunscreen lotions, and various nanocomposite materials.

In the beginning of nanotechnology, growth was based on the revelation of new phenomena and the intersection of disciplines from the basic ideas of control of matter at nanoscale to novel science principles which leads to bridging the gap between reality and imagination due to all these developments (Jiang et al., 2010; Dubiel et al., 2011). At the global level, nanotechnology, and Nanoscience are gaining momentum in the domain of engineering and science, specifically for their possible industrial applications. The difference between nanoscience and nanotechnology is given in Table 4.1.

TABLE 4.1 Difference between Nanotechnology and Nanoscience

Sl. No.	Nanotechnology	Nanoscience
1	It is engineering and science of modeling and manipulation of matter on an atomic molecular and supramolecular scale.	It is the atomic physics which merges with the physics and chemistry of complex systems.
2	Nanotechnology is used in multi-disciplinary field which enclose understanding and control of matter about 1–100 nm.	It involves fundamental interaction of systems confined to its properties and nanoscale dimensions.
3	It is the science and technology of matter at the nanoscale level.	Interpretating the fundamental interactions of system confined to nanoscale dimension and their properties.

4.2 TYPES OF NANOPARTICLES

NMs are categorized based on the source, chemical composition and also on their size. Normally, there are four different material-based divisions.

4.2.1 CARBON-BASED NANOMATERIALS (NMS)

These materials possess carbon, where they are found in morphologies such as ellipsoids, spheres or hollow tubes. Carbon nanotubes (CNTs), carbon black and grapheme (Gr), fullerenes (C60), carbon nanofibers would likely to be incorporated in the category of carbon-based NMs. Chemical vapor deposition (CVD), arc discharge and Laser ablation are the most supreme production methods for those carbon-based material manufacturing (Yan et al., 2011).

4.2.2 INORGANIC-BASED NANOMATERIALS (NMS)

These types of NMs mainly consist of metal and metal oxide nanoparticles (Liu & Swihart, 2014). These materials can be produced as metals likely gold or silver nanoparticles (AgNPs), metal oxides (MOs) such as titanium dioxide (TiO_2) and zinc oxide nanoparticles (ZnONPs) and also semiconductors namely silicon and ceramics (Ningthoujam et al., 2007) (Figure 4.3).

4.2.3 ORGANIC-BASED NANOMATERIALS (NMS)

Organic-based NMs are often made up of organic matter, except inorganic-based NMs and carbon-based materials. The usage of non-covalent (weak) interactions for the design of molecules and for self-assembly helps to convert the organic NMs into desired specific structures probably as micelles, polymer Nanoparticles, dendrimers, and liposomes (Mody et al., 2014) (Figure 4.3).

4.2.4 COMPOSITE-BASED NANOMATERIALS (NMS)

This type of NMs is usually multiphase nanoparticles with one phase on the nanoscale dimension which can either combine with another nanoparticles that fuse with larger or with bulk-type materials (e.g., hybrid nanofibers) or more complex structures like metal organic frameworks (Shi et al., 2018).

These composites can be of any compositions of metal-based, carbon-based, organic-based with any suitable form of ceramic, metals, and polymer bulk materials. NMs are synthesized by different morphologies as specified in Figure 4.3 looking on the need properties for the corresponding application (Wang et al., 2002).

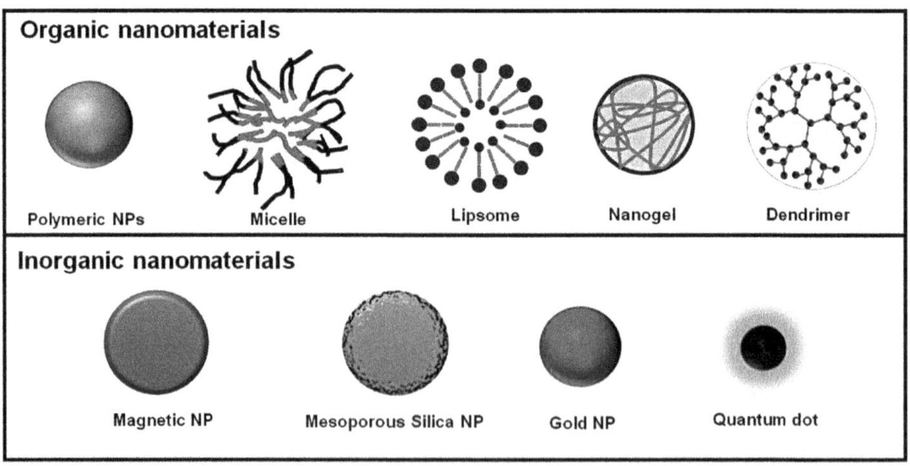

FIGURE 4.3 Classification of nanoparticles based on their composition and structure.

Source: Adapted with permission from: Shi et al. (2018). © 2017 Elsevier B.V.

4.3 APPLICATIONS OF NANOTECHNOLOGY

In recent years, nanotechnology applications have been extended to various sectors. The details are given in subsections.

4.3.1 NANOTECHNOLOGY FOR THE AGRICULTURAL SECTOR APPLICATIONS

The estimated world population as of 2020 is 7.8 billion that implies the need for new food, water, and energy systems to ensure good health. The European Commission (EC) has recognized nanotechnology as one of its six "key enabling technologies" which contributes to sustainable competitiveness and growth in various fields of industrial applications underlying the shift to a greener economy (Marchiol, 2018). Agricultural production is greatly affected by various forms of plant diseases and insect pests, causing significant economic loss. Nanotechnology can be applied in agriculture through the accurate and controlled release of fertilizers, pesticides, insecticides, insect repellents, herbicides, and antimicrobial compounds, the rapid identification of pathogens, increasing production, smart delivery systems and use of nano-based kits (Robles-Garcia et al., 2016).

4.3.1.1 PLANT GROWTH

In recent years, the effects of NMs on plant germination and growth have been studied with the goal of promoting its use for agricultural applications (Khodakovskaya et al., 2009) Concluded that by increasing the absorption of seed water, MWCNTs can penetrate tomato seeds and increase the germination rate (Figure 4.4). The MWCNTs raised seed germination in 20 days and plant biomass to 90% (compared to 71% in control) (Shah & Belozerova, 2009) have shown that nanoparticles (Pd, Au at low concentrations; Si, Cu at higher concentrations, and Au and Cu combination) have a positive effect on seed germination, increase seedling and enhance root-to-root ratio.

Nanoparticles can have a positive or negative influence on the plants (Monica & Cremonini, 2009). Their phytotoxicity is one of the most important points in the application of NMs for germinating seed. The phytotoxicity degree may be dependent on the kind of nanomaterial and its application. For instance, the applicability of silica nanoparticles labeled with fluorescein

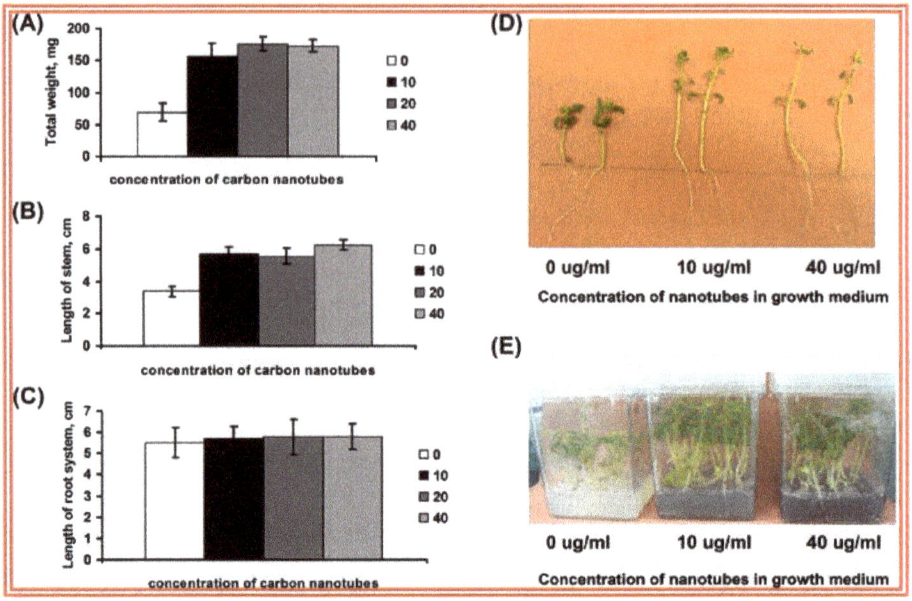

FIGURE 4.4 Effect of CNTs on growth and development of tomato seedlings.

Note: Results are shown as average ±SE of measurements of 10 plants per each condition. Around 27-day-old seedlings were used for all measurements. (A) Weight of total fresh biomass of tomato seedlings growing on medium with and without CNTs. (B) Length of stem of tomato seedlings growing on medium with and without CNTs. (C) Length of root system of tomato seedlings growing on medium with and without CNTs. (D) Phenotypes of 27-day-old tomato seedlings growing on medium with and without CNTs. And (E) phenotypes of 25-day-old tomato seedlings growing on medium without and with CNTs (10 and 40 μg/mL).

Source: Adapted with permission from: Khodakovskaya et al. (2009). Copyright 2009, American Chemical Society.

isothiocyanate (FTIC) and photostable Cadmium-Selenide (CdSe) quantum dots was tested to boost seed germination.

They claimed that FTIC-labeled silica nanoparticles promote seed germination in rice while CdSe quantum dots inhibit germination (Nair et al., 2011; Lin & Xing, 2007) nanomaterial assessed phytotoxicity (MWCNTs, Aluminum oxide-Al_2O_3, ZnO, Al, and Zn) and its effect on radish, rape canola, ryegrass, lettuce, corn, and cucumber germination levels and they stated that nano-sized Zn (35 nm) and ZnO (~20 nm) higher concentrations (2,000 mg/L) prevented germination in ryegrass and maize, respectively. The root length of the species studied was also affected by the use of nano-Zn and ZnO at 200 mg/L. Nano-Al and Al_2O_3 phytotoxicity greatly increased the root stretching of ryegrass and maize, respectively, while nano-Al helped the radish. Environmental Protection Agency (EPA) was allowed to use nano-silver in agriculture (Bergeson, 2010), and more than 100 pesticides contain nano-Ag because of their anti-microbial properties.

4.3.1.2 PLANT PROTECTION

Nanopesticides can be described as very small particles of pesticide active components or other small, engineered structures with useful pesticide properties. Nanopesticides can increase agricultural formulations' dispersion and wettability (i.e., decrease in organic solvent runoff) and harmful movement of pesticides.

NMs and biocomposites show good properties such as stiffness, biocompatability, crystallinity, solubility, and thermal stability, and permeability (Bouwmeester et al., 2009; Bordes et al., 2009) which are important for nanopesticides formulation. Nanopesticides also have a broad specific surface area that increases target affinity. Reported the launch of such modified sodium dodecyl sulfate (SDS) photocatalytic TiO_2/Ag dimethomorph nanomaterial joint (DMM), a pesticide widely used in agricultural production. Updated formulation of 96 nm average granularity, increased dispersiveness and breakdown of the pesticide in soil while enhancing its effect in (cabbage and cucumber) vegetable seedling studies.

The SDS altered nano-formulation of Ag/TiO_2 imidacloprid was developed via microencapsulation process that used chitosan (CS) and alginate. This was used on soybean plants transplanted to soil with 3.1% dry matter content and 6.2 pH. Within the first eight days, the formulation residues in soil and plants deteriorated faster and after 20 days were minimal to undetectable. The SDS has been used in the above applications to increase the photodegradation of nanoparticles in soil.

Liu et al. (2008) say that the stable nanopesticide (bifenthrin) has been successfully manufactured using polymer stabilizers such as poly(acrylic acid)-b-poly(butylacrylate) (PAA-b-PBA), polyvinylpyrrolidone (PVP), and polyvinyl alcohol (PVOH). Using a flash nano-precipitation technique 60–200 nm of bifenthrin nanoparticles were prepared.

Another important research item could be the development of NMs that can be used as a protective layer to allow traditional pesticides and fertilizers to be released slowly. For instance, Corradini et al. (2010) discovered the possibility of using CS nanoparticles as a highly degradable antibacterial material for slow release of NPK fertilizer.

4.3.2 NANOTECHNOLOGY IN BIOMEDICAL SECTOR APPLICATIONS

Nanotechnology is an incredible asset for testing developments in pharmaceutical science and novel plans. Nanoparticles have macromolecular materials and can be utilized restoratively as drug carriers in which the vital agent is disintegrated. Different kinds of nanopharmaceuticals are presently being utilized; like nanoemulsions, nanosuspensions, nanospheres, nanoshells, nanotubes, nanocapsules, lipid nanoparticles, dendrimers, and nanofluids to improve medicate conveyance into malignancy cells and tissues.

4.3.2.1 DRUG DELIVERY

Drug delivery based on nanoparticles has many benefits, such as enhancing the drug therapy performance and pharmacological characteristics. These nanoparticles improve the solubility of poorly water-soluble drugs, increase drug half-life by reducing immunogenicity, modify pharmacokinetics, improve bioavailability, increase specificity towards the target cell or tissue (therefore reducing side effects), delivery of two or more drugs simultaneously for combination therapy enable a more controllable release of therapeutic compounds and diminish drug metabolism (Sanvicens et al., 2008).

Implementing nanotechnology to drug delivery need to attain the following advantages:

- Develop the ability to deliver poorly water-soluble drugs.
- Including site-specific targeting to minimize accumulation of drugs within healthy tissue.
- Helps in the retention of drug in the body long enough for effective treatment.
- Outgrowth of drug bioactivity through biological environmental protection.
- Enable drugs to be transported through epithelial and endothelial barriers.
- Combine therapeutic and diagnostic methods into one agent (Sahoo & Labhasetwar, 2003; Suri et al., 2007; Safari & Zarnegar, 2014; Gu et al., 2013; Liu et al., 2012).

Cancer reportedly ranks among the world's leading causes of death (Bray et al., 2018). Current therapies may include chemotherapy, radiation, and surgery, but the results of these procedures may affect the tumor tissue as well as normal tissue (Prise et al., 1997). Weinberg and Hanahan have identified a set of six cancer characteristics that may help to differentiate between normal tissue and tumor tissue and may suggest alternative better treatments (Hanahan & Weinberg, 2011). These characteristics include the maintenance of proliferative signaling, the avoidance of growth suppressors, the activation of invasion and metastasis, replicative immortality, angiogenesis, and cell death resistance (Pickup et al., 2014; Mesri et al., 2014).

Nanotechnology has provided significant improvement in the diagnosis, treatment, and prevention of cancer treatment (Siddiqui et al., 2012). It's much more efficient and advanced therapeutic approaches that have provided latest methods and potential for disease treatment are minimal, due to the cancer cell determination is not significant (Baeuerle & Reinhardt, 2009).

Nanoparticles are used as guided tumor drug delivery that can enhance local contact with tumor cells and help increase biomarker preparation (Gray et al., 2013). Nanomedicines are used for accurate imaging of cancer (Ju et al., 2017). Iron oxide nanoparticles are uniformized as one of the cancer imaging techniques due to the specific coatings of iron nanoparticles, which are especially binding to the tumor cells and are used as an indicator of cancer detection due to cancer biomarkers (Perera & Coppens, 2019). During the early stage of cancer, levels of biological markers are used as imaging agents during MRI scans for their magnetic properties to provide a very high resolution to precise visualization of lesions (Sun et al., 2008).

4.3.3 NANOTECHNOLOGY IN ENERGY APPLICATIONS

4.3.3.1 SOLAR ENERGY APPLICATION

The photocatalytic water electrolysis is defined as the solar energy is required to break water molecules into oxygen and hydrogen (Takanabe, 2017). In brief, hydrogen is stored in solar cells (Roy et al., 2019). It is well known that artificial photosynthesis method is a leading research field, motivated by a demand for cheap hydrogen, which is expected to rise in the new hydrogen economy (Tachibana et al., 2012). Nanotechnology based solar energy production is an important area due to their high efficiency, low cost, environmentally friendly (Zach et al., 2006).

Many semiconductor nanoparticles (SNPs) are used for catalysis applications, including CdS, SiC, $CulnSe_2$ and TiO_2, etc., Among all TiO_2 expensively used, since it fulfills the above-mentioned requirements (Wang et al., 2014). But, this technology is still in the initial stage, because of their cost and low efficiency (Kouro et al., 2015). When nano-sized TiO_2 catalyst is used, the recombination of hydrogen and oxygen to produce water takes place. Moreover, the band gap of TiO_2 is around 3.2 eV, hence UV light is used as a source. Finally, a rapid recombination of photogenerated electron/hole pairs has also been reported in the literature. Consequently, a lot of work is being carried out to reduce these drawbacks shown by TiO_2 nanoparticles. Leung and co-workers, summarizes the different approaches can be classified into two broad categories of photocatalytic water splitting using TiO_2. With the addition of chemical additives, we will modify the photocatalyst physical and chemical properties. The added chemical additives contain electron donors, which are avoiding the electron hole recombination, which will be increasing the quantum efficiency of the electron/hole separation. Although inorganic ions including S_2/SO_3^{2-} Ce^{4+}/Ce^{3+} or IO_3/I are used as most common electron donors. Nada et al. (2005) has been categorized the hydrogen production capability in the following order: EDTA, methanol, ethanol, and, finally, lactic acid. Moreover, he observed that suppresses the backward reaction, while using the carbonate salts as chemical additives, because the carbonate species consume the photo generated holes in adequate quantities. Thus, considerably increase both hydrogen and oxygen production.

Platinum is the most common metal, but is expensive, other lower cost metals are Ni, Ag, Au, Cu, Pd, and Rh. Khan et al. (2008) have been synthesized platinum ion-exchanged titania nanotubes: (i) which shows high catalytic activity on the stoichiometric hydrogen production water and aqueous methanol; (ii) Ion doping, including transition metal and rare earth metal ions as well as anions like sulfur that can expand the photo catalytic response to visible region; (iii) dye sensitization and the coupling of semiconductors; and (iv) metal-ion implantation. These techniques are the most effective techniques of metal-ion implantation and dye sensitization. Moreover, in dye sensitization techniques, the additional use of sacrificial agents can enhance the hydrogen production rate. As an example, Nakato and co-workers have synthesized a thin film electrode made of a polycrystalline-Si/doped TiO_2 semiconductor. The system consists of a particulate doped TiO_2 thin film on the surface on inexpensive polycrystalline Si. Thus allowing to the absorption of short- and long-wavelength parts of the solar light, respectively. Consequently, this combination can yield a high solar-to-chemical conversion efficiency of more than 10%, which results in a very promising approach to efficient and low-cost solar energy conversion.

4.3.3.2 SUPERCAPACITORS

Electrochemical capacitors (ECs), also called supercapacitors and ultracapacitors, which store electrical energy like batteries but following a different mechanism (Simon & Gogotsi, 2009). While batteries do it chemically, ultracapacitors store electricity by separating the positive and negative charges (Schindall, 2007). Since the discovery of ECs, these devices have attracted considerably less attention than batteries as energy storage devices (Simon & Gogotsi, 2012). Nevertheless, thanks to both the contribution of nanotechnology and the better understanding of charge storage mechanisms (ion behavior in small pores) the interest on ECs has noticeably increased recently (Yang et al., 2007).

The drawbacks of classical capacitors can be summarized in three main points: (i) high cost of high performance electrodes due to the miniaturization; (ii) the larger requirements for long cycle life (around 105 cycles) (Markandya & Wilkinson, 2007); and (iii) their low efficiency-materials that combine both high surface and with a low resistivity are required.

The main determining factor for maximum power output and power density are the surface area of each electrode, which makes up the capacitor (El-Kady et al., 2012). The use of nanostructured materials significantly increases the surface area (e.g., up to 1,000 m^2/g of carbon) (Siegel, 1993). Moreover, in contrast to capacitors (Ma et al., 1999), supercapacitors utilize a small volume of electrolyte (Kandalkar et al., 2008) which interacts with the surface of each electrode to store charge. Consequently, they offer a unique combination of high power and high energy performance parameters.

The NMs typically used are metal-based nanocomposites and conductive polymers (Bogue, 2011; Jayathilaka et al., 2018), carbon-based nanostructures and hybrid inorganic/organic nanocomposites (Zhu & Diao, 2011), respectively. Electric double layer capacitors (EDLCs) are the most common devices currently. Recent technologies are mainly based on porous materials with a conductive addictive. A transition from activated carbon electrodes to carbon-based nanostructures is being done in order to improve the efficiency of these devices (Jiang et al., 2013). CNTs form an open mesoporous network, which allows the solvated ions for the interface of the double layer (Frackowiak & Béguin, 2002). Similar results have been observed for carbon nanofibers.

High conductivity can be achieved by the blending of activated carbon with less quantities of CNTs (Li et al., 2018), which can be further improved by doping of MOs or conductive polymers (Ishaq et al., 2019), but currently the cost is too high (Feng & Berntsson, 1997). Another approach includes the use of carbon aerogels (Gu & Yushin, 2013) which could lead to an improvement in capacitance and cyclability, because of their low electronic charging and iconic resistance (Marley et al., 2015). Similar characteristics have been proved for fullerene-like carbon nanoparticles (Harris et al., 2000). Finally, the use of nanotemplates (i.e., mesoporous silica, zeolites, etc.) (Kim et al., 2009) to synthesize nanosized carbon has also been explored for carbon electrodes.

NMs used for the nanostructuration of the cathode in Li-ion batteries are good candidates for redox supercapacitors (Kumar et al., 2018; Zuo et al., 2017). In this sense, transition-MOs such as V_2O_5, RuO_2, MnO_2 and transition metal nitrides can be used in the form of nanoparticles or nano-films (Zhao et al., 2019). The synthesis of nanotubes and nanowires of vanadium oxide can improve the efficiency since the agglomeration of nanoparticles is avoided with these morphologies. Shi et al. (2013) have been reported the synthesis of two-dimensional MnO_2 nanosheet used as cathode and graphene nanosheet as anode. First time they have achieved a high specific capacitance of 774 F g-1 even after 10,000 cycles. Interestingly, they have shown the energy storage devices and a Chinese panda supercapacitor is capable of red light emitting diode (Figure 4.5) (Shi et al., 2013).

Finally, hybrid supercapacitors based on nanotechnology combine the advantages of Li-ion batteries with those typically found in EDLCs, they are composed by a negative electrode (Liu et al., 2016), which is a battery-like electrode acting as energy source and a positive electrode made of carbon-based nanostructures (Mariappan et al., 2018), similar to that used in EDLCs, acting as a power source. Both electrodes are in the same cell with an electrolyte (Iwahara, 1990), which could be either aqueous or organic. For the capacitor electrode, i.e., the positive one (Raymundo-Piñero et al., 2006), NMs typically CNTs can be used in the development of ultracapacitors, which will have important uses for many walks of life-from communication to transportation-since they significantly increase this surface area (i.e., up to 103 m^2/g of carbon) (Nicole et al., 2001). Regarding the negative electrode based on nanotechnology, we have categorized into two types: (i) electrodes consisting on an inorganic nanomaterial capable of lithium (i.e., $Li_4Ti_5O_{12}$) (Wang et al., 2008) and (ii) pseudocapacitive electrodes with MOs NMs (Singh & Chandra, 2015). Consequently, their characteristics and alternatives are the same that already pointed out in batteries and supercapacitors (Pandolfo & Hollenkamp, 2006). The main contributions of nanostructured materials in the near future to hybrid supercapacitors should focus on increasing the energy density of these devices (Serrano et al., 2009).

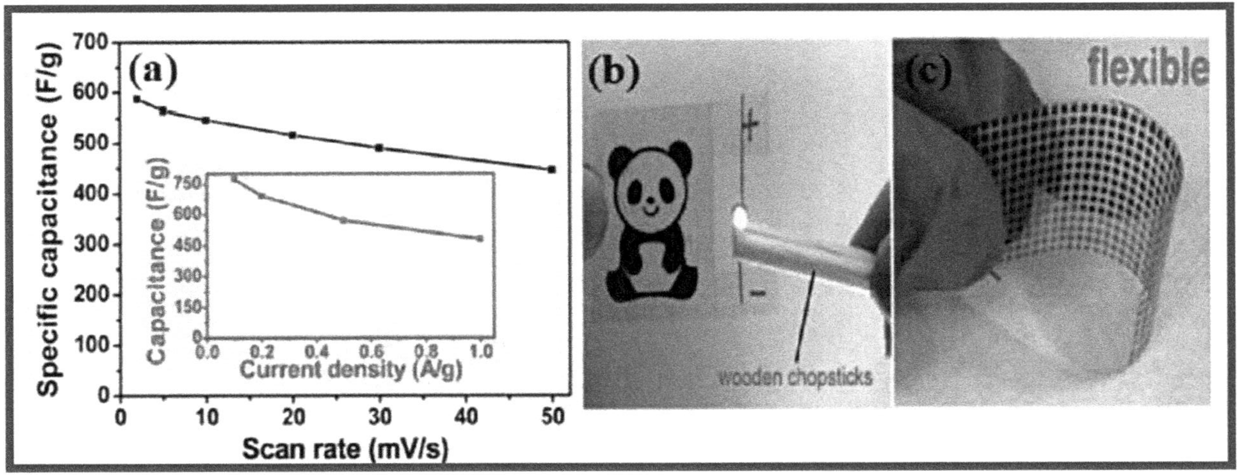

FIGURE 4.5 (a) Specific capacitance vs. scan rate of the 2D MnO_2 nanosheets; the inset is the capacitance versus discharging current density. Photographs of the (b) "panda" asymmetric MS/GA supercapacitor lighting up a red light-emitting diode. And (c) electrodes demonstrating mechanical flexibility.

Source: Adapted with permission from: Shi et al. (2013). Copyright 2013, Springer Nature. https://creativecommons.org/licenses/by/4.0/

4.3.4 NANOTECHNOLOGY IN THE SENSOR APPLICATIONS

4.3.4.1 GOLD NANOPARTICLES (AUNPS)

Gold nanoparticles (AuNPs) are conductive materials, which have an enormous surface area and show novel optical properties (Jans & Huo, 2012). In AuNPs, the surface plasmon is limited offering ascend to localized surface plasmon resonance (LSPR) (Zou et al., 2016). In this manner their color changes from red to yellow and their size become 100 nm (Mullen, 1985). AuNPs experience oscillations which are practically equivalent to their metallic lattice (Huang & El-Sayed, 2010). Depending on the shape of AuNPs, the heat and light dispersing emerging from the surface plasmonic decay may be affected. Henceforth, color changes in nanospheres are less articulated than GNRs. Another factor is an aggregation of nanoparticles. This boundary can be saddled for creating biosensors and optical immunoassays. Multifunctional AuNPs are presently being broadly utilized for identifying different biomarkers for cancer (Kanwar et al., 2012), neurological issues (Méndez et al., 2010), diabetes mellitus (Guoyao, 2009), nucleic acid (Wang et al., 2013), amino-acids (Tiwari et al., 2013), hemoglobin (Otto et al., 2005), and a variety of pathogens (Ascenzi et al., 2005). Chaichi and Ehsani immobilized GOx on CS shells with Fe_3O_4 cores and coordinated it with a luminal-chemiluminescence framework to build up an optical sensor for glucose (Zhu et al., 2019). They utilized AuNPs to catalyze the luminol reaction just as the reaction among GOx and glucose that lead to the generation of H_2O_2. The linear range of the neoteric was observed to be 1×10^{-4} to 8.5×10^{-7} M and with the detection limit of 4.3×10^{-7} M. Guo and collaborators announced a comparable enzymatic sensor with an electrochemical detection mechanism.

The sensor was accounted for to be particularly against electroactive non-specific biomolecules (Cheng et al., 2006), such as acetaminophen, UA, and AA, and had a life expectancy of as long as 90 days (Xishan et al., 2014). The sensor gave a direct amperometric response from 0.05 to 15 mM. AuNPs are likewise materials of decision for pathogen quantification. A fluorescence resonance energy transfer (FRET)-based immunoassay utilized AuNPs for the serious quenching of CdTe quantum dots fluorescence to detect outer membrane protein W (OmpW) of Vibrio cholerae. The AuNPs were functionalized with 11-mercaptoundecanoic corrosive (11-MUA) and afterward conjugated with OmpW. When OmpW reacts with its antibody, the separation among the two nanoparticles lessens under 10 nm. Lowering the emission intensity was observed because of the energy transfer from CdTeQDs to AuNPs. The fluorescence quenching increments with a rise in the AuNPOmpQ concentration by FRET. The sensor permitted the quantification of OmpW with a linear range from 2 and 10 nm. In another study, an enzyme-free electrochemical sensor was created utilizing a nanocomposite of CuZr MOFs with aptamer as the signaling probe for the detection of *Pseudomonas aeruginosa* (*P. aeruginosa*). A highly conductive gold working electrode deposited with Super P, a stable form of carbon black also AuNPs to acquire an improved and stable signal. This has improved the sensitivity.

Sabouri and collaborators built up a chemiluminescence sandwich immunosensor for the identification of hepatitis B utilizing functionalized AuNPs as capture probes. The AuNPs were at first permitted to respond with 11-MUA before conjugation with luminol and cAb. The identification test comprised of dAb. The predominant ability has shown by AuNPs, and it was harnessed to obtain a CL reaction somewhere in the range of 0.12 and 30 $ng.mL^{-1}$ of hepatitis B surface antigen (HBsAg) with a detection limit of 14 $pg.mL^{-1}$. An investigation for zika virus (ZKV) identification utilized silsesquioxane polyelectrolyte (SiPy) as a steady help for AuNPs. To develop the impedimetric biosensor, ssDNA was immobilized on an oxidized glassy carbon electrode (oxGCE) that was changed with AuNPs-SiPy. Good linearity was observed from 1.0×10^{-12} to 1.0×10^{-6} M with a detection limit of 0.82 pM.

4.3.4.2 MOLECULARLY IMPRINTED POLYMERIC NANOPARTICLES

Molecularly imprinted polymers (MIPs) are artificial materials synthesized by the template molecule with the polymerization of monomers and cross-linking agents (Sellergren & Hall, 2012). After the removal of the template, cavities have been generated, which corresponds with the shape, size of the template molecule (Wulff, 1995). MIPs thus provided the high selectivity and good physical and mechanical stability than other probes. MIPs can be sustained in harsh environments and their synthesis and functionalizations are easy.

MIP-based nanostructures have gained momentum in the field of sensor and medical diagnosis (Figure 4.6) (Qiao et al., 2011). For example, nano-MIP-based SPR sensor-based vancomycin detection in milk by Altintas (2018). In this study, itaconic acid (ITA) was polymerized using trimethylolpropane trimethacrylate and ethylene glycol dimethacrylate as

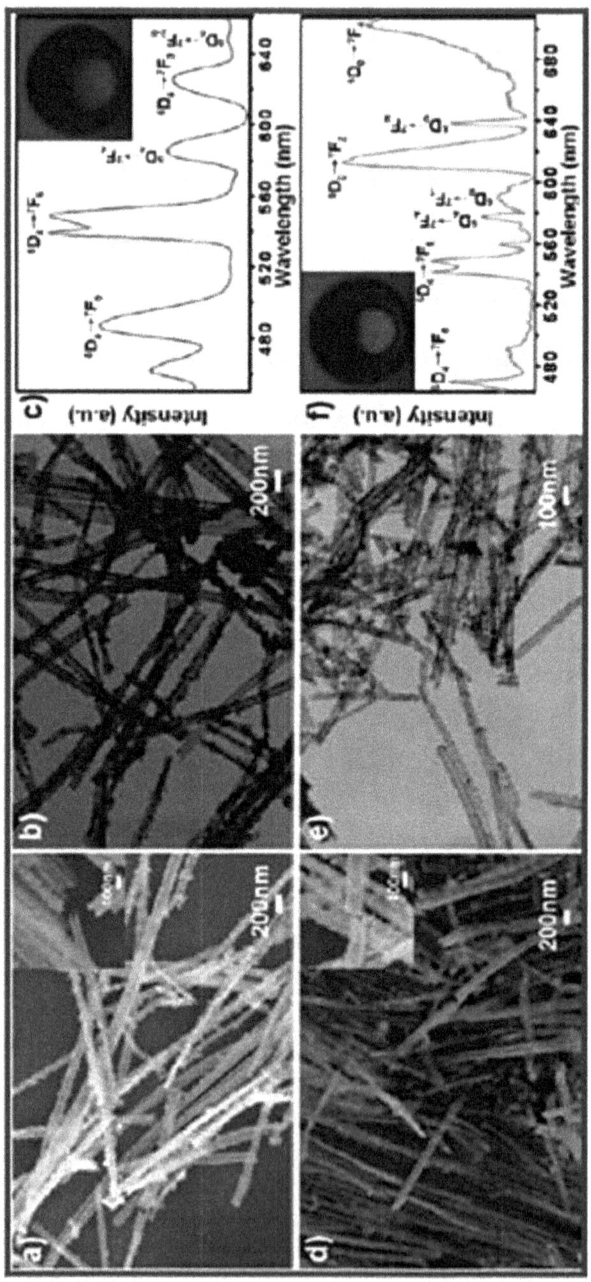

FIGURE 4.6 Silica nanotubes that are obtained from Tb3p-cholate (3.0 mM/1.5 mM) hydrogel (a–c) and Tb3p/Eu3p-cholate (3.0 mM/1.5 mM, Tb3p/Eu3p = 99/1) hydrogel (d–f). FE-SEM images ((a) and (d)); TEM images ((b) and (e)); and room temperature emission spectra ((c) and (f)) of the silica nanotubes (λex = 370 nm).

Source: Adapted with permission from: Qiao et al. (2011). Copyright 2011, American Chemical Society

cross-linkers. The nanoMIPs were immobilized on an 11-MUA functionalized Au sensor chip. The sensor has shown high specificity and affinity for the antibiotic. A linear range from 10 to 125 ng/ml was obtained and the LOD was 4.1 ng/ml.

Chen et al. (2014) developed a method for lysozyme (LZM) detection using surface imprinting strategy. The LZM template was immobilized on SiNP, followed by N-(4-vinylbenzyl) iminodiacetic acid (VBIDA) was polymerized in the presence of methylenebisacrlamide N-isopropylacrylamide, and acrylamide. VIBDA chelated with Cu^{2+} ions and recognized LZM, because of metal chelate substantivity. The enzyme binding kinetics have been improved because of their surface imprinted nanoparticles (Zhang, 2019). The detection range in this case was 0.1–1.0 mg/ml and LOD was 0.1 mg/ml. A similar technique was followed for the determination of bilirubin, where bilirubin imprinted Fe_3O_4-hydroxyapatite (HAP)-PPy Nanoparticles was immobilized on a magnetic glassy carbon electrode (MGCE) and applied for sensor application.

Yun & co-workers (2018) have developed for the sensor of amantadine drug using MIPs with RGO and AuNPs. Denizli's group prepared a MIP based QCM sensor for the detection of synthetic cannabinoids, in biological samples.

They have used 2-mthacryloyl-(L)-phenyllalanine (MAPA) monomer. A good linearity was observed from 0.0005 to 1.0 mg/ml with LODs of 0.2 to 0.45 pg/ml for four different cannabinoids. Altintas and co-workers reported a protein imprinting and achieved more stable surface conformation of the neuron specific enolase (NSE) protein in the sensing media. Altintas et al. also reported the sensitive detection of *E. coli* endotoxins using ITA as a functional monomer (Abdin et al., 2015).

4.4 CONCLUSION AND TRENDS

Nanotechnology provides to escort in a technological revolution due to their fascinating physical, chemical, and mechanical properties and nanometer size. In particular, metal and metal oxide nanoparticles having surface plasmon resonance and semiconducting properties, respectively. Therefore, nanotechnology is extensively used in several fields. In this chapter, I have provided the details about the basics of nanoscience and nanotechnology, and I hope that it will be very useful to the early-stage researchers who are starting to carry out the research. Then, I have well discussed the applications of nanotechnology in agriculture sector in particularly nanotechnology applications in plant protection, biomedical applications, drug delivery, solar cells and supercapacitor applications and sensor applications.

Nanotechnology, even though it is one of the emerging fields, but still there is an urgent need to develop new technologies for the direct societal applications. For example, COVID-19 virus drug delivery and vaccine development are needed today. Further, the nanotechnologies also have to concentrate on developing a new technology for the cheap cost kit for HIV, COVID-19 viruses using nanoscience. On the other hand, the portable kit preparation is important today. Vasimalai and co-workers have developed smart phone aided paper based kit for onsite monitoring of toxins in environmental water (Maruthupandi et al., 2020). This type of technology is urgently needed today to protect the living beings from their hazards. A various range of applications are there in nanotechnology, but utilizing it in a commercial aspect is much more important so that everyone can be benefited. Change in perspective is also important because many people focus on only one type of usage of nanoparticles for example up conversion nanoparticles are now widely being researched in the field of detection sensors and many researchers are currently working on it but up conversion nanoparticles can also be utilized in drug delivery and many other cases. Similarly, many types of applications by a single type of material are not taken into the research, so if we focus on multiple applications of a single compound, tremendous growth can be reached in the field of nanotechnology and also be very much useful to human society.

KEYWORDS

- **biomedical applications**
- **energy**
- **nanoscience**
- **nanotechnology**
- **plant nanotechnology**
- **sensor**

REFERENCES

Abdin, M. J., Altintas, Z., & Tothill, I. E., (2015). In silico designed nanoMIP based optical sensor for endotoxins monitoring. *Biosens. Bioelectron., 67*, 177–183.

Altintas, Z., (2018). Surface plasmon resonance-based sensor for the detection of glycopeptide antibiotics in milk using rationally designed nanoMIPs. *Sci. Rep., 8*, 1–12.

Andreescu, S., Njagi, J., Ispas, C., & Ravalli, M. T., (2009). JEM Spotlight: Applications of advanced nanomaterials for environmental monitoring. *J. Environ. Monit., 11,* 27–40.

Ascenzi, P., Bocedi, A., Visca, P., Altruda, F., Tolosano, E., Beringhelli, T., & Fasano, M., (2005). Hemoglobin and heme scavenging. *IUBMB Life, 57*, 749–759.

Baeuerle, P. A., & Reinhardt, C., (2009). Bispecific T-cell engaging antibodies for cancer therapy. *Cancer Res., 69*, 4941–4944.

Bamrungsap, S., Zhao, Z., Chen, T., Wang, L., Li, C., Fu, T., & Tan, W., (2012). Nanotechnology in therapeutics: A focus on nanoparticles as a drug delivery system. *Nanomedicine, 7,* 1253–1271.

Battal, D., Akgönüllü, S., Yalcin, M. S., Yavuz, H., & Denizli, A., (2018). Molecularly imprinted polymer-based quartz crystal microbalance sensor system for sensitive and label-free detection of synthetic cannabinoids in urine. *Biosens. Bioelectron., 111*, 10–17.

Bergeson, L. L., (2010). Nanosilver: US EPA's pesticide office considers how best to proceed. *Manag. Environ., 19*, 79–85.

Bogue, R., (2011). Nanocomposites: A review of technology and applications. *Assembly Autom, 31*, 106–112.

Bohr, M. T., (2002). Nanotechnology goals and challenges for electronic applications. *IEEE Trans. Nanotechnol., 1*, 56–62.

Bordes, P., Pollet, E., & Averous, L., (2009). Nano-biocomposites: Biodegradable polyester/nanoclay systems. *Prog. Polym. Sci., 34*, 125–155.

Bouwmeester, H., Dekkers, S., Noordam, M. Y., Hagens, W. I., Bulder, A. S., De Heer, C., & Sips, A. J. A. M., (2009). Review of health safety aspects of nanotechnologies in food production. *Regul. Toxicol., 53*, 52–62.

Bray, F., Ferlay, J., Soerjomataram, I., Siegel, R. L., Torre, L. A., & Jemal, A., (2018). Global cancer statistics 2018: GLOBOCAN estimates of incidence and mortality worldwide for 36 cancers in 185 countries. *CA-Cancer J. Clin., 68*, 394–424.

Chao, J., Zhu, D., Zhang, Y., Wang, L., & Fan, C., (2016). DNA nanotechnology-enabled biosensors. *Biosens. Bioelectron., 76*, 68–79.

Chen, H., Kong, J., Yuan, D., & Fu, G., (2014). Synthesis of surface molecularly imprinted nanoparticles for recognition of lysozyme using a metal coordination monomer. *Biosens. Bioelectron., 53,* 5–11.

Cheng, M., Cuda, G., Bunimovich, Y., Gaspari, M., Heath, J., Hill, H., & Thundat, T., (2006). Nanotechnologies for biomolecular detection and medical diagnostics. *Curr. Opin. Chem. Biol., 10*, 11–19.

Dubiel, E. A., Martin, Y., & Vermette, P., (2011). Bridging the gap between physicochemistry and interpretation prevalent in cell– surface interactions. *Chem. Rev., 111*, 2900–2936.

Duncan, T. V., (2011). Applications of nanotechnology in food packaging and food safety: Barrier materials, antimicrobials and sensors. *J. Colloid Interface Sci., 363*, 1–24.

El-Kady, M. F., Strong, V., Dubin, S., & Kaner, R. B., (2012). Laser scribing of high-performance and flexible graphene-based electrochemical capacitors. *Science, 335*, 1326–1330.

Feng, X., & Berntsson, T., (1997). Critical COP for an economically feasible industrial heat-pump application. *Appl. Therm. Eng., 17*, 93–101.
Frackowiak, E., & Béguin, F., (2002). Electrochemical storage of energy in carbon nanotubes and nanostructured carbons. *Carbon, 40*, 1775–1787.
Fu, Y., (2014). *Physical Models of Semiconductor Quantum Devices*. Springer: Netherlands.
Gardner, J., (2015). Nanotechnology in medicine and healthcare: Possibilities, progress and problems: Education and training. *SA Journal of Bioethics and Law, 8*, 50.
Gordon, R., Losic, D., Tiffany, M. A., Nagy, S. S., & Sterrenburg, F. A. S., (2009). The glass menagerie: Diatoms for novel applications in nanotechnology. *Trends Biotechnol., 27*, 116–127.
Gray, B. P., Li, S., & Brown, K. C., (2013). From phage display to nanoparticle delivery: Functionalizing liposomes with multivalent peptides improves targeting to a cancer biomarker. *Bioconjugate Chem., 24*, 85–96.
Gu, W., & Yushin, G., (2013). Review of nanostructured carbon materials for electrochemical capacitor applications: Advantages and limitations of activated carbon, carbide-derived carbon, zeolite-templated carbon, carbon aerogels, carbon nanotubes, onion-like carbon, and grapheme. *Wires Energy Environ., 3*, 424–473.
Gu, W., Wu, C., Chen, J., & Xiao, Y., (2013). Nanotechnology in the targeted drug delivery for bone diseases and bone regeneration. *Int. J. Nanomedicine., 8*, 2305.
Guoyao, W. J., (2009). Amino acids: Metabolism, functions, and nutrition. *Amino Acids, 37*, 1–17.
Hanahan, D., & Weinberg, R. A., (2011). Hallmarks of cancer: The next generation. *Cell, 144*, 646–674.
Harris, P. J. F., Vis, R. D., & Heymann, D., (2000). Fullerene-like carbon nanostructures in the Allende meteorite. *Earth Planet. Sci., 183*, 355–359.
Heath, J. R., (1998). A defect-tolerant computer architecture: Opportunities for nanotechnology. *Science, 280*, 1716–1721.
Horton, M., Khan, A., & Maddison, S., (2006). Delivering nanotechnology to the healthcare, IT and environmental sectors-a perspective from the 'London Centre for nanotechnology.' *BT Technol. J., 24*, 175–183.
Huang, W., Ling, S., Li, C., Omenetto, F. G., & Kaplan, D. L., (2018). Silkworm silk-based materials and devices generated using bio-nanotechnology. *Chem. Soc. Rev., 47*, 6486–6504.
Huang, X., & El-Sayed, M. A., (2010). Gold nanoparticles: Optical properties and implementations in cancer diagnosis and photothermal therapy. *J. Adv. Res., 1*, 13–28.
Ishaq, S., Moussa, M., Kanwal, F., Ehsan, M., Saleem, M., Van, T. N., & Losic, D., (2019). Facile synthesis of ternary graphene nanocomposites with doped metal oxide and conductive polymers as electrode materials for high performance supercapacitors. *Sci. Rep., 9*, 1–11.
Iwahara, H., (1990). High temperature solid electrolyte fuel cells using perovskite-type oxide based on $BaCeO_3$. Gold nanoparticle-enabled biological and chemical detection and analysis. *J. Electrochem. Soc., 137*, 462.
Jans, H., & Huo, Q., (2012). Gold nanoparticle-enabled biological and chemical detection and analysis. *Chem. Soc. Rev., 41*, 2849–2866.
Jayathilaka, W. A. D. M., Qi, K., Qin, Y., Chinnappan, A., Serrano-García, W., Baskar, C., & Ramakrishna, S., (2018). Significance of nanomaterials in wearables: A review on wearable actuators and sensors. *Adv. Mater.,* 1805921.
Jiang, H., Lee, P. S., & Li, C., (2013). 3D carbon-based nanostructures for advanced supercapacitors. *Energy Environ. Sci., 6*, 41–53.

Jiang, J., Yu, J., & Corma, A., (2010). Extra-large-pore zeolites: Bridging the gap between micro and mesoporous structures. *Angew. Chem., 49*, 3120–3145.
Johnson, C. J., Zhukovsky, N., Cass, A. E. G., & Nagy, J. M., (2008). Proteomics, nanotechnology and molecular diagnostics. Proteomics, *8*, 715–730.
Ju, Y., Zhang, H., Yu, J., Tong, S., Tian, N., Wang, Z., Wang, X., et al., (2017). *ACS Nano, 11*, 9239–9248.
Kandalkar, S. G., Gunjakar, J. L., & Lokhande, C. D., (2008). Preparation of cobalt oxide thin films and its use in supercapacitor application. *Appl. Surf. Sci., 254*, 5540–5544.
Kanwar, J. R., Sun, X., Punj, V., Sriramoju, B., Mohan, R. R., Zhou, S. F., & Kanwar, R. K., (2012). Nanoparticles in the treatment and diagnosis of neurological disorders: Untamed dragon with fire power to heal. *Nanomed: Nanotechnol., 8*, 399–414.
Khan, M. A., Akhtar, M. S., Woo, S. I., & Yang, O. B., (2008). Enhanced photoresponse under visible light in Pt ionized TiO_2 nanotube for the photocatalytic splitting of water. *Catal. Commun., 10,* 1–5.
Khodakovskaya, M., Dervishi, E., Mahmood, M., Xu, Y., Li, Z., Watanabe, F., & Biris, A. S., (2009). Carbon nanotubes are able to penetrate plant seed coat and dramatically affect seed germination and plant growth. *ACS Nano, 3*, 3221–3227.
Kim, Y. H., Yoon, D. K., & Jung, H. T., (2009). Recent advances in the fabrication of nanotemplates from supramolecular self-organization. *J. Mater. Chem., 19*, 9091–9102.
Kouro, S., Leon, J. I., Vinnikov, D., & Franquelo, L. G., (2015). Grid-connected photovoltaic systems: An overview of recent research and emerging PV converter technology. *IEEE Ind. Electron. M., 9*, 47–61.
Kumar, K. S., Choudhary, N., Jung, Y., & Thomas, J., (2018). Recent advances in two-dimensional nanomaterials for supercapacitor electrode applications. *ACS Energy Lett., 3*, 482–495.
Li, X., Tang, Y., Song, J., Yang, W., Wang, M., Zhu, C., & Lin, Y., (2018). Self-supporting activated carbon/carbon nanotube/reduced graphene oxide flexible electrode for high performance supercapacitor. *Carbon, 129*, 236–244.
Lin, D., & Xing, B., (2007). Phytotoxicity of nanoparticles: Inhibition of seed germination and root growth. *Environ. Pollut., 150*, 243–250.
Liu, W., Li, J., Feng, K., Sy, A., Liu, Y., Lim, L., & Yu, A., (2016). Advanced Li-ion hybrid supercapacitors based on 3D graphene–foam composites. *ACS Appl. Mater., 8*, 25941–25953.
Liu, X., & Swihart, M. T., (2014). Heavily doped colloidal semiconductor and metal oxide nanocrystals: An emerging new class of plasmonic nanomaterials. *Chem. Soc. Rev., 43*, 3908–3920.
Liu, Y., Tan, J., Thomas, A., Ou-Yang, D., & Muzykantov, V. R., (2012). The shape of things to come: Importance of design in nanotechnology for drug delivery. *Ther. Deliv., 3*, 181–194.
Liu, Y., Tong, Z., & Prud'homme, R. K., (2008). Stabilized polymeric nanoparticles for controlled and efficient release of bifenthrin. *Pest Manag. Sci., 64*, 808–812.
Lobenberg, (2003). *Proceedings International Conference on MEMS, NANO and Smart Systems*. 0-7695-1947-4/03.
Ma, R., Liang, J., Wei, B., Zhang, B., Xu, C., & Wu, D., (1999). Processing and performance of electric double-layer capacitors with block-type carbon nanotube electrodes. *Bull. Chem. Soc. Jpn., 72*, 2563–2566.
Marchiol, L., (2018). *New Visions in Plant Science*. BoD–Books on Demand.

Mariappan, V. K., Krishnamoorthy, K., Pazhamalai, P., Sahoo, S., Nardekar, S. S., & Kim, S. J., (2018). Nanostructured ternary metal chalcogenide-based binder-free electrodes for high energy density asymmetric supercapacitors. *Nano Energy, 57*, 307–316.
Markandya, A., & Wilkinson, P., (2007). Electricity generation and health. *Lancet, 370*, 979–990.
Marley, P. M., Horrocks, G. A., Pelcher, K. E., & Banerjee, S., (2015). Transformers: The changing phases of low-dimensional vanadium oxide bronzes. *Chem. Comm., 51*, 5181–5198.
Maruthupandi, M., Thirupathi, D., & Vasimalai, N., (2020). One minute synthesis of green, fluorescent copper nanocluster: The preparation of smartphone aided paper-based kit for on-site monitoring of nanomolar level mercury and sulfide ions in environmental samples. *J. Hazard. Mater., 392*, 122294.
McNeil, S. E., (2010). *Unique Benefits of Nanotechnology to Drug Delivery and Diagnostics* (pp. 3–8). Springer.
Méndez, J. D., Xie, J., Aguilar-Hernández, M., & Méndez-Valenzuela, V., (2010). Molecular susceptibility to glycation and its implication in diabetes mellitus and related diseases. *Mol Cell Biochem., 344*, 185–193.
Mesri, E. A., Feitelson, M. A., & Munger, K., (2014). Human viral oncogenesis: A cancer hallmarks analysis. *Cell Host Microbe, 15*, 266–282.
Mody, N., Tekade, R. K., Mehra, N. K., Chopdey, P., & Jain, N. K., (2014). Dendrimer, liposomes, carbon nanotubes and PLGA nanoparticles: One platform assessment of drug delivery potential. *AAPS Pharm. Sci. Tech., 15*, 388–399.
Monica, R. C., & Cremonini, R., (2009). Nanoparticles and higher plants. *Caryologia, 62*, 161–165.
Mullen, K. T., (1985). The contrast sensitivity of human color vision to red-green and blue-yellow chromatic gratings. *J. Appl. Physiol., 359*, 381–400.
Munir, A. B., & Yasin, S. H. M., (2007). Nanotechnology in healthcare: Are existing laws adequate? *Eur. J. Health Law., 14*, 261–272.
Nada, A., Barakat, M., Hamed, H., Mohamed, N., & Veziroglu, T., (2005). Studies on the photocatalytic hydrogen production using suspended modified photocatalysts. *Int. J. Hydrog. Energy, 30*(7), 687–691.
Nair, P. M. G., Park, S. Y., & Choi, (2011). Expression of catalase and glutathione S-transferase genes in *Chironomus riparius* on exposure to cadmium and nonylphenol. *J. Comp Biochem Phys. C., 154*, 399–408.
Neethirajan, S., & Jayas, D. S., (2010). Nanotechnology for the food and bioprocessing industries. *Food Bioprocess Tech., 4*, 39–47.
Nicole, J., Tsiplakides, D., Pliangos, C., Verykios, X., Comninellis, C., & Vayenas, C., (2001). Electrochemical promotion and metal–support interactions. *J. Catal., 204*, 23–34.
Ningthoujam, R. S., Sudarsan, V., Godbole, S. V., Kienle, L., Kulshreshtha, S. K., & Tyagi, A. K., (2007). SnO_2: Eu^{3+} nanoparticles dispersed in TiO_2 matrix: Improved energy transfer between semiconductor host and Eu^{3+} ions for the low temperature synthesized samples. *Appl. Phys. Lett., 90*, 173113.
Otto, B. R., Sijbrandi, R., Luirink, J., Oudega, B., Heddle, J. G., Mizutani, K., & Tame, J. R. H., (2005). Crystal structure of hemoglobin protease, a heme binding autotransporter protein from pathogenic *Escherichia coli. Int. J. Biol. Chem., 280*, 17339–17345.
Ozin, G. A., Manners, I., Fournier-Bidoz, S., & Arsenault, A., (2005). Dream nanomachines. *Adv. Mater., 17*, 3011–3018.

Pandolfo, A. G., & Hollenkamp, A. F., (2006). Carbon properties and their role in supercapacitors. *J. Power Sources, 157*, 11–27.
Perera, A. S., & Coppens, M. O., (2019). Re-designing materials for biomedical applications: From biomimicry to nature-inspired chemical engineering. *Philos. Trans. Royal Soc. A, 377*, 20180268.
Pickup, M. W., Mouw, J. K., & Weaver, V. M., (2014). The extracellular matrix modulates the hallmarks of cancer. *EMBO Reports, 15*, 1243–1253.
Porter, A. L., Youtie, J., Shapira, P., & Schoeneck, D. J., (2007). Refining search terms for nanotechnology. *J. Nanoparticle Res., 10*, 715–728.
Prasad, M., Lambe, U. P., Brar, B., Shah, I., Ranjan, M. J. K., & Prasad, G., (2018). Nanotherapeutics: An insight into healthcare and multi-dimensional applications in the medical sector of the modern world. *Biomed Pharmacother., 97*, 1521–1537.
Prise, E. L., Etienne, P. L., Meunier, B., Maddern, G., Hassel, M. B., Gedouin, D., & Launois, B., (1994). A randomized study of chemotherapy, radiation therapy, and surgery versus surgery for localized squamous cell carcinoma of the esophagus. *Cancer, 73*, 1779–1784.
Pustovalov, V. K., & Babenko, V. A., (2004). Optical properties of gold nanoparticles at laser radiation wavelengths for laser applications in nanotechnology and medicine. *Laser Phys. Lett., 1*, 516–520.
Qiao, Y., Chen, H., Lin, Y., Yang, Z., Cheng, X., & Huang, J., (2011). Photoluminescent lanthanide-doped silica nanotubes: Sol−gel transcription from functional template. *J Phys Chem C, 115,* 7323–7330.
Raffa, V., Vittorio, O., Riggio, C., & Cuschieri, A., (2010). Progress in nanotechnology for healthcare. *Minim. Invasiv. Ther., 19,* 127–135.
Rai, M., & Ingle, A., (2012). Role of nanotechnology in agriculture with special reference to management of insect pests. *Appl. Microbiol. Biotechnol., 94*, 287–293.
Raymundo-Piñero, E., Kierzek, K., Machnikowski, J., & Béguin, F., (2006). Relationship between the nanoporous texture of activated carbons and their capacitance properties in different electrolytes. *Carbon, 44,* 2498–2507.
Robles-García, M. A., Rodríguez-Félix, F., Márquez-Ríos, E., Aguilar, J. A., Barrera-Rodríguez, A., Aguilar, J., & Del-Toro-Sánchez, C. L., (2016). Applications of nanotechnology in the agriculture, food, and pharmaceuticals. *J. Nanosci. Nanotechno., 16*, 8188–8207.
Roy, N., Suzuki, N., Terashima, C., & Fujishima, A., (2019). Recent improvements in the production of solar fuels: From CO_2 reduction to water splitting and artificial photosynthesis. *Bull. Chem. Soc. Jpn., 92,* 178–192.
Saeedi, M., Eslamifar, M., Khezri, K., & Dizaj, S. M., (2019). Applications of nanotechnology in drug delivery to the central nervous system. *Biomed. Pharmacother., 111*, 666–675.
Safari, J., & Zarnegar, Z., (2014). Advanced drug delivery systems: Nanotechnology of health design: A review. *J. Saudi Chem. Soc., 18*, 85–99.
Sahoo, S. K., & Labhasetwar, V., (2003). Nanotech approaches to drug delivery and imaging. *Drug Discov. Today, 8*, 1112–1120.
Sanvicens, N., & Marco, M. P., (2008). Multifunctional nanoparticles–properties and prospects for their use in human medicine. *Trends Biotechnol., 26*(8), 425–433.
Sawhney, A. P. S., Condon, B., Singh, K. V., Pang, S. S., Li, G., & Hui, D., (2008). Modern applications of nanotechnology in textiles. *Text. Res. J., 78*, 731–739.
Schindall, J., (2007). The charge of the ultracapacitors. *IEEE Spectrum, 44*(11), 42–46.
Sellergren, B., & Hall, A. J., (2012). Molecularly imprinted polymers. *Supramolecular Chemistry: From Molecules to Nanomaterials.*

Serrano, E., Rus, G., & García-Martínez, J., (2009). Nanotechnology for sustainable energy. *Renew. Sust. Energ. Rev., 13*(9), 2373–2384.
Shah, V., & Belozerova, I., (2008). Influence of metal nanoparticles on the soil microbial community and germination of lettuce seeds. *Water Air Soil Pollut., 197*(1–4), 143–148.
Shi, S., Xu, C., Yang, C., Chen, Y., Liu, J., & Kang, F., (2013). Hierarchically structured Co_3O_4@Pt@MnO_2 nanowire arrays for high-performance supercapacitors. *Sci. Rep., 3*, 1–8.
Shi, Z., Xu, C., Guan, H., Li, L., Fan, L., Wang, Y., & Zhang, R., (2018). Magnetic metal organic frameworks (MOFs) composite for removal of lead and malachite green in wastewater. *Colloids Surf. A Physicochem. Eng. Asp., 539*, 382–390.
Siddiqui, I. A., Adhami, V. M., Christopher, J., & Mukhtar, H., (2012). Impact of nanotechnology in cancer: Emphasis on nanochemoprevention. Impact of nanotechnology in cancer: Emphasis on nanochemoprevention. *Int. J. Nanomedicine, 7*, 591.
Siegel, R. W., (1993). *Proceedings of the First International Conference on Nanostructured Materials Nanostructured Materials, 3*, 1–18.
Simon, P., & Gogotsi, Y., (2008). Materials for electrochemical capacitors. *Nature Materials, 7*, 845–854.
Simon, P., & Gogotsi, Y., (2012). Capacitive energy storage in nanostructured carbon–electrolyte systems. *Acc. Chem. Res., 46*, 1094–1103.
Singh, A., & Chandra, A., (2015). Significant performance enhancement in asymmetric supercapacitors based on metal oxides, carbon nanotubes, and neutral aqueous electrolyte. *Sci. Rep, 5*.
Sonvico, F., Dubernet, C., Colombo, P., & Couvreur, P., (2005). Metallic colloid nanotechnology, applications in diagnosis and therapeutics. *Curr. Pharm. Des., 11*, 2091–2105.
Sparrow, R., (2007). Revolutionary and familiar, inevitable and precarious: Rhetorical contradictions in enthusiasm for nanotechnology. *Nanoethics, 1*, 57–68.
Subramani, K., (2006). Applications of nanotechnology in drug delivery systems for the treatment of cancer and diabetes. *Int. J. Nanotechnol., 3*, 557.
Suehiro, J., (2010). Fabrication and characterization of nanomaterial-based sensors using dielectrophoresis. *Biomicrofluidics, 4*, 022804.
Sun, C., Jerry. S. H. L., & Miqin, Z., (2008). Magnetic nanoparticles in MR imaging and drug delivery. *Adv. Drug Deliv. Rev., 60*(11), 1252–1265.
Sun, Q., Wang, J., & Guo, H., (2005). Quantum transport theory for nanostructures with Rashba spin-orbital interaction. *Phys. Rev. B Condens. Matter., 71*.
Suri, S., Fenniri, H., & Singh, B., (2007). Nanotechnology-based drug delivery systems. *J. Occup. Med. Toxicol., 2*, 16.
Tachibana, Y., Vayssieres, L., & Durrant, J. R., (2012). Artificial photosynthesis for solar water-splitting. *Nat. Photon., 6*, 511–518.
Takanabe, K., (2017). Photocatalytic water splitting: Quantitative approaches toward photocatalyst by design. *ACS Catal., 7*, 8006–8022.
Tiwari, B. K., Pandey, K. B., Abidi, A. B., & Rizvi, S. I., (2013). Markers of oxidative stress during diabetes mellitus. *J. Biomarkers*, 1–8.
Toumey, C., (2009). Plenty of room, plenty of history. *Nat. Nanotechnol., 4*, 783–784.
Vashist, S. K., Venkatesh, A. G., Mitsakakis, K., Czilwik, G., Roth, G., Von, S. F., & Zengerle, R., (2012). Nanotechnology-based biosensors and diagnostics: Technology push versus industrial/healthcare requirements. *Bionanoscience, 2*, 115–126.

Venugopal, J. R., & Ramakrishna, S., (2016). Nanotechnology: 21st century revolution in restorative healthcare. *Nanomedicine, 11,* 1511–1513.
Wang, D. W., Fang, H. T., Li, F., Chen, Z. G., Zhong, Q. S., Lu, G. Q., & Cheng, H. M., (2008). Aligned titania nanotubes as an intercalation anode material for hybrid electrochemical energy storage. *Adv. Funct. Mater., 18,* 3787–3793.
Wang, K. H., Choi, M. H., Koo, C. M., Xu, M., Chung, I. J., Jang, M. C., & Song, H. H., (2002). Morphology and physical properties of polyethylene/silicate nanocomposite prepared by melt intercalation. *J. Polym. Sci. Pol. Phys., 40,* 1454–1463.
Wang, M., Chen, J., Liao, X., Liu, Z., Zhang, J., Gao, L., & Li, Y., (2014). Highly efficient photocatalytic hydrogen production of platinum nanoparticle decorated SiC nanowires under simulated sunlight irradiation. *Int. J. Hydrogen Energ., 39,* 14581–14587.
Weiwei, W., Zhenlong, W., Zhaolai, D., Ying, Y., Junjun, W., & Guoyao, W., (2013). Glycine metabolism in animals and humans: Implications for nutrition and health. *Amino Acids, 45,* 463–477.
Whitesides, G., (2005). Nanoscience, nanotechnology, and chemistry. *Small, 1,* 172–179.
Wulff, G., (1995). Molecular imprinting in cross-linked materials with the aid of molecular templates—A way towards artificial antibodies. *Angew. Chem., 34,* 1812–1832.
Xishan, G., Bo, L., Jinming, J., Yelei, Z., & Xuesong, Y., (2014). Glucose biosensor based on a platinum electrode modified with rhodium nanoparticles and with glucose oxidase immobilized on gold nanoparticles. *Microchim Acta, 181,* 519–525.
Yan, L., Zhao, F., Li, S., Hu, Z., & Zhao, Y., (2011). Low-toxic and safe nanomaterials by surface-chemical design, carbon nanotubes, fullerenes, metallofullerenes, and graphenes. *Nanoscale, 3,* 362–382.
Yang, C., Irwin, P. C., & Younsi, K., (2004). The future of nanodielectrics in the electrical power industry. *IEEE Trans. Dielectr. Electr. Insul., 11,* 797–807.
Yang, H., Wang, Y., Lai, S., An, H., Li, Y., & Chen, F., (2007). Application of atomic force microscopy as a nanotechnology tool in food science. *J. Food Sci., 72,* R65–R75.
Yun, Y., Pan, M., Fang, G., Gu, Y., Wen, W., Xue, R., & Wang, S., (2018). An electrodeposited molecularly imprinted quartz crystal microbalance sensor sensitized with AuNPs and rGO material for highly selective and sensitive detection of amantadine. *RSC Adv., 8,* 6600–6607.
Zach, M., Hagglund, C., Chakarov, D., & Kasemo, B., (2006). Nanoscience and nanotechnology for advanced energy systems. *Curr. Opin. Solid State Mater Sci., 10,* 132–143.
Zhang, H., (2019). Molecularly imprinted nanoparticles for biomedical applications. *Adv. Mater.,* 1806328.
Zhao, W. W., Bothra, P., Lu, Z., Li, Y., Mei, L. P., Liu, K., & Cui, Y., (2019). Improved oxygen reduction reaction activity of nanostructured CoS_2 through electrochemical tuning. *ACS Appl. Energy Mater.,* 8605–8614.
Zhu, M., & Diao, G., (2011). Review on the progress in synthesis and application of magnetic carbon nanocomposites. *Nanoscale, 3,* 2748.
Zhu, N., Gu, L., Wang, J., Li, X., Liang, G. X., Zhou, J., & Zhang, Z., (2019). Novel and sensitive chemiluminescence sensors based on 2D-MOF nanosheets for one-step detection of glucose in human urine. *J. Phys. Chem. C., 123,* 9388–9393.
Zou, J., Si, Z., Cao, Y., Ran, R., Wu, X., & Weng, D., (2016). Localized surface plasmon resonance assisted photothermal catalysis of CO and toluene oxidation over Pd–CeO_2 catalyst under visible light irradiation. *J Phys Chem C., 120,* 29116–29125.
Zuo, W., Li, R., Zhou, C., Li, Y., Xia, J., & Liu, J., (2017). Battery supercapacitor hybrid devices: Recent progress and future prospects. *Adv. Sci., 4,* 1600539.

CHAPTER 5

A New Dawn for Human Health: Bio/Nanomaterials

TUERXUN DUOLIKUN,[1,2] CHIN WEI LAI,[2] BEY FEN LEO,[2,3] and MOHD. RAFIE BIN JOHAN[2]

[1]*Department of Mechanical Engineering, Faculty of Engineering, University of Malaya, Kuala Lumpur, Malaysia*

[2]*Nanotechnology and Catalysis Research Center (NANOCAT), Institute for Advanced Studies, University of Malaya, Kuala Lumpur, Malaysia*

[3]*Faculty of Medicine, University of Malaya, Kuala Lumpur, Malaysia*

ABSTRACT

Nano/biomaterials are as a pioneer of the fast-developed field of nanomedicine, due to its unique dimension-based properties make this material glossy in the field. In this chapter, a summary of the short history of biomaterials, advantages of nanomaterials (NMs) in the medicine, recent trends, and commercialization prospects have been discussed.

5.1 INTRODUCTION

Biomaterials have a long history, and it was a concern to repair unhealthy or wounded tissues or organs. The first generation of biomaterials, only investigating repair organs. Then benefited from the industrial revolution early of 20th century. After that, promising polymeric materials helped to the transformation and elaboration, as well as used in the sophisticated living systems.

Advanced Nanomaterials for Biological, Nutraceutical, and Medicinal Applications.
Sreeraj Gopi, Augustine Amalraj, and T. R. Sreeraj (Eds.)

The history of biomaterials can start referring to ancient *Roman civilization*, where already used gold as the dental filling materials and framing of the broken teeth. In the *Arabian civilization*, Dr. Rhazes (865–932 AD Persia) paid particular attention to tooth repair (Migonney, 2014).

Around 40 years ago, polyaryletherketones were played a vital role as biomaterials for trauma, orthopedic, and spinal implants (Kurtz et al., 2007). Nowadays, following the direct/indirect reasons, medicine also has been "zoom in" ed as nanomedicine, and high quality of advanced approach has been parallelly developed to satisfy the clinical demand. Absolutely sure, bionanomaterials (NMs) make nanomedicine succeed by its huge portion affords.

The biomaterials discipline involves knowledge from multidisciplinary fields such as materials science, chemistry, biology, mechanical, and medicine. Biomaterials applied in the various area such as orthopedics (joint replacements, bone plates, bone cement, artificial ligaments), cardiovascular (blood vessel prostheses, heart valves), dental fillers for tooth fixation, ophthalmic (contact lenses), wound healing and skin repair devices (Choudhary et al., 2019).

5.2 POLYMER IN MEDICAL APPLICATION

Biomaterials are used widely in a variety of fields, such as applications on food, medical, along with personal care. Biomaterials are also used with many other similar polymers through the whole of disciplines (Shoichet et al., 2009). Polymers (both natural and synthetic) and their composites are broadly researched for a variety of biomedical applications which includes tissue regeneration, wound therapeutics, and synthetic skins (Choudhary et al., 2019).

Due to natural polymers' easily available, non-toxic, non-inflammatory, biocompatible, and biodegradable qualities, they and their derivatives are commonly used in biomedical applications. Namely polysaccharides (cellulose, starch, alginate, pectin, carrageen gums), proteins, and some microbes-based polyester (polyhydroxyalkanaotes) and hyaluronate polysaccharide. Depending on the structure, the natural polymer is divided into four categories, such as polysaccharides, polypeptides, and proteins, polynucleotide, and polyester (Hemamalini et al., 2018). Bioresorbable polymers together with nanosize based structures can shed light on to new outlooks in the self-assembly of nanosized biomaterials for medical applications with tunable mechanical, thermal, and electrical properties (Armentano et al., 2010).

5.2.1 BIOCOMPATIBLE POLYMERS

While many polymers have been studied for medical applications, biocompatible nanoparticles for biophysical and biomedical applications share specific properties that are fundamental to their use as part of current materials for biomedicine. To be used in tissue engineering (TE), they must be biocompatible, non-toxic, and noninflammatory, which is particularly crucial when designing biodegradable polymers as the biomedical products should meet those requirements.

5.2.2 BIODEGRADABLE POLYMERS

Temporary structures are often a chosen design for polymeric scaffolds which have the desired geometry and the physical, chemical, and mechanical properties required for implantation. The biodegradability and biocompatibility of aliphatic polyesters as polylactide, poly(glycolides), poly(3-caprolactone), and poly(p-dioxanone) have triggered further interest for researching the applicability in the biomedicine, direct in the living cell environments (Armentano al., 2010).

Biopolymeric scaffolds have played a critical role in the TE fields, in particular bone scaffolds. Tissue-engineering systems have attempted to work as temporary ECMs as it is highly mimicking the main functions through replacing damaged cells with biopolymeric artificial scaffolds. Following the damaged bone scaffold is healed, the biodegradable polymeric ECMs will completely be absorbed by the body (Sabir et al., 2009).

5.2.3 BIOSTABLE POLYMERS

Biostable polymers have the potential to be used in encapsulated cell therapy, where cells are protected in a polymeric membrane from the host immune system to support an immune-privileged environment in which transplanted cells produce therapeutically relevant molecules for extended periods.

An implant usually stays longer in the human body. Biostability is connected with the polymers integrity aspect to stay in the living organism, where biopolymers are protected from corrosion. In polymers, different degradation can occur and also be accelerated through changes in pH and stress in the integrated interconnected lines. Polydimethylsiloxane is an effective biostable polymer which belongs to the group of silicone elastomers with minimal inflammatory reaction in the living organism, extensive use in application to personal care. As well as biocompatible and bio-stable makes it the most studied implantable polymers (Victor et al., 2019).

5.2.4 SMART POLYMERS

Smart polymers also called intelligent polymers; it can respond to a single/multiple stimuli such as:

i. **Physical:** Temperature, ionic strength, solvents, radiation (UV, visible), electric field, mechanical stress, high pressure, sonic radiation, magnetic field.

ii. **Chemical:** pH, specific ions, chemical agents.

iii. **Biochemical:** Enzyme substrates and affinity ligands (Hoffman et al., 2013).

Mostly smart polymers include peptides, polymer hydrogels useable in nanomedicine, such as drug and gene delivery system, new substrates for cell cultures or scaffolds for TE, and wound dressing. The inflammation-responsive polymeric systems also help in cancer-related diseases (Aguilar et al., 2019).

The polymer-biomolecule conjugates have been attracting worldwide research, to developing some sensitiveness of creating materials for much fascinating application in medicine (Hoffman et al., 2004). Besides, polyurethane-based shape memory polymers (SMP) are lightweight, have a high strain/shape recovery ability, are easy to process, to be biocompatible, non-toxic, and non-mutagenic. And required properties can be tailored for a variety of medical applications such as self-deployable vascular and coronary devices (Sokolowski et al., 2007). The electroactive smart polymers (ESPs) have been studied for biomedical applications. Thus, these ESPs have advantages in speeding up the regeneration mechanism in muscles and bones (Palza et al., 2019). As a carrier, smart polymeric drug delivery systems (SPDDS) can respond to environmental changes and consequently, alter their properties reversibly enabling an efficient and safe drug delivery (Heggannavar et al., 2019). Different significant routes of administration and targeting strategies are represented in Figure 5.1.

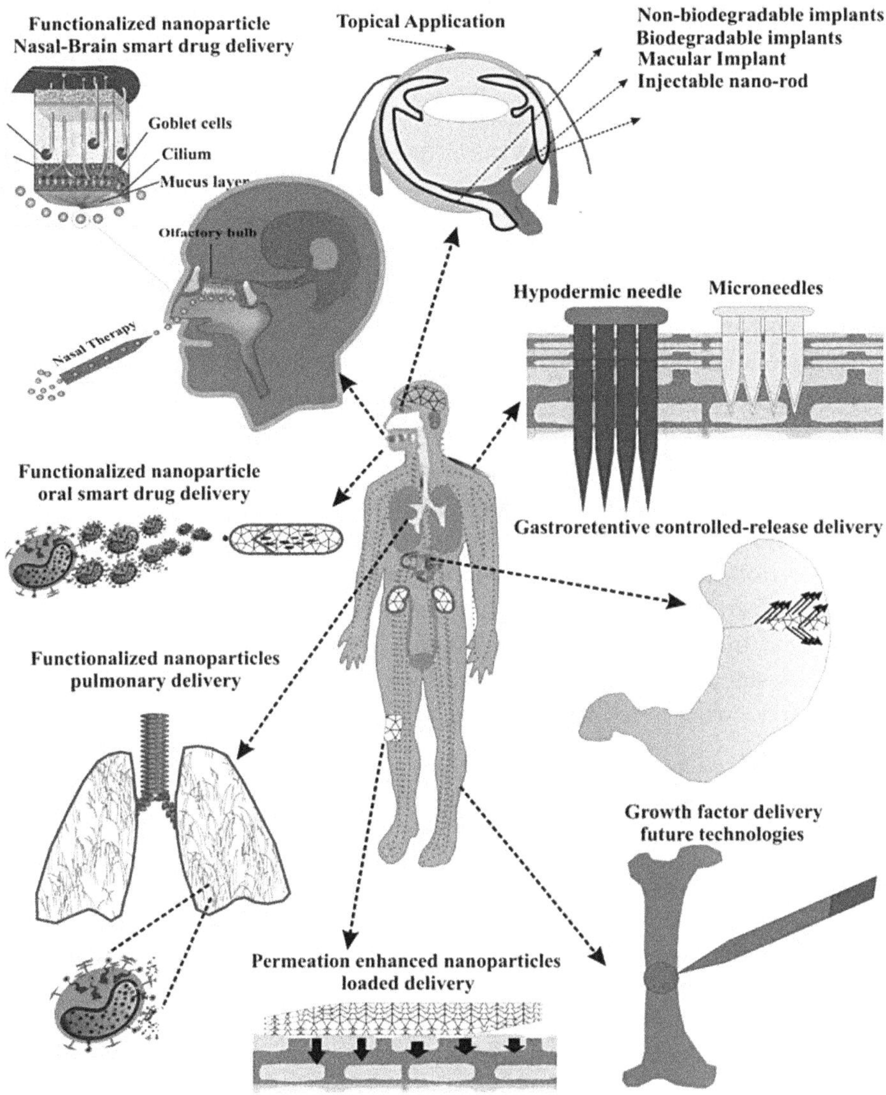

FIGURE 5.1 Schematic diagrams represent the recent developments of various significance routes of administration and targeting strategies (Bennet et al., 2014).

Source: Reprinted with permission from Bennet et al., 2014.

5.2.5 HYDROGEL

The hydrogel is a promising material with antibacterial functions. The hydrogel is a typical biomaterial synthesized via a natural water-soluble polymer that turns into gel through changing the parameters like temperature, ionic strength, or pH, etc. Due to its good hydrophilicity, unique 3D network, excellent biocompatibility, mimics physical properties of human tissue and cell adhesion, the hydrogel is one of the suitable biomaterials for drug delivery in antimicrobial areas (Chen et al., 2018).

The hydrogel is mainly used as nano-antimicrobials (NAMs) which are inherent antimicrobial activity or NMs that improve the efficacy and safety of antimicrobial drugs. In an aqueous medium for instance in body fluids hydrogel can swell rapidly, while keeping its structure intact and controlling the drug release (Yang et al., 2018).

Extracellular matrix (ECM) hydrogels can be derived from various tissue sources and used to treat ischemic injuries and organ replacement (Spang et al., 2018). The boronic ester-based hydrogels exhibited excellent self-healing property and dual pH/sugar responsiveness (Chen et al., 2018). The hydrogel-tissue material which can be probed and manipulated in new ways, in organisms and tissues across biology (including in cancer diagnostics, HIV infection, developmental biology, addiction, and mental health-related issues and fear/anxiety disorders) (Gradinaru et al., 2018). Natural polymeric biomaterial hydrogels (agarose, alginate, fibrin, and collagen chitosan) to mimic the satellite cell niche, which would allow for and promote an enormous regeneration potential (Pollot et al., 2018).

5.3 OTHER MAIN TYPES OF NANOMATERIAL USED IN MEDICAL APPLICATION

5.3.1 SEMICONDUCTOR NANOMATERIALS (NMS)

Developing the nanoscale semiconductor devices as new biomaterials will elicit a deeper understanding of electrical activity in a biological context., such as molecular and cellular biophysics studies (Zimmerman et al., 2014). Electrically conductive π-conjugated materials are attracting interest in a variety of biomedical applications ranging from neural electrode coatings to biosensors, cell, and TE, and artificial muscles (Tovar et al., 2013). Semiconductor quantum dots are exceptionally good for biological imaging and beneficial to conventional dyes. The advent of high-quality, solution-grown NMs that may be manipulated as a typical reagent while still maintaining their photoluminescent properties has allowed new applications, notably in bioimaging. For an instant, CdSe/ZnS core-shell materials have been reported to use in cancer imaging. Others, such as CdTe/CdSe suitable for optical coding applications (Green et al., 2004). Figure 5.2 represents microbeads containing various levels of CdSe/ZnS quantum dots.

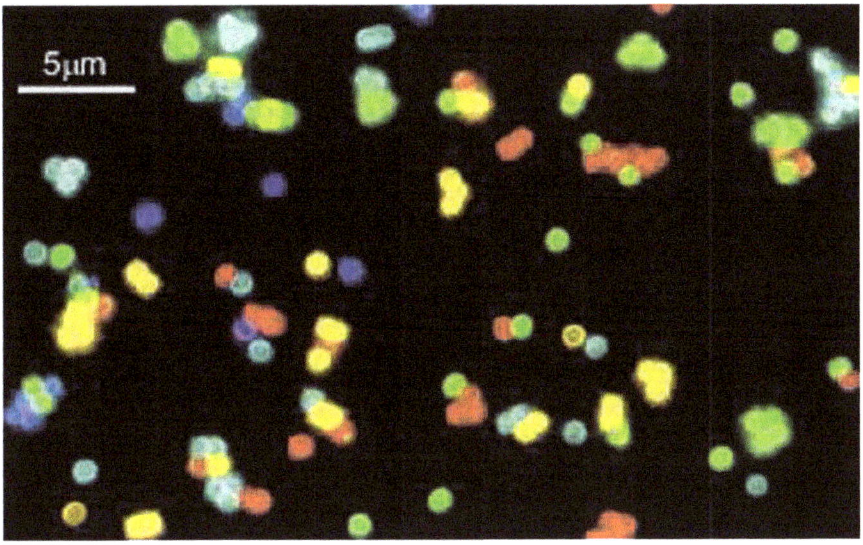

FIGURE 5.2 Microbeads containing various levels of CdSe/ZnS quantum dots (Han et al., 2001).

Source: Reprinted with permission from Han et al., 2001. Copyright 2001 Nature Publishing Group.

SNPs based thin films are a good potential for biomedical applications, coating SNPs with SiO_2 and organic polymers can render them sufficiently biocompatible to allow *ex-vivo* and animal studies. Improved biocompatibility of CdTe/PDDA multilayers has been studied of interactions between SNP/polymer composites and living tissues (Sinani et al., 2003).

5.3.2 MAGNETIC NANOMATERIALS (NMS)

In the recent past, magnetic nanoparticles (MNPs) have been progressively applied in biomedicine, more specifically in TE and drug delivery. Magnetite (Fe_3O_4) and maghemite (γ-Fe_2O_3) are the main MNPs which practical in biomedicine as a result of their low toxicity and high magnetization. Magnetic responsive surfaces that could control better the cellular behavior, implantable constructs with the shape-controlled by magnetic fields, magnetic beads for stem cell expansion, and magnetic hydrogel composites as remotely activated microfluidic devices (Gil et al., 2014).

Developing the methodology of biomolecular recognition elements that reveal the reproducible improvement and sensitive enhancement is crucial in proteomics and clinical diagnostics. Electrochemical immunosensors hold high potentials as the upgraded detection strategy which take advantages from their excellent sensitivity, easy miniaturization and cost-effective. The magnetic mesoporous $NiCo_2O_4$ nanosheet, an interlayer of Nafion/thionine organic molecules and a nanogold layer show excellent adsorption properties for the attachment of horseradish peroxidase-labeled secondary anti-CEA antibody (HRP-anti-CEA). With a sandwich-type immunoassay format, the functional bionanomaterials present unique analytical features to facilitate and modulate the way it was integrated onto the electrochemical immunosensors (Li et al., 2011). Besides, magnetic hydroxyapatite (HAP)/γ-Fe_2O_3/Au nanocomposite can be used for the simple, sensitive, and reusable piezoelectric immunosensor (Zhang e al., 2008).

The photothermal effect of clustered Fe_3O_4 magnetic nanoparticles was reported for the photothermal ablation of tumors, both *in vitro* in cellular systems but also *in vivo* study. It has shown with long irradiation time was more effective in killing A549 tumor cells (Shen et al., 2015).

5.3.3 METAL NANOPARTICLES

Biomedical metals (BMs) are bio-functionalized materials with bioactivity, anti-tumor, and anti-microbial, anticoagulation, and drug loading functions. For example, stainless steel, cobalt alloy, gold, Titanium, and its alloy have been widely used for clinical applications, especially for the dental, orthopedics, and other hard tissue replacement, such as dentures and related kits, internal fracture fixation, artificial hip joint, artificial cardiac valves, intravascular stent, and cardiac pacemaker box (Xiao et al., 2017). Recently, photosensitizers integrated PCN-224-Pt, which is metal-based nodes (metal ions or clusters) with organic linkers (organic ligands) have been developed for enhanced photodynamic therapy (PDT) (Zhang et al., 2018).

Silver nanoparticles (AgNPs) and gold nanoparticles (AuNPs) are highly remarkable BMs that used in biomedicine. AgNPs have been shown to possess fast, broad-spectrum antibacterial activity. The AgNPs based biomaterials are actively utilized for the treatment of all types of chronic wounds, and this combination plays an essential role in wound healing applications due to its non-cytotoxic and safe for patients (Kumar et al., 2018).

AuNPs have attracted tremendous interests from different fields of biomedicine, including sensing, bioimaging, photothermal therapy, due to high X-ray absorption coefficient, ease of synthetic manipulation, controllable physicochemical features, unique optoelectronic properties (Elahi et al., 2018). As cancer is one of the leading causes of mortality worldwide, the AuNPs for cancer diagnosis, photothermal, and PDT are great potential for improved cancer treatment and reduced side effects (Guo et al., 2017).

Corrosion is an essential factor in the design and selection of metals and alloys for service *in vivo*. The corrosion resistance of biomaterial is a primary factor that affects biocompatibility. Ion toxicity levels, possibly leading to hypersensitivity and cancer, may be attained even at corrosion rates that are insignificant concerning the physical performance of the implant (Eliaz et al., 2019).

5.3.4 CARBON NANOMATERIALS (CNMS)

Carbon nanomaterials (CNMs) are a class of low-dimensional carbonaceous materials that have aroused a great deal of interest in the past few decades. CNMs such as fullerenes, carbon nanotubes (CNTs), carbon quantum dots and graphene in materials sciences, nano-electronics, and photonics, their contributions to biology and biomedicine, in particular, biosensing, bioimaging, and as novel biomimetic materials (Teradal et al., 2017).

One of the good synthetic scaffold materials from CNMs should include the functionalized multi-walled carbon nanotubes (MWNTs) for bone regeneration. The results indicated MWNTs based biomaterials enhance osteoblast viability, adhesion, and proliferation at the desired level with comparable favors (Gutiérrez-Hernández et al., 2017). For the targeted drug delivery, CNTs are functionalized with a tumor-targeting ligand so that they can easily distinguish tumor cells from surrounding healthy cells (Kumar et al., 2017).

CNMs in the field of cancer phototherapy, namely photothermal therapy, PDT, and combined phototherapies attracted much attention over the last decades due to their strong absorption in the visible NIR regions, which require relatively lower energy and laser intensity for photo induction. Besides that, the appropriate size and large surface area of CNMs are also promising in the development of new generations of anticancer systems (Jiang et al., 2019).

CNMs also used in the intracellular translocation of biomacromolecules, in particularly, CNTs as intracellular shuttles to be applicable in this field, due to cellular uptake capability and protecting them from enzyme digestion (Mohajeri et al., 2019). Different CNMs and their potential applications are depicted in Figure 5.3.

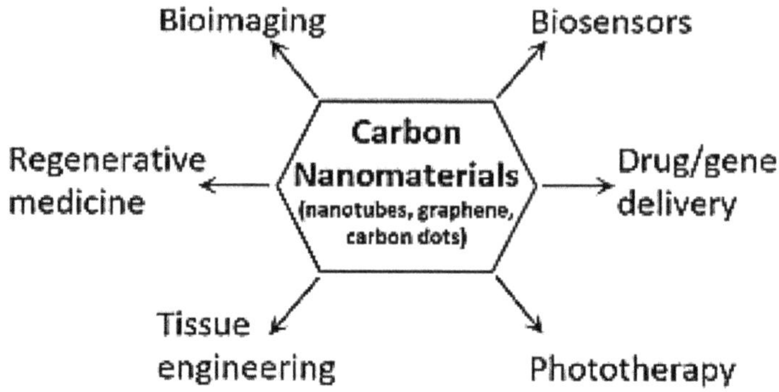

FIGURE 5.3 Carbon nanomaterials and potential applications (Teradal et al., 2017).

Source: Reprinted with permission from Teradal et al., 2017.

CNMs are applicable in cardiac tissue scaffolds, due to excellent mechanical and electrical properties. The enhanced cardiac tissue constructs have the potential to approach drug discovery and tissue regeneration, improving the current protocol on heart diseases treatment (Dozois et al., 2017).

5.3.5 HYDROGEL NANOCOMPOSITES

Nanocomposite hydrogels (NCHs) are hybrid material for biomedical applications with good properties such as highly porous nature and excellent water retention. NCHs are good potential in TE-related applications, like skeletal and electroactive tissues (Mehrali et al., 2017). These biomaterials benefit from retaining a large amount of water, effective mass transfer, similar to natural tissues and the ability to form different shapes. The first natural polymers exploited for the fabrication of hydrogels are biodegradable materials such as fibrin, collagen, hyaluronic acid (HA) and alginate, or rather, hydrogels can be made of synthetic polymers such as poly(ethylene glycol), poly(acrylamide) and poly(vinyl alcohol) (Vedadghavami et al., 2017).

Reinforced NCHs also great potential for resembling the ECM that useful in cartilage regeneration application (Asadi et al., 2018).

Hydrogels with reinforced mechanical properties that can repair themselves when damaged are of broad interest for biomedical applications due to a better stimulation of biological tissues and an improvement of materials' lifetime. Meanwhile, Stimulus-responsive NC hydrogels have been particularly impressive in TE, drug delivery and wound healing applications because they can undergo reversible physical and chemical changes after being exposed to external stimuli, for an instant, temperature, electric, and magnetic fields (Motealleh et al., 2017).

5.3.6 LIPOSOMES

Liposomes are spherical structures that are formed by one or several concentric lipid bilayers surrounding discrete aqueous spaces. Liposomes present several distinctive characteristics compared with other drug delivery systems including biocompatibility, no immunogenicity, the ability for self-assembly, improve the agents' solubility, ability to carry large drug payloads and protect the encapsulated agents from the external media (Gutiérrez-Hernández et al., 2017; Kumar et al., 2017; Jiang et al., 2019; Mohajeri et al., 2019), ability to reduce the toxicity of the encapsulated agent and the exposure of sensitive tissues to toxic drugs coupled with the strength of site-specific targeting and improving penetration into tissues. Indeed, for drug formulation, liposomes are the most versatile and sophisticated nanoparticle type since they have the capacity to deliver several biologically active compounds and macromolecules (DNA, peptides, and proteins), both in their lipid bilayer and in their lumen. Several liposomes based nanomedicine applications have been approved for clinical use (Vahed et al., 2017).

Stimuli-responsive liposomes with the use of various triggers such as pH, temperature, light, enzyme, and redox potential, etc., are emerging tactics for the effective localization of the bioactive at the tumor site with desired molecular manifestations (Jain et al., 2018). The dual-functional drug liposomes (DFDL) are taking control of the dual-function that drug efficacy, as well as the drug carrier. DFDL are first defined as the drug-containing phospholipid bilayer vesicles that possess a dual-function of providing primary therapeutic efficacy of the drug and the extended effect of the drug carrier (Mu et al., 2017). Their application in the destruction of vasculogenic mimicry channels by regulating the signaling pathways are illustrated in Figure 5.4.

The dual-functional epirubicin liposomes by consisting of celecoxib to destroy vasculogenic mimicry channels. Their mechanism study revealed that the drug liposomes were able to downregulate the vasculogenic mimicry channels related proteins, including matrix metalloproteinase-9 (MMP-9), vascular endothelial cadherin (VE-Cad), hypoxia-inducible factor-1α (HIF-1α), focal adhesion kinase (FAK), and ephrin type-A receptor 2 (EphA2), etc. (Mu et al., 2017).

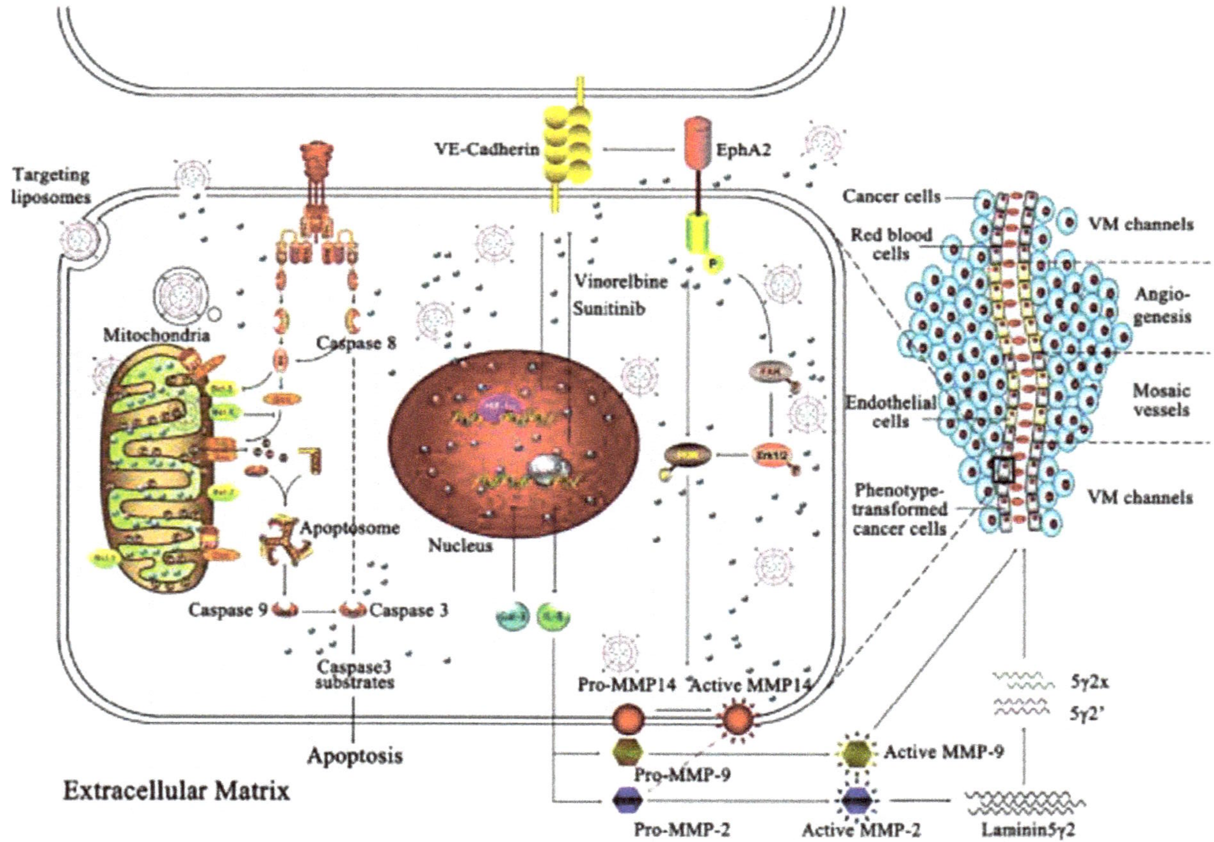

FIGURE 5.4 Schematic representation for the destruction of vasculogenic mimicry channels by regulating the signaling pathways using dual-functional drug liposomes (Mu et al., 2017).

Source: Reprinted with permission from Mu et al., 2017. Copyright © 2017 Elsevier B.V.

In addition, the possibility of liposome-supported enzymatic peritoneal dialysis (LSEPD) is determined to increase the dialysis functions for the model indications of acute alcohol poisoning (Pratsinis et al., 2017).

5.3.7 DENDRIMERS

Dendrimers are 3D macromolecular structures that demonstrated some unique features in molecular uniformity, weight distribution, size, and shape. Overall, these physicochemical features expanded related applications to pro-drugs and vehicles for encapsulation, complexation, delivery, and contrast agents (Wu et al., 2015).

Nowadays, nuclear medicine has become one of the key tools in the diagnosis and treatment of various diseases such as targeted delivery of multiple radionuclides in specific manners to improve the effectiveness of diagnosis. The unique molecular structures support adaptable platforms in radiolabeled NPs for nuclear medicine applications (Zhao et al., 2017).

Dendrimers have a high degree of branching, monodispersed, multivalence, and globular structures in the field of novel drug delivery method. The poly(propyleneimine) and poly(amidoamine) dendrimers are the popular dendrimers that concerning their excellent properties, synthesis, as well as toxicity (Kaur et al., 2016).

5.3.8 NANOCELLULOSE (NC)

Cellulose is a fascinating biopolymer and sustainable raw material. Nanocellulose (NC) is a general term for cellulose-based nanostructures, which have been increasingly studied in recent years, mainly due to their flexibility regarding chemical modifications, physical properties and the possible myriad applications. NC includes nanofibrillated cellulose (NFC), nanocrystalline cellulose (NCC) and bacterial nanocellulose (BNC). NC applications based on the production of face masks, bandages, skin replacements for implants, burns, cuffs for nerve surgery, artificial blood vessels, cell carriers and support matrices for the immobilization of enzymes (Seabra et al., 2018).

The BNC is focused as scaffolds for cell cultivation, drug delivery and medical implants. Developed BNC-based medical products can be found from the market:

- SYNTHECEL® Dura Repair (implant) produced by DePuy Synthes, USA;
- epicite^{hydro} (wound dressing) provided by JeNaCell GmbH, Germany;
- Celmat (wound dressing), supplied by BOWIL, Poland.

Figure 5.5 represents the biomedical applications of cellulose nanoparticles.

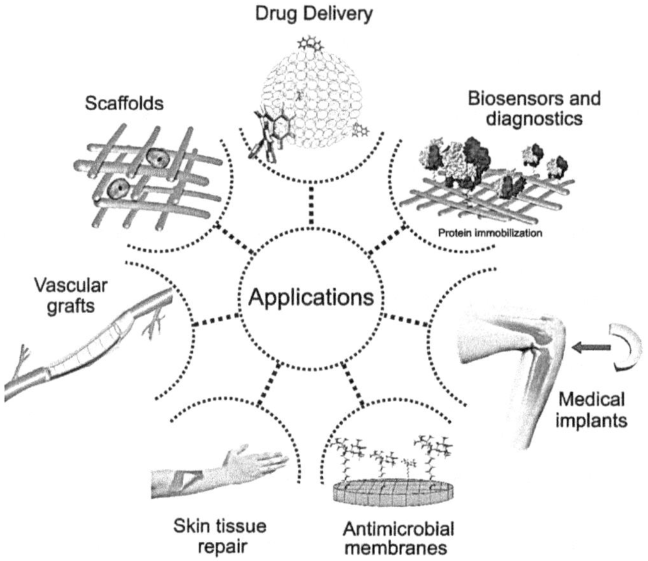

FIGURE 5.5 Biomedical applications of cellulose nanoparticles (Rojas et al., 2015).

Source: Reprinted with permission from Rojas et al., 2015. http://creativecommons.org/licenses/by/3.0

For NFC, the main large-scale applications include papermaking, composites, and rheological modifiers (food and paint), as well as the electronic, cosmetics, and nanofiller devices. NCC is most destined as interface stabilizers, rheological modifiers, coatings, and enhancing agents in composites, as well as high potential in biomedical devices (Klemm et al., 2018). NCC also determined as new aerogel material for bone regeneration, as the high porosity was shown to support the proliferation of the cells (Osorio et al., 2019).

Most recently, 3D printing techniques have been applied NC in combination with polymers is being studied to develop regenerative therapy. The injectability and printability of NC based hydrogels owing to their shear-thinning behavior and the possible support to living cells make NC the tremendous potential (Sultan et al., 2017).

5.4 MAIN RESEARCH STREAM IN NANOMEDICINE

The possible future trend in bio-nanomaterials for nanomedicine should be listed as biocompatibility of NMs, biosensor, targeted drug delivery, MD, nano-safety for NMs in medical use.

Early this year, the researchers from Tel Aviv University has been demonstrated their new approach to 3D-print thick, vascularized, and perfusable cardiac patches that match entirely the immunological, cellular, biochemical, and anatomical properties of the patient is reported (Noor et al., 2019). 3D printed heart model can be supporting the experimental and theoretical studies of heart-related disasters, and this cannot be achieved without advanced printable bio-nanomaterials conjunctions. Also, research on artificial livers even playing a critical role in the clinical treatments for liver failure (Li et al., 2019). That artificial tissue and organs paramount concern on the future development of biomedicine for better health conditions of a human being in all aspects of medicine. A 3D printing of personalized thick and perfusable cardiac patches and hearts are given in Figure 5.6.

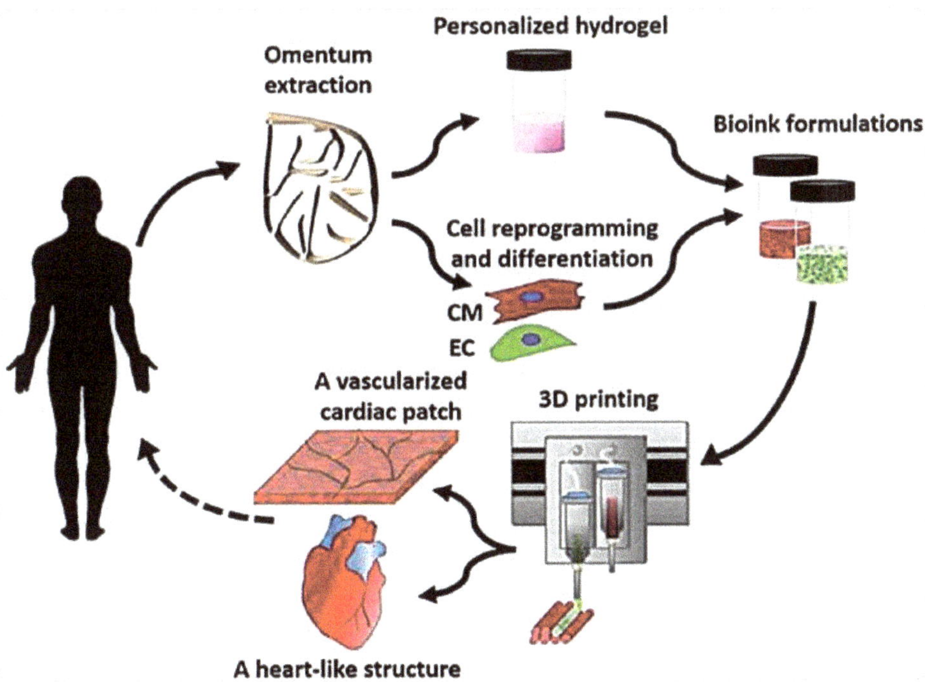

FIGURE 5.6 Schematic representation of 3D printing of personalized thick and perfusable cardiac patches and hearts (Noor et al., 2019).

Source: Reprinted from Noor et al., 2019. https://creativecommons.org/licenses/by/4.0/

5.5 RECENT DEVELOPMENTS

5.5.1 TISSUE ENGINEERING (TE)

Tissue engineering (TE) is a comprehensive field that mainly concentrates on the fundamental understanding of physiochemical, mechanical, engineering, and clinical sciences to resolve the tissue-related biomedical issues (Armentano et al., 2010).

The primary trend of biomaterials in the TE field focused on wound healing, bone/nerve tissue regeneration and cancer therapy.

5.5.2 ECM SCAFFOLD

Biological scaffold materials were applied in a different TE-related biomedical with favorable constructive remodeling properties such as angiogenesis, and stem cell recruitment (Huleihel et al., 2016). ECM scaffolds are typically used in regenerative medicine strategies for tissue and organ regeneration. The methods of scaffold decellularization have already been used in several tissues or organs, such as blood vessels, liver, bladder, lungs, and trachea. In this regard, the researcher has been successfully developed the ECM scaffold for kidney regeneration (Guan et al., 2015).

The living organism has the ability to self-healing small defects. Severe injuries or significant defects are significantly difficult or even not possible for the body to self-healing. However, ECM based biomaterials widely reported on the ability to use in regenerative medicine (Hinderer et al., 2016). The study on the applicability of the chondrogenic ECM incorporated collagen/chitosan (CS) scaffold in cartilage regenerative therapy has been succeeded (Ravindran et al., 2015).

5.5.3 MEDICAL DEVICES (MDS)/IMPLANTS

Medical devices (MDs) include a wide range from bedpans to sophisticated programmable pacemakers with micro-chip technology and laser surgical devices. Furthermore, MDs covers *in vitro* diagnostic products, such as general-purpose lab equipment along with test kits. As the Food and Drug Administration (FDA), has stated, an MD is "an instrument, apparatus, implement, machine, contrivance, implant, *in vitro* reagent, or other similar or related article, including a component part…" [Is the Product a Medical Device? 2018 (cited 2019 August 19); Available from: https://www.fda.gov/medical-devices/classify-your-medical-device/product-medical-device].

At this time, biopolymeric materials are fast developing in MDs preparation as a favor of long-term implantable MDs. Sterilization of biopolymeric implantable MDs is important and the sterilization method on the different MDs should be taken into the considerations (Tipnis et al., 2018).

The nanocomposites (PVA–PAA–PbO_2) have been investigated to be a one of the novel pressure-sensitive nanocomposites, due to high sensitivity for the pressure, low cost, easy processing, and lightweight (Hashim et al., 2018). Recently, polymeric heart valves (PHV) have been investigated for an emerging health issue. PHV materials included: Silicone, PTFE, PUs (polyester urethane: PEU, PCU), PVA, SIBS, PDMS–PHMO and POSS–PCU nanocomposite (Ghanbari et al., 2009).

Any MD by selecting biomaterials should address critical concerns its acceptability by the body. There have three factors in identifying the success rate properly, which (i) physiochemical and biological aspect of biomaterial for MD; (ii) MD's biocompatibility; and (iii) convenience of patient health condition (Pandey et al., 2016). Figure 5.7 shows the areas in the human body, which are suitable for implants.

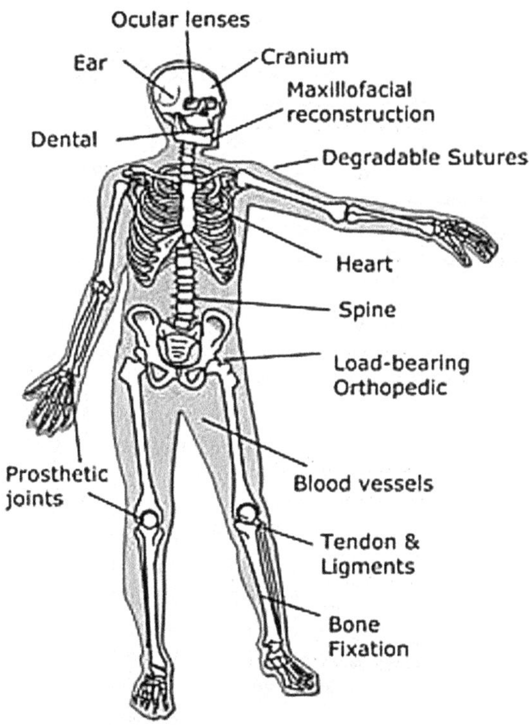

FIGURE 5.7 Human body areas suitable for implants (Pandey et al., 2016).

In 1808, Sir Humphry Davy produced initial metallic magnesium MDs. After two centuries, the high-quality corrosion-resistant biodegradable magnesium-based metal implants are becoming the breakthrough (Witte et al., 2010). As a new generation of the biomaterial, magnesium-based material gets attention because of its biodegradability and without concern of removal operation. Even though several clinical cases were reported 200 years ago. However, improved features of magnesium-based MDs were succeeded in Germany for bone implants (Zhao et al., 2017).

5.6 APPLICATIONS

Nano/biomaterials have been developed for many applications, especially in the drug delivery system, MD/implants, biosensor, cosmetics, and wound dressing products. Thus, the more and more potential use still awaiting the researcher who has innovated mind to make a breakthrough in the field. Underdeveloped applications/products will be possible in the market near future, which includes:

- Fluorescent biological labels;
- Gene delivery;
- Targeted drug delivery;
- Pathogens detections;
- Protein detections;
- DNA probing;
- Wound healing;
- Tumor destruction via;
- Separation of cells;
- MRI contrast enhancement;
- Phagokinetic studies.

Different possible configurations of nano-biomaterials for medical or biological applications are represented in Figure 5.8.

5.7 CONCLUSION

This chapter has aimed to critically review the potential, recent advances, challenges, and future research directions about bio-nanomaterials in nanomedicine as a new dawn for human health concern. In summary, it can be concluded that these discussed NMs carry the tremendous potential to be used as biomaterials for applications in TE, targeted drug/gene delivery, MDs/implants and diagnose/detection by the biosensor, as well as the enhanced

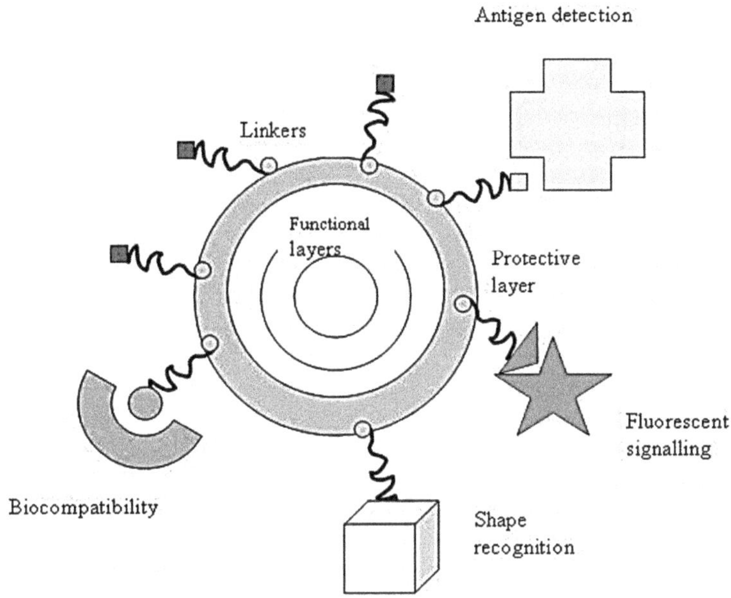

FIGURE 5.8 Typical configurations utilized in nano-biomaterials applied to medical or biological problems (Salata et al., 2004).

actual killing of cancer cells. Therefore, the biocompatibility, biodegradability, bioactivity, and mechanical, chemical, as well as biological properties of bio-nanomaterials have been summarized together with developed/ undergo applications.

The desired properties of those materials make them unique in the potential application within the nanomedicine field. The newly developing artificial organ model together with 3D techniques open the new door for future investigations and innovations for organ-on-chip to carefully studies on any failure, regeneration, targeted delivery, micro-mechanism and fluid aspect of human tissue and organs very critical ways. Future of bio-nanomaterials are mysterious, but we will keep curiosity for discovering and our innovative mind for maximizing the real concern on health issue as we afford.

ACKNOWLEDGMENT

This study was supported by the University of Malaya Research Fund Assistance RU001-2018, TOP100PDNANOCAT and UMRG program – AET ITRC (RP045D-17AET), are acknowledged.

KEYWORDS

- **biomaterial**
- **innovation**
- **nanomedicine**
- **nanoparticles**
- **polymers**

REFERENCES

Aguilar, M. R., & San, R. J., (2019). Introduction to smart polymers and their applications. In: Aguilar, M. R., & Román, J. S., (eds.), *Smart Polymers and Their Applications* (pp. 1–11). Elsevier.

Armentano, I., Dottori, M., Fortunati, E., Mattioli, S., & Kenny, J. M., (2010). Biodegradable polymer matrix nanocomposites for tissue engineering: A review. *Polym. Degrad. Stabil., 95*, 2126–2146.

Asadi, N., Alizadeh, E., Salehi, R., Khalandi, B., Davaran, S., & Akbarzadeh, A., (2018). Nanocomposite hydrogels for cartilage tissue engineering: A review. *Artif. Cells Nanomed. Biotechnol., 46*, 465–471.

Bennet, D., & Kim, S., (2014). Polymer nanoparticles for smart drug delivery. In: Sezer, A. D., (ed.), *Application of Nanotechnology in Drug Delivery* (pp. 257–310). Intech-Open: UK.

Chen, Y., Diaz-Dussan, D., Wu, D., Wang, W., Peng, Y. Y., Asha, A. B., Hall, D. G., et al., (2018). Bioinspired self-healing hydrogel based on benzoxaborole-catechol dynamic covalent chemistry for 3D cell encapsulation. *ACS Macro Lett., 7,* 904–908.

Choudhary, R., Saraswat, M., & Venkatraman, S. K., (2019). A fundamental approach toward polymers and polymer composites: Current trends for biomedical applications. In: Sadasivuni, K. K., Ponnamma, D., Rajan, M., Ahmed, B., & Al-Maadeed, M. A. S. A., (eds.), *Polymer Nanocomposites in Biomedical Engineering* (pp. 1–28). Springer: Cham.

Dozois, M. D., Bahlmann, L. C., Zilberman, Y., & Tang, X. S., (2017). Carbon nanomaterial-enhanced scaffolds for the creation of cardiac tissue constructs: A new frontier in cardiac tissue engineering. *Carbon, 120*, 338–349.

Elahi, N., Kamali, M., & Baghersad, M. H., (2018). Recent biomedical applications of gold nanoparticles: A review. *Talanta, 184*, 537–556.

Eliaz, N., (2019). Corrosion of metallic biomaterials: A review. *Mater., 12*, 407.

Ghanbari, H., Viatge, H., Kidane, A. G., Burriesci, G., Tavakoli, M., & Seifalian, A. M., (2009). Polymeric heart valves: new materials, emerging hopes. *Trends Biotechnol., 27*, 359–367.

Gil, S., & Mano, J. F., (2014). Magnetic composite biomaterials for tissue engineering. *Biomater. Sci., 2*, 812–818.

Gradinaru, V., Treweek, J., Overton, K., & Deisseroth, K., (2018). Hydrogel-tissue chemistry: Principles and applications. *Ann. Rev. Biophys*.

Green, M., (2004). Semiconductor quantum dots as biological imaging agents. *Angew. Chem. Int. Ed. Engl., 43*, 4129–4131.

Guan, Y., Liu, S., Liu, Y., Sun, C., Cheng, G., Luan, Y., Li, K., et al., (2015). Porcine kidneys as a source of ECM scaffold for kidney regeneration. *Mater. Sci. Eng. C, 56*, 451–456.

Guo, J., Rahme, K., He, Y., Li, L. L., Holmes, J. D., & O'Driscoll, C. M., (2017). Gold nanoparticles enlighten the future of cancer theranostics. *Int. J. Nanomed., 12*, 6131.

Gutiérrez-Hernández, J. M., Escobar-García, D. M., Escalante, A., Flores, H., González, F. J., Gatenholm, P., & Toriz, G., (2017). *In vitro* evaluation of osteoblastic cells on bacterial cellulose modified with multi-walled carbon nanotubes as scaffold for bone regeneration. *Mater. Sci. Eng. C, 75*, 445–453.

Han, M., Gao, X., Su, J. Z., & Nie, S., (2001). Quantum-dot-tagged microbeads for multiplexed optical coding of biomolecules. *Nature Biotech., 19*, 631.

Hashim, A., & Hadi, A., (2018). Novel pressure sensors made from nanocomposites (biodegradable polymers–metal oxide nanoparticles): Fabrication and characterization. *Ukr. J. Phys., 63*, 754–754.

Heggannavar, G. B., Achari, D., Fernandes, C., Mitchell, G. R., Morouço, P., & Kariduraganavar, M. Y., (2019). Smart polymers in drug delivery applications. *Appl. Mech. Mater., 890*, 324–339.

Hemamalini, T., & Dev, V. R. G., (2018). Comprehensive review on electrospinning of starch polymer for biomedical applications. *Int. J. Biol. Macromol., 106*, 712–718.

Hinderer, S., Layland, S. L., & Schenke-Layland, K., (2016). ECM and ECM-like materials—Biomaterials for applications in regenerative medicine and cancer therapy. *Adv. Drug Deliv. Rev., 97*, 260–269.

Hoffman, A. S., & Stayton, P. S., (2004). Bioconjugates of smart polymers and proteins: Synthesis and applications. *Macromol. Symp., 207*, 139–152.

Hoffman, A. S., (2013). Stimuli-responsive polymers: Biomedical applications and challenges for clinical translation. *Adv. Drug Deliv. Rev., 65*, 10–16.

Huleihel, L., Hussey, G. S., Naranjo, J. D., Zhang, L., Dziki, J. L., Turner, N. J., Stolz, D. B., & Badylak, S. F., (2016). Matrix-bound nanovesicles within ECM bioscaffolds. *Sci. Adv., 2*, e1600502.

Is The Product a Medical Device? (2018). Available from: https://www.fda.gov/medical-devices/classify-your-medical-device/product-medical-device (accessed on 13 February 2022).

Jain, A., & Jain, S. K., (2018). Stimuli-responsive smart liposomes in cancer targeting. *Curr. Drug Targets, 19*, 259–270.

Jiang, B. P., Zhou, B., Lin, Z., Liang, H., & Shen, X. C., (2019). Recent advances in carbon nanomaterials for cancer phototherapy. *Chem. Eur. J., 25*, 3993–4004.

Kaur, D., Jain, K., Mehra, N. K., Kesharwani, P., & Jain, N. K., (2016). A review on comparative study of PPI and PAMAM dendrimers. *J. Nanopart. Res., 18*, 146.

Klemm, D., Cranston, E. D., Fischer, D., Gama, M., Kedzior, S. A., Kralisch, D., Kramer, F., et al., (2018). Nanocellulose as a natural source for groundbreaking applications in materials science: Today's state. *Mater. Today, 21*, 720–748.

Kumar, S. S. D., Rajendran, N. K., Houreld, N. N., & Abrahamse, H., (2018). Recent advances on silver nanoparticle and biopolymer-based biomaterials for wound healing applications. *Int. J. Boil. Macromol., 115*, 165–175.

Kumar, S., Rani, R., Dilbaghi, N., Tankeshwar, K., & Kim, K. H., (2017). Carbon nanotubes: A novel material for multifaceted applications in human healthcare. *Chem. Soc. Rev., 46*, 158–196.

Kurtz, S. M., & Devine, J. N., (2007). PEEK biomaterials in trauma, orthopedic, and spinal implants. *Biomater., 28*, 4845–4869.

Li, Q., Zeng, L., Wang, J., Tang, D., Liu, B., Chen, G., & Wei, M., (2011). Magnetic mesoporous organic– inorganic $NiCo_2O_4$ hybrid nanomaterials for electrochemical immunosensors. *ACS appl. Mater. Interfaces, 3*, 1366–1373.

Li, Y., He, M., Zou, Z., Bian, X., Huang, X., Yang, C., Wei, S., & Dai, S., (2019). Artificial liver research output and citations from 2004 to 2017: A bibliometric analysis. *PeerJ., 6*, e6178.

Mehrali, M., Thakur, A., Pennisi, C. P., Talebian, S., Arpanaei, A., Nikkhah, M., & Dolatshahi-Pirouz, A., (2017). Nanoreinforced hydrogels for tissue engineering: Biomaterials that are compatible with load-bearing and electroactive tissues. *Adv. Mater., 29*, 1603612.

Migonney, V., (2014). History of biomaterials. *Biomater.*, 1–10.

Mohajeri, M., Behnam, B., & Sahebkar, A., (2019). Biomedical applications of carbon nanomaterials: Drug and gene delivery potentials. *J. Cell. Physiol., 234*, 298–319.

Motealleh, A., & Kehr, N. S., (2017). Nanocomposite hydrogels and their applications in tissue engineering. *Adv. Healthc. Mater., 6*, 1600938.

Mu, L. M., Ju, R. J., Liu, R., Bu, Y. Z., Zhang, J. Y., Li, X. Q., Zeng, F., & Lu, W. L., (2017). Dual-functional drug liposomes in treatment of resistant cancers. *Adv. Drug Deliv. Rev., 115*, 46–56.

Noor, N., Shapira, A., Edri, R., Gal, I., Wertheim, L., & Dvir, T., (2019). 3D printing of personalized thick and perfusable cardiac patches and hearts. *Adv. Sci., 6*, 1900344.

Osorio, D. A., Lee, B. E., Kwiecien, J. M., Wang, X., Shahid, I., Hurley, A. L., Cranston, E. D., & Grandfield, K., (2019). Cross-linked cellulose nanocrystal aerogels as viable bone tissue scaffolds. *Acta Biomater., 87*, 152–165.

Palza, H., Zapata, P. A., & Angulo-Pineda, C., (2019). Electroactive smart polymers for biomedical applications. *Mater., 12*, 277.

Pandey, E., Srivastava, K., Gupta, S., Srivastava, S., & Mishra, N., (2016). Some biocompatible materials used in medical practices: A review. *Int. J. Pharm. Sci. Res., 7*, 2748–2755.

Pollot, B. E., Rathbone, C. R., Wenke, J. C., & Guda, T., (2018). Natural polymeric hydrogel evaluation for skeletal muscle tissue engineering. *J. Biomed. Mater. Res. Part B: Appl. Biomater., 106*, 672–679.

Pratsinis, A., Zuercher, S., Forster, V., Fischer, E. J., Luciani, P., & Leroux, J. C., (2017). Liposome-supported enzymatic peritoneal dialysis. *Biomaterials, 145*, 128–137.

Ravindran, S., Kotecha, M., Huang, C. C., Ye, A., Pothirajan, P., Yin, Z., Magin, R., & George, A., (2015). Biological and MRI characterization of biomimetic ECM scaffolds for cartilage tissue regeneration. *Biomaterials, 71*, 58–70.

Rojas, J., Bedoya, M., & Ciro, Y., (2015). Current trends in the production of cellulose nanoparticles and nanocomposites for biomedical applications. In: Poletto, M., (ed.), *Cellulose – Fundamental Aspects and Current Trends* (pp. 193–228). In-Tech: UK.

Sabir, M. I., Xu, X., & Li, L., (2009). A review on biodegradable polymeric materials for bone tissue engineering applications. *J. Mater. Sci., 44*, 5713–5724.

Salata, O. V., (2004). Applications of nanoparticles in biology and medicine. *J. Nanobiotechnology, 2*, 3.

Seabra, A. B., Bernardes, J. S., Fávaro, W. J., Paula, A. J., & Durán, N., (2018). Cellulose nanocrystals as carriers in medicine and their toxicities: A review. *Carbohydr. Polym., 181*, 514–527.

Shen, S., Wang, S., Zheng, R., Zhu, X., Jiang, X., Fu, D., & Yang, W., (2015). Magnetic nanoparticle clusters for photothermal therapy with near-infrared irradiation. *Biomater., 39*, 67–74.

Shoichet, M. S., (2009). Polymer scaffolds for biomaterials applications. *Macromol., 43*, 581–591.

Sinani, V. A., Koktysh, D. S., Yun, B. G., Matts, R. L., Pappas, T. C., Motamedi, M., Thomas, S. N., & Kotov, N. A., (2003). Collagen coating promotes biocompatibility of semiconductor nanoparticles in stratified LBL films. *Nano Lett., 3*, 1177–1182.

Sokolowski, W., Metcalfe, A., Hayashi, S., & Raymond, J., (2007). Medical applications of shape memory polymers. *Biomed. Mater., 2*, S23.

Spang, M. T., & Christman, K. L., (2018). Extracellular matrix hydrogel therapies: *In vivo* applications and development. *Acta Biomater., 68*, 1–14.

Sultan, S., Siqueira, G., Zimmermann, T., & Mathew, A. P., (2017). 3D printing of nano-cellulosic biomaterials for medical applications. *Curr. Opin. Biomed. Eng., 2*, 29–34.

Teradal, N. L., & Jelinek, R., (2017). Carbon nanomaterials in biological studies and biomedicine. *Adv. Healthcare Mater., 6*, 1700574.

Tipnis, N. P., & Burgess, D. J., (2018). Sterilization of implantable polymer-based medical devices: A review. *Int. J. Pharm., 544*, 455–460.

Tovar, J. D., (2013). Supramolecular construction of optoelectronic biomaterials. *Acc. Chem. Res., 46*, 1527–1537.

Vahed, S. Z., Salehi, R., Davaran, S., & Sharifi, S., (2017). Liposome-based drug co-delivery systems in cancer cells. *Mater. Sci. Eng. C., 71*, 1327–1341.

Vedadghavami, A., Minooei, F., Mohammadi, M. H., Khetani, S., Kolahchi, A. R., Mashayekhan, S., & Sanati-Nezhad, A., (2017). Manufacturing of hydrogel biomaterials with controlled mechanical properties for tissue engineering applications. *Acta Biomater., 62*, 42–63.

Victor, A., Ribeiro, J., & Araújo, F. F., (2019). Study of PDMS characterization and its applications in biomedicine: A review. *JMEB, 4*, 1–9.

Witte, F., (2010). The history of biodegradable magnesium implants: A review. *Acta Biomater., 6*, 1680–1692.

Wu, L. P., Ficker, M., Christensen, J. B., Trohopoulos, P. N., & Moghimi, S. M., (2015). Dendrimers in medicine: Therapeutic concepts and pharmaceutical challenges. *Bioconjug. Chem., 26*, 1198–1211.

Xiao, M., Chen, Y. M., Biao, M. N., Zhang, X. D., & Yang, B. C., (2017). Bio-functionalization of biomedical metals. *Mater. Sci. Eng. C., 70*, 1057–1070.

Yang, K., Han, Q., Chen, B., Zheng, Y., Zhang, K., Li, Q., & Wang, J., (2018). Antimicrobial hydrogels: Promising materials for medical application. *Int. J. Nanomed., 13*, 2217.

Zhang, Y., Wang, F., Liu, C., Wang, Z., Kang, L., Huang, Y., Dong, K., et al., (2018). Nanozyme decorated metal–organic frameworks for enhanced photodynamic therapy. *ACS Nano, 12*, 651–661.

Zhang, Y., Wang, H., Yan, B., Zhang, Y., Li, J., Shen, G., & Yu, R., (2008). A reusable piezoelectric immunosensor using antibody-adsorbed magnetic nanocomposite. *J. Immunol. Meth., 332*, 103–111.

Zhao, D., Witte, F., Lu, F., Wang, J., Li, J., & Qin, L., (2017). Current status on clinical applications of magnesium-based orthopaedic implants: A review from clinical translational perspective. *Biomaterials, 112*, 287–302.

Zhao, L., Zhu, M., Li, Y., Xing, Y., & Zhao, J., (2017). Radiolabeled dendrimers for nuclear medicine applications. *Molecules, 22*, 1350.

Zimmerman, J., Parameswaran, R., & Tian, B., (2014). Nanoscale semiconductor devices as new biomaterials. *Biomater. Sci., 2*, 619–626.

CHAPTER 6

An Insight into the Nanomaterials Fabricated Nanotechnology in the Nutraceutical Field

ASMITA DURAGKAR,[1] SHWETA MISHRA,[2] and ANIL M. PETHE[3]

[1]*Nikalas Mahila Mahavidyalaya, Panapratap Nagar, Nagpur, Maharashtra, India*

[2]*Shobhaben Pratapbhai Patel School of Pharmacy and Technology Management, SVKM's NMIMS, V.L. Mehta Road, Vile Parle (W), Mumbai, Maharashtra, India*

[3]*School of Pharmacy and Technology Management, SVKM's NMIMS, B-4, TSIIC, Polepally SEZ, Jadcherla, Hyderabad Campus, Telangana, India*

ABSTRACT

This contribution comprehensively summarizes the nanotechnological breakthrough in the area of scientific inventions by means of extremely small particle dimension, thereby imparting novel properties and increasing efficacy of the material. These contemporary qualities of nanotechnology are extended in the escalating field of nutraceuticals to conquer the restrictions posed by inherent complex nature of bioactives and to elevate pharmacokinetic and pharmacodynamic potential, hence, boosting their bioavailability. Nanonutraceuticals is rapidly burgeoning field encompassing health benefits, wellness and preventive healthcare approach. Nanodelivery systems for traditional nutraceuticals have shown improved efficacy, maintaining good health as well as in the management of many diseases like Alzheimer's,

Advanced Nanomaterials for Biological, Nutraceutical, and Medicinal Applications.
Sreeraj Gopi, Augustine Amalraj, and T. R. Sreeraj (Eds.)

cardiovascular disease, cancer, arthritis. Novel nanocarriers like nanoparticles, solid lipid nanoparticles, niosomes, nanoemulsions, nanotubes, phytosomes exhibit advantages of increased permeability and absorption, high bioavailability, controlled and site specific targeted release, and improved physical stability. However, assured human and environmental risk assessment and balanced regulatory guidelines for nanonutraceuticals is the need of an hour for a safe future.

6.1 INTRODUCTION

Nanotechnology is an integrated frontier of science and engineering that manipulates matter at the nanometer scale. This new dimension made it a key-enabling technology offering infinite possibilities in diverse field of science like electronics, information technology, green energy production, environmental science, pharmaceuticals, food and dietary supplement industry and many others (Bouwmeester et al., 2014). Nanotechnological research has taken a great leap in real sense in the 21st Century and over the last few years, this powerful sector is getting huge attention by researchers, scientists and investigators to enable breakthroughs for health and wellness of human kind. Nanosized materials possess novel properties due to their very small size compared to their bulk forms. This small size enables them to interact with the body at molecular scale. The field started growing rapidly from environmental to various medical applications. Such immense potential of nanotechnology has touched the food industry and exponentially expanding to revolutionize nutraceutical field (Singh & Singh, 2017).

Heightened economic development leads to altered modus vivendi, which has brought its own luggage of a variety of lifestyle diseases known to take negative toll on human health. This initiated awareness about maintaining overall good health with nutraceuticals, being recognized as a preventive healthcare approach to tackle lifestyle disorders (Dang et al., 1998). Nutraceuticals are primarily diet supplements. Nutraceuticals are defined as any material that may be considered as component of food which offers health benefits in addition to providing nutrition and also includes prophylaxis and treatment of disease (Brower et al., 1998; Kalra et al., 2003; Biesalski et al., 2002; Keservani et al., 2017). Use of nutraceuticals for enhancement of health and treatment of diseases has very old history. Very few civilizations, e.g., Indians, Egyptians, Chinese and Sumerians have documented the effective use of food as medicines (Radhika et al., 2011; Gupta et al.,

2010; Kannappan et al., 2011). Nutraceuticals are being used for wellness by administering known bioactive compound from natural foods in the required amount than that derived from normal consumption of food products (Sohaimy et al., 2012). Nutraceuticals encompass a wide range of herbal products, isolated nutrients, dietary fiber, antioxidants, polyunsaturated fatty acids, proteins, vitamin, minerals, prebiotics, probiotics and phytochemicals. The naturally occurring chemical compounds in plants like polyphenols, flavonoids, carotenoids, lycopene exhibit significant health benefits resulting in improved immunity playing vital role in treatment of lifestyle diseases like obesity, type II diabetes, metabolic disorders, and also effective in chronic diseases like Alzheimer's disease, cancers, cardiovascular diseases. However, poor physicochemical properties and intrinsic complexity of these phytochemicals results in poor absorption and bio-distribution. This necessitates the emergence of multidisciplinary nanotechnological principles to improve delivery and bioavailability of nutraceuticals. Nanotechnology has tremendous potential to remodel many allied disciplines such as food, pharmaceuticals, nutraceuticals, and cosmeceuticals.

6.1.1 NUTRACEUTICALS: BACKGROUND

Today's disorderly way of life has led to the emergence of various lifestyle diseases. The repercussions of this is the entrance of sedulous multiple drug therapies associated with innumerable side effects. Thus, with an objective to maintain and enhance human health, a new health promoting dietary regimen is receiving great demand from people worldwide. This has led to the evolution of the concept of nutraceuticals. Even though harnessing nutraceuticals by people around the world has a long history, only recently health benefits of regular consumption of nutritional bio-actives in combating lifestyle diseases has been documented.

"Nutraceuticals" term was coined by Stephen De Felice, Founder, and Chairman of the Foundation for Innovation in Medicine (FIM), Cranford, NJ in 1989 and it is made up of two words which is "nutrition" and "pharmaceutical" (Dillard & German, 2000). There exist several definitions for nutraceutical, but According to DeFelice, nutraceutical can be defined as, "a food (or a part of food) that provides medical or health benefits, including the prevention and or treatment of a disease." Functional foods are defined according to the International Food Information Council (IFIC) as "foods or dietary components that may provide health benefits beyond basic nutrition"

(Bagchi et al., 2016). Nutraceuticals and functional foods are tailored to provide nutrition, enhance the immune mechanism of the body, promote wellness, prevent malignant processes and control symptoms (Dureja et al., 2003). With advanced research in the dietary supplement segment, boundary between nutraceutical and functional food has almost blurred. Functional food with bioactive nutrient need to be consumed in large potions whereas nutraceuticals can be considered as concentrated food bio-actives, required in small amount for its health promoting effect. Nutraceuticals involves a wide range of products including minerals, fatty acids, dietary fibers, vitamins, phytonutrients, peptides, antioxidants, etc.

Nutraceuticals are classified by either of the ways, the first one being on the basis of source or on the basis of type (Keservani et al., 2015). Nutraceuticals can be derived from plant (e.g., ascorbic acid, quercetin, cellulose, and lutein), animal (e.g., lecithin, choline, co-enzyme Q10) and microbial source (e.g., *Saccharomyces boulardii* (yeast), *Lactobacillus acidophilus* (LC1), *B. longum*).

Nutraceuticals are categorized by different types:

1. **Functional Food** (Probiotics Fortified Foods, Omega Fatty Acid Fortified Food, Branded Ionized Salts, Branded Wheat Flour Market, and Other Functional Food);
2. **Functional Beverages (**Fruits and Vegetable Juices & Drinks, Dairy & Dairy Alternative Drinks, Non-Carbonated Drinks (Bottled water, Tea, and Coffee, and Others (Herbal tea, Sports drinks, Energy drinks);
3. **Dietary Supplements (**Proteins & Peptides, Vitamins & Minerals, Herbals (Ayurveda extracts, plant extracts, algal extracts, phytochemicals), and Others (Fatty acids, Fiber); and
4. **Personal Care:** All organic food also comes under Nutraceuticals category and the awareness and choice of such food is majorly on the increase (Keservani et al., 2015; Hilton, 2017).

Nutraceuticals are versatile in nature as they are used in numerous industries such as pharmaceutical, food & beverages, animal feed additives, and personal care. Recent innovations in novel delivery of nutraceuticals ensuring its high efficacy and bioavailability are expected to create new opportunities in this booming sector, the global market is going to witness assured growth over the coming years (Andlauer et al., 2002).

The global nutraceutical market, which was initially ruled by North America (U.S., Mexico, Rest of North America) and Europe (UK, France, Germany, Rest of Europe) segment, is now being taken over by Asia-Pacific (India, China, Japan, Rest of Asia-Pacific) nutraceutical market (Keservani et al., 2015). This region has high potential and is fastest-growing nutraceutical market in the world (Keservani et al., 2015). The LAMEA (Latin America, Middle East, and Africa) Nutraceuticals market is an emerging market owing to increasing health and wellness awareness. Among all the segments, Functional Foods & Beverages currently dominate the nutraceuticals market and will continue with decent growth in the future (Keservani et al., 2015).

The primary key drivers are the developed & health-conscious nations. However, emerging market consumers from developing nations also contribute to the augmented growth of the nutraceutical market due to increased consciousness towards a holistic approach towards life and raised disposable income of the general population. Several other forces that promote rapid growth in the industry are increasing awareness of diet-disease relationships, proactive approach in attaining higher well-being, demographic transitions and focus on preventive healthcare system.

6.1.2 BIOPHARMACEUTICAL PERFORMANCE CONCERN

Nutraceuticals, potent functional foods, have always been regarded essentially safe and green alternatives for providing health benefits beyond basic nutrition. They can ameliorate the effects of poor diet habits, substandard food quality and adverse effects due to multidrug regimens, thereby promoting health and wellness (Espín et al., 2007). Several nutraceuticals, with many therapeutic benefits, are being consumed by people for example polyvitamins, omega-3 fatty acids, carotenoids, polyphenols (anthocyanins, proanthocyanidins, flavanones, isoflavones, and ellagic acid) (Zaki et al., 2014). Amongst all these varied bio-actives, at present, antioxidants and probiotics are dominating the global nutraceutical market.

The majority of these nutraceutical products are formulated for oral administration in the form of tablets, capsules, powders, pills, however, it is quite challenging to achieve satisfactory bioavailability and their site-specific delivery. The bioavailability is often intimidated by poor solubility, chemical instability and low permeability of the bioactive in the GIT. The labile nature, low absorption rate and lack of targeted delivery are the major fundamental causes for poor nutraceuticals oral delivery (Yi et al., 2014).

The bioavailability of the lipophilic bioactive compounds is hampered by its low aqueous solubility and slow dissolution rate in GI fluids. The absorption process gets more complicated by high molecular weight of these bio-actives (Rao et al., 2013; Waraho et al., 2011). The biggest challenge in the formulation of functional foods and nutraceuticals is the vulnerability of lipophilic and hydrophilic bio-actives leading to fast degradation under normal storage condition (Mcclements et al., 2000). This problem is especially seen with omega-3 fatty acid, lycopene, carotenoids, curcumin and many others (Boon et al., 2010; Gupta et al., 2013). Oral administration of curcumin usually leads to low serum levels due to apparent rapid metabolism and short circulation half-life (Axelos et al., 2017). One of the rapidly growing health promoting approach is administration of probiotic bacteria as nutraceuticals. Bioavailability and efficacy of orally administered probiotics are hindered by the low stability of live bacteria at GIT conditions resulting in loss of viability, which is attributed to the high acid and bile salt concentrations present (Anand et al., 2007).

This burning issue of bioavailability of key ingredients evoked researchers to design delivery vehicles that will not only prevent the degradation of nutraceuticals, but also increase their bioavailability. Many promising delivery systems have already been developed to enhance the therapeutic efficiency of nutraceuticals (Cook et al., 2012; Panwar et al., 2015; Raja et al., 2016; De Vos et al., 2010). Advent of nanotechnology enabled engineering of advanced nutraceutical formulation designs at nanoscale to stabilize and enhance nutraceuticals' functionality in food products and to ensure maximum bioavailability and safety (Hafez et al., 2013).

6.2 NANOTECHNOLOGY: IMPACT ON NUTRACEUTICALS

Nanotechnology focuses on characterization, fabrication and manipulation of functional materials of the order of 1–100 nm. Nanoscaling imparts these molecules with very novel and unique properties which can be endlessly exploited in every domain of life. The potential benefits of nanoscience have been realized by scientific fraternity and consequently stimulated extensive research for developing highly innovative and smart materials for the betterment of mankind.

Currently, with the rising number of health issues, many traditionally used natural ingredients are being re-explored for their beneficial effect as well as newly discovered phytonutrients finding their space in the daily diet for good physical health. Micronutrients like vitamins, minerals, antioxidants, living cells- probiotics as well as lipids, peptides and fatty acids are available in various forms claiming to provide health benefits in the prevention and treatment of a range of diseases. However, oral administration of these compounds in their free form results in low bioavailability which is mainly due to chemical instability to food processing conditions (i.e., temperature, oxygen, or presence of light) or in the gastrointestinal (GI) tract (i.e., pH and presence of enzymes and other agents), insufficient gastric residence time, minimal solubility in GIT and less permeability through biological membrane (Ruiz et al., 2016). All of which hampers bioaccessibility of nutrients in the human body, thereby lowering the potency of nutraceuticals (Jampilek et al., 2019; Nedovic et al., 2011; McClements et al., 2009).

Inventions and discoveries in nanotechnology have begun influencing the fast developing industry of food and Nutraceuticals, wellness promoting functional foods and nutraceuticals are gaining increasing attention. The advent of nanoencapsulation technology rendered new dimension to the research in the nutraceutical field by opening infinite avenues to deal with problems like instability, solubility, degradation, permeability and bioavailability (Singh & Singh, 2017; Nedovic et al., 2011). The delivery of bioactive agent can be improved by entrapping a compound within a carrier substance (Singh et al., 2017; Nedovic et al., 2011). Encapsulation enables retardation of the degradation process, better solubility, protecting and enhancing its activity, improved bioavailability or permitted targeted delivery (Mc Clements et al., 2009; Esfanjani et al., 2016). The solubility of lipophilic bio-actives (lipids) can be enhanced by nanoencapsulating a lipophilic compound in a hydrophilic carrier or vice versa. The stability of labile compounds (i.e., vitamins) can be enhanced by a barrier film of encapsulating material between the reactive molecule and the environment. Also protecting evaporation and degradation of volatile substances (i.e., aromas), as well as masking unpleasant sensorial properties (i.e., bitter taste, astringency, or smell) is possible with nanotechnology (Cerqueira et al., 2013).

The beneficial effects of nanodelivery systems are due to enormously increased surface area per unit volume (Chau et al., 2007). Due to their subcellular size, they can efficiently cross bio-membrane. (i.e., epithelial membrane) as well as exhibit prolonged GI transit time, leading to higher biological activity than their larger counterparts with the same chemical composition (Chau et al., 2007). A study demonstrated the improved absorption and oral bioavailability of many essential minerals (i.e., iron, zinc, calcium, selenium, and magnesium) (Misra et al., 2011). Nanosized self-assembled emulsion developed by Garti et al. 2004; Garti et al., 2007) exhibited 7–20 fold superior solubility properties for highly lipophilic lycopene, phytosterol and lutein (Garti et al., 2004; Garti et al., 2007). Curcumin loaded PLGA nanoparticles observed to improve clinical management of leukemia (Neethirajan et al., 2011). Development in nanoencapsulation technologies offers viable methods for solving compatibility issues (i.e., aggregation and phase separation) with the food matrix affecting aesthetics, stability, or flavor of the functional food product and allows control release at the desired site (i.e., nutrients delivery in part of GIT where they are maximally absorbed) (Yao et al., 2015; Rajam et al., 2015).

To optimize the formulation design and performance of the delivery system for nutraceutical purpose, sound knowledge of characteristics of carrier polymer and bioactive compounds to be used, as well as possibility of physical and chemical interactions at the molecular level, is crucial. Presently, many established pharmaceutical nanotechnological methods are being extended for food industry which created need for finding bio-based, biodegradable food-grade alternatives that are certified as "Generally Recognized As Safe" (GRAS) materials for food applications to replace nonfood-grade polymers and materials (Mcclements et al., 2014; Goncalves et al., 2018) and basic safety requirements by governmental agencies. Biodegradable, barrier forming materials like Polysaccharides (i.e., starch and their derivatives, alginate, pectin, dextran, and chitosan) and Proteins (i.e., gelatin, whey proteins, and zein) are most widely used materials. Lipids (i.e., medium chain triglycerides, tristearin, and corn oil) and low molecular weight surfactants are also suitable for encapsulation due to their biodegradability, biocompatibility and non-toxicity (Gunasekaran, et al., 2014). Nutraceutical delivery systems can be divided in two groups: (i) lipid and surfactant-based carriers, which include emulsions, liposomes, solid lipid particles, nanostructured lipid carriers and self-dispersing lipid formulations and (ii) biopolymer-based carriers (i.e., polysaccharide and protein-based carriers), which include hydrogels, protein-polysaccharide complex and polymeric micelles.

The naturally derived functional ingredients occur in various molecular and physical forms, with different polarities, molecular weight and chemical properties. Features of both, bioactive and encapsulating material, influence delivery system characteristic like density, polarity, charge, particle size, entrapment efficiency, and release characteristics (Joye et al., 2014; Aditya et al., 2017). Customized approach should be followed to design a nanosystem suitable for a specific nutraceutical. Generally, the criteria for tailor made design are physicochemical properties of nutraceuticals (e.g., solubility, chemical interactions and stability) and final application (e.g., amount of nutraceutical required, storage conditions, type of food matrix, physiological activity and its shelf-life) (Augustin et al., 2009).

Nanotechnology employed in nutraceutical field is evolving exponentially with major advances in delivery system design. Different encapsulation methodologies are being employed to achieve desired formulation properties as none of them can be considered as universally applicable due to the varied nature of bioactive and encapsulating material (Martins et al., 2018; Ezhilarasi et al., 2013). The nanoencapsulation techniques are based on the principle of either top-down or bottom-up approaches for development of nanomaterials. Commonly used methods are emulsification, emulsification-solvent evaporation, spray drying, spray cooling, freeze drying, nanoprecipitation, coacervation, inclusion complexation, and supercritical fluid technique (Cerqueira et al., 2017).Top-down approaches involve force assisted size reduction and shaping of the structure for the desired application of the nanomaterial. Techniques such as emulsification and emulsification-solvent evaporation are based on top-down approach. Bottom-up approach involves self-assembling and self-organization of molecules in a constructive way. This process is influenced by factors like pH, temperature, concentration, and ionic strength (Sanguansri et al., 2006). Supercritical fluids, inclusion complexation, coacervation, and nanoprecipitation makes use of bottom-up approach (Aghbashlo et al., 2012).

Spray drying is fast, relatively cheap, and reproducible and therefore frequently used technique. The principle of this approach involves the atomization of dispersion of bioactive and biopolymer under heated air, producing dried product with rapid evaporation of solvent. The limitation of this method involves restricting the use of thermo labile, volatile functional active (Ingale et al., 2018). On the contrary, atomization at low temperatures is employed in spray cooling technique, thereby making it suitable for thermo labile compounds. Immobilization of enzymes, flavor and protein hydrolysates are executed by spray cooling technique. Very high value probiotic bacteria are preserved by encapsulation using a freeze drying technique under vacuum (Fathi et al., 2014). Emulsification can be achieved by inducing electrostatic, or hydrophobic, or hydrogen bonding between the bioactive compounds and an encapsulating material and the surfactant addition, to promote formation of bilayers, micelles and reverse micelles encapsulating bio-active (Doublier et al., 2017). Coacervation is one of the most easily used techniques for nanosystems. Formulators can select simple coacervation between oppositely charged bioactive and biopolymer or complex coacervation with carrier system formed by electrostatic complexation of oppositely charged biopolymers (positively charged chitosan and negatively charged alginate or pectin) for encapsulation of lipophilic compounds, like oils as well as some hydrophilic bio nutrients. Though Inclusion complexation, supercritical fluid process, nanoprecipitation, emulsification-solvent evaporation techniques are employed in the functional food industry, more flexible, less expensive, stability imparting spray drying and freeze drying has always been the technique of choice in food industry (Pinto Reis et al., 2006; Guterres et al., 2009).

6.3 NANOFORMULATIONS FOR NUTRACEUTICALS DELIVERY

Nutraceuticals covering sizable majority of preventive health care domain encounter loss of efficiency due to above stated biopharmaceutical problems. Bio-based nanoscale delivery systems are looked upon as a potential approach to overcome bioavailability issues. As shown in Figure 6.1, several delivery formulations such as nanoparticles, solid lipid nanoparticles (SLN), phytosome, niosome, nano-liposomes, nanoemulsions, nanofibers, nanodispersions, and carbon nanotubes (CNTs) and furnish the intelligent deliverance of nutraceuticals (Ajazuddin et al., 2010).

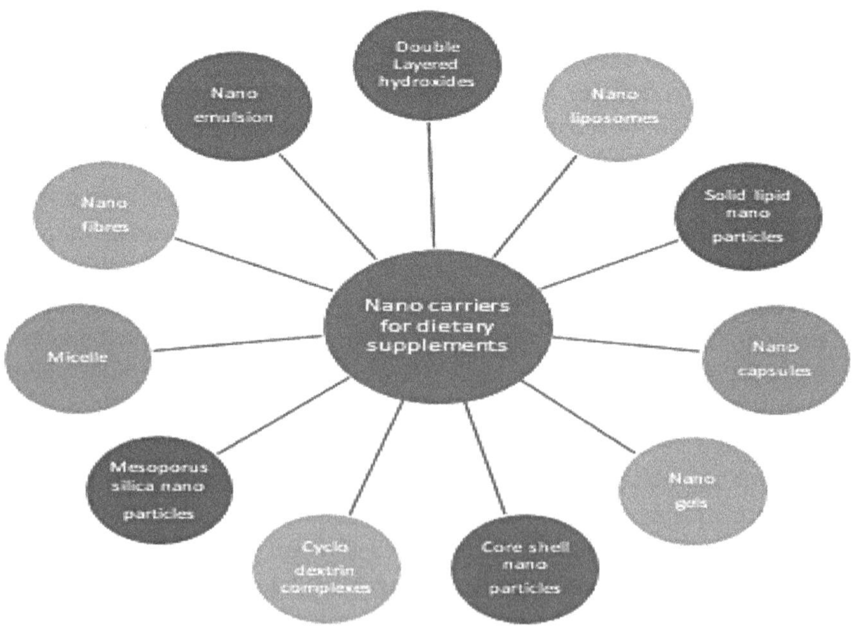

FIGURE 6.1 Nano-delivery systems for nutraceuticals.

6.3.1 NANOPARTICLES

To combat restrictions of conventional forms of nutraceuticals like low absorption rate, less permeability, poor bioavailability, nanoparticles, the sub-micronized delivery systems, were devised by scientists for encapsulation of bioactives that also serve to protect their intrinsic physicochemical nature (Alvarez-Román et al., 2001). Polymeric nanoparticles, size ranging between 10–1000 nm, are formulated using either biodegradable on non-biodegradable polymers. They can further be categorized as nanospheres and nanocapsules. Nanospheres represent the matrix arrangement of polymer chains, whereas reservoir type vesicular structure is exhibited by nanocapsules (Couvreur et al., 2002; Soppimath et al., 2001). The encapsulation mechanism involves dispersion, entrapment, or dissolution of bioactive within the polymeric nanoparticle or even it can get adsorbed on the surface of nanoparticle (Lopes et al., 2000; Cruz et al., 2006a; Cruz et al., 2006b).

Nanoparticles (NPs) can be formulated by employing natural or synthetic polymers. Gelatin, chitosan, albumin, sodium alginate are perpetually used natural polymers whereas, polyglycolides (PGA), polylactides (PLA), poly (lactideco-glycolides) (PLGA), polycyanoacrylates, polycaprolactone, polyglutamic acid belongs to synthetic polymers (Nagavarma et al., 2012; Lemes et al., 2016). With an objective of incorporating functional foods and nutraceuticals in various food products, a need is created for exploitation of food grade as well as GRAS status polymers for encapsulation. Unique qualities like low toxicity, easy digestibility, capability of incorporation of diverse range of bioactives and most prominently biocompatibility makes them ideal encapsulating materials in the blooming nutraceutical industry (Nagavarma et al., 2012; Lemes et al., 2016).

Different methods of preparation of polymeric NPs include mechanical methods like fluidization, homogenization, dispersion, emulsification process implementing the top-down approach, whereas chemical methods like precipitation-condensation, sol-gel synthesis, are based on bottom-up approach. Nanocapsules reported to be manufactured using emulsification-diffusion method and same method excluding oil can be utilized for preparation of nanospheres (Kumari et al., 2010). Currently revolutionizing approach of green synthesis of NPs, is being implemented by researchers (Peñalva et al., 2015).

Physical and chemical properties of engineered NPs such as surface area, surface charge and surface chemistry are essential parameters for stability and efficacy of nutraceuticals. Submicron size of NPs, associated with enormous surface are, found to be responsible for enhanced permeability and improved bioaccessibility of bioactive. Surface charge, a key feature, of nanoparticle influences absorption and biodistribution of nutrient in the body. Hydrophobicity, feature of paramount importance, plays crucial role in biological processes like bio-membrane permeation, protein adsorption and cellular uptake (Moinard-Checot et al., 2006).

The bioactive embedded polymer matrix, nanospheres, can be amorphous or crystalline in nature. They are extensively employed in enzyme encapsulation leading to enhanced thermal and ultrasonic stability of the enzyme. Nanoparticles with external polymeric membrane and inner liquid core, i.e., nanocapsules can be prepared by ionic pregelation/coacervation technique (Singh et al., 2017; Shima et al., 2013). The method involves cross linking of polyelectrolyte in presence of cation or anion (Madadlou et al., 2010). Other methods involved are dispersion of preformed polymers, nanoprecipitation and use of supercritical fluids (Nedovic et al., 2011). Nanoparticles dispense extensive applications in nutraceuticals due to numerous features like elevated bioavailability, improved mechanical and thermal stability, superior shelf life, feasibility of controlled release and reduced amount of nutrient as compared to a conventional formulation (Bharimalla et al., 2017). Owing to these diverse features, many nutraceuticals have been formulated as nanoparticles. Curcumin, poorly aqueous soluble phytonutrient, exhibits various pharmacotherapeutic actions like antioxidant, anti-inflammatory, and anticancer activity. Curcumin resulted in the quick systemic clearance when administered in conventional formulation. Thus PLGA nanoparticles loaded with curcumin were manufactured using high pressure emulsification-solvent evaporation method. These NPs showed extended retention time and also ensued blood-brain barrier passage (Rao et al., 2011). Patra et al. (2018) explained the formulation of a dual (magnetic and thermal) responsive nanoparticles using a novel nano-technique for potent delivery and increased efficacy of curcumin. Researchers found PLGA nanoparticles to be an effective carrier for potent nutraceuticals curcumin and quercetin, where phytonutrient moieties may either get adsorbed on PLGA surface or get entrapped inside (Tsai et al., 2011; Luo et al., 2013). Similarly, oleoyl alginate ester and vitamin D3 nanoparticles revealed improved cell membrane permeability, with excellent anti-rickets activity, can be employed as a potential nanonutraceutical delivery system (Bennet et al., 2012; Khalil et al., 2013; Sun et al., 2012). A green tea catechin, Epigallocatechin-3-gallate (EGCG), possesses strong antioxidant property, but it is unstable in plasma and GIT, eventually leads to low bioavailability and reduced potency. EGCG was incorporated in Gallic acid (GA) grafted chitosan (CS, GA-g-CS for GA grafted CS) and caseinophosphopeptides (CPP) polymeric nanoparticles, which revealed potent antioxidant activity with augmented anticancer potential against Caco-2 colon cancer cells (Ernest et al., 2018).

Similarly, a powerful antioxidant and anticancer agent, Ferulic acid (FA), was entrapped in chitosan-tripolyphosphatepentasodium (CS-TPP) nanoparticles to augment its solubility and permeability with refined cytotoxic activity against ME-180 cell lines (Paul et al., 2020).

The use of polymeric and organic nanoparticles presents many limitations like vulnerability to microbial contamination, organic solvent residue, and low chemical stability. So, ample of inorganic nanocarriers are being exploited to augment the bioavailability and clinical efficacy of phytonutrients and potent bioactives in the nutraceutical field (Panwar et al., 2016). Inorganic NPs employ noble metals like gold, silver, and platinum and also silica is being utilized. These nanocarriers possess unique biological and physicochemical properties than their bulk counterparts. Seong-Yun Jeong synthesized murine monoclonal anti-EGFR ß-antibody conjugated gold nanoparticles of lapachone, a quinine containing natural anticancer molecule, for systemic administration to conquer poor aqueous solubility and non-specific distribution issues. Silica based gold nanoshells in a temperature sensitive hydrogel were able to successfully deliver insulin and lysozyme upon irradiation (Bikram et al., 2007). Mesoporous ß-silica nanoparticles effectively achieved controlled release of curcumin in cancer treatment (Gangwar et al., 2013). TiO_2 NP finds its use as food coloring and sun screening agent Thus, nutraceuticals efficacy can be improved and augmented with judicious use of nanoparticle system (Weir et al., 2012).

6.3.2 NANODISPERSIONS

Nutraceuticals in conventional form known to possess numerous drawbacks and nanoscience proved to be a boon for the nutraceutical industry. Customarily employed nanocarriers for the nutraceuticals like liposomes, polymer based encapsulation or protein-carbohydrate based systems have certain limitations like toxicity quotient, limited availability of appropriate biopolymer and thermal and chemical overexposure during processing (Fathi et al., 2012; Tamjidi et al., 2013). Lipid based carriers, mostly nanodispersions, are getting attention from research fraternity owing to its enhanced encapsulation capacity, low toxicity and compatibility with biological systems. Nanodispersions which comprises of microemulsions, nanoemulsions and nanosuspensions, possess globules /particles whose size have been always less than 1000 nm in which bioactive is dispersed and is stabilized by surface active agent.

Microemulsions are isotropic, thermodynamically stable, non-viscous systems stabilized by a surface active agent (Kumar et al., 2006; Spernath et al., 2006). The colloidal dispersion consists of small spheroidal particles ranging from 20 to 200 nm in diameter (Agboola et al., 2020). Microemulsion, comprised of oil, water, nutrient moiety, surfactant and preferably co-surfactant, is self-assembling system utilizing low energy for the process (Rosano et al., 1987). This nanodispersion has been explored in the nutraceutical industry with an aim to improve solubilization of phytonutrients like lycopene, lutein and phytosterols using PG, alcohol, (R+)-D-limonene stabilized by Tween 80 and Tween 60. Oral bioavailability of Vitamin A palmitate was found to be enhanced when it was formulated in the form of o/w microemulsion (Garti et al., 2004; Dizaj et al., 2015).

Despite numerous benefits offered by microemulsion, nanoemulsions are favored due to less proportion of surfactant needed, thus minimizing the toxicity factor. Other factors furnishing the escalating acceptance for nanoemulsions in the nutraceutical domain are ease of formulation, elevated solubilization capacity, simple production, high absorption, and flexibility of encapsulating lipophilic and hydrophilic molecules concurrently (Blasiak et al., 2006; Dixon et al., 2008). Nanoemulsions are nanoscaled dispersions with droplet size ranging between 10–100 nm. The nanometric size range of droplets imparts optical transparency to this formulation making it an ideal carrier for nutrients, especially in beverages (Izquierdo et al., 2002). This ideal nature is further complimented by nanoemulsion's unique ability to retain the polydispersity index even diluted with water (Grishkewich et al., 2017). Even though nanoemulsion is not as thermodynamically stable as microemulsions, it accomplishes long term kinetic stability (McClements et al., 2005). Bioactive molecule can be integrated within the dispersed phase droplet or in the continuous medium or even can be concentrated in the interfacial region. The chemical stability of bioactive can be preserved by altering the interfacial layer properties and overall stability can be improved by achieving the small droplet size (Livney et al., 2014).

Nanoemulsions can be formulated using two approaches, high-energy and low energy approach. In high-energy approach, larger droplets of conventional emulsion are brought down to nanosize by mechanical energy using high pressure homogenizer, sonicator or microfluidizer and thus nanoemulsion is achieved. Low energy process adopts either dilution of macroemulsion or reduction of temperature until the inversion point of the system, which yields nanoemulsion with small droplet size. Both the methods suffer from their own limitations like the high production cost in high energy approach and large amount of surfactants in low-energy approach (Aboalnaja et al., 2016; Chowdhury et al., 2018).

The nutraceuticals sector is employing this nanocarrier system to boost the bioavailability of therapeutically potent phytonutrients. Lycopene-enriched tomato extract was formulated into nanoemulsion using emulsification-evaporation method. Smaller droplets size imparted better protection to bioactive against environmental degradation (Ha et al., 2015). Parthasarathi et al. established correlation between nanorange droplet size and improved stability by a formulation vitamin E nanoemulsion which yielded droplet size of 277 nm whereas conventional emulsion with droplet size of 1285 μm exhibited poor stability (Parthasarathi et al., 2016). Conventional formulation of natural antioxidant, Resveratrol, produces low biological availability due to first pass metabolism and poor aqueous solubility. Sessa et al. formulated a resveratrol nanoemulsion using high pressure homogenization technique and achieved protection against degradation and improved permeability (Davidov-Pardo and McClements, 2015). Davidov-Pardo and McClements (2015) used GRAS status and biocompatible ingredients to formulate resveratrol nanoemulsion maintaining chemical stability. Researchers made use of distinctive property of nanoemulsions of encapsulating more than one bioactive moiety either hydrophilic or lipophilic in the same structure by formulating nanoemulsion with resveratrol and curcumin, that protected them from degradation (Nasr et al., 2016). ß-carotene, effective bioactive moiety in reducing risk of chronic diseases like cancer and cardiovascular diseases, displays aqueous insolubility and slight solubility in oils. This results in precipitation in GIT leading to very low bioavailability. This restriction is overcome by incorporating ß-carotene in nanoemulsion using digestible carrier corn oil and improving its bioavailability (Rao et al., 2013). Mahanasar formulated a nanoemulsion with Hyaluronic acid and Vitamin E, which facilitated the partitioning of lipophilic vitamin E from the stratum corneum to epidermis and improved diffusion of hydrophilic Hyaluronic acid through highly lipophilic skin layers, stratum corneum. Owing to the numerous benefits offered and feasibility of industrial scalability due to simple formulation methods, nanoemulsion form of Vitamin A, D, E and K has been launched in the market with the brand name of Vitalipid as parenteral Nutraceuticals (Nasr et al., 2016). Diverse varieties of nanoemulsions are being tested for efficient nutraceutical delivery. Novel nanostructured multiple emulsions (O/W/O and W/O/W) allows different phytonutrients, that might counter-interact, to be incorporated in two separate and distant phases enabling successful delivery of multiple therapeutic agents in a single product. Another novel variant under nanoemulsion is nanostructured multilayer emulsion consisting of oil droplet enveloped by polyelectrolyte layer whose thickness measured in nanometer. It is prepared by layer-by-layer (LbL) electrostatic deposition and utilizes food grade ingredients (Choudhary et al., 2018; Bhusnure at al., 2015).

Another form of nanodispersion which got established as an effective nutraceutical delivery system is nanosuspension. Nanosuspensions are colloidal dispersions of nanosized drug particles and are stabilized by surfactants. Poorly water-soluble biocomponent suspended in dispersion without any matrix material comprises Submicron colloidal dispersions. Bioactive molecules possessing poor solubility and low permeability characteristics can be delivered as nanosuspension, which resolves the aforementioned problems and yields formulation with elevated bioavailability as well as amending pharmacokinetics of nutrients, achieving high plasma concentration and augmented bionutrient efficacy.

Nanosuspensions are prepared using high pressure homogenization where poorly-water soluble nutrients are passed through the narrow aperture and homogenized under high pressure. High-shear media mills or pearl mills are used to prepare nanocrystals where homogenization is effected in non-aqueous media. Deep-freeze homogenization is employed to formulate nanopure in non-aqueous media. Particle size reduction to nanosize can also be achieved using nanojet technology.

Marine bioproduct, algae, rich in natural vitamins, carbohydrates, mineral-trace elements and essential fatty acids, are a potential nutraceutical bioactive. Formulation of algal extract in nanosuspension form offers ease of manufacturing helps to combat solubility issues due to the sticky nature of bioactive and improve bioavailability. Even though, nanodispersipersions aforethought as promising nutraceutical delivery systems, it endured certain restrictions such as requiring expensive methodology, stability is questioned due to Ostwald's ripening on long-term storage and need of the high amount of surfactant and co-surfactant for stable formulation.

6.3.3 LIPID NANOPARTICLES

In 1991 Rainer H. Müller instigated a novel nanoformulation, lipid nanoparticles, where oil used in nanoemulsion formulation was replaced by solid lipid leading to the formation of solid lipid nanoparticles (Muller et al., 2000).

Solid lipid nanoparticles (SLN) are composed of highly organized crystalline solid lipid core with the nutraceutical moiety embedded within it, which is further stabilized by a peripheral layer of surfactant (Katouzian et al., 2017). Presence of solid lipid in SLN offers improved stability, supplementing the innate attributes of liposomes and emulsions, imparting sustained release of embedded bioactive (Tamjidi et al., 2013). The attributes to be focused while selecting lipid matrix for the formulation of SLN include solubility of bioactive, which influences the encapsulation efficiency. Further biocompatibility, thermal and chemical stability of lipid and nature and surfactant blend to be adsorbed at interface strongly influence the stability of SLN (Gramdorf et al., 2008). Diverse methods are employed for SLN manufacturing like high pressure homogenization, high shear homogenization with ultrasonicaton, microemulsion technique and solvent evaporation/emulsification process. The choice of method depends on the desired particle size, administration route of product and instruments available (Weiss et al., 2008). Anticancer bioactive, resveratrol exhibited improved stability when incorporated into SLNs. In another study anti-malarial activity of curcuminoid has been shown to be augmented by incorporating it into SLN by virtue of improved chemical stability and bioavailability (Mehnert et al., 2001). Combination of topical and conventional oral delivery of certain nutraceuticals has been anticipated to offer the additional advantage by surpassing degradation via GIT transport. Lipophilic SLN effectively crosses the stratum corneum resulting in lipid exchange and successful bioactive delivery (Reis et al., 2013). Nutraceutical based topical delivery of curcuminoid has been studied by Tiyaboonchai et al. (2007) via formulating curcuminoid SLN using a blend of stearic acid and glycerylmonostearate (GMS) and poloxamer 188 as a surfactant. The cream containing curcuminoid SLN was found to improve texture and wrinkled condition of skin when tested on human volunteers (Zamarioli et al., 2015; Tiyaboonchai et al., 2007). Owing to high lipophilicity and small size, SLN easily crosses blood brain barrier by endocytosis and therefore are employed for brain targeting through nasal application and found to exhibit increased nasal retention time due to adhesive property. SLN containing Rosmarinic acid for nuroprotective effect and resveratrol for anti-tumor effect has been studied (Singh et al., 2012).

Disadvantages faced by SLN like gelation tendency and lipid recrystallization resulting in bioactive expulsion lead to invention of second generation colloidal lipid carriers designated as nanostructured lipid carrier (NLC). In contrast to highly organized SLN due to presence of solid lipid, NLC has less organized crystalline structure and thus exhibit improved stability (Bhatt et al.,

2015). The less organized structure of NLC is attributed to partial replacement of solid lipid in SLN with liquid lipid component, which is responsible for improved solubility of bio-actives in lipid core. Frequently used liquid lipids/oils are medium chain triglycerides, oleic acid and natural edible oils, whereas glycerides, fatty acids and waxes may contribute as solid lipid to the lipid core of NLC (Jose et al., 2014). NLCs are composed of lipid blend and peripheral layer of the surfactant molecule. The solid lipid is incorporated in the form of its oil constituent, which eventually transits into a solid matrix at body temperature (Muller et al., 2007). NLC manufacturing can be accomplished by micro emulsions and hot and cold homogenization methods, precipitation, and ultrasound dispersion technique (Berge et al., 1977; Manjunath et al., 2005). NLC containing Coenzyme Q10 were formulated using hot homogenization and ultra-sonication in combination at 10,000 rpm and demonstrated high encapsulation efficiency (98%) and high bioavailability as well. Quercetin NLC demonstrated maximum bio-accessibility as compared to its SLN formulation and can be comfortably dispensed in food products owing to transparency and ultrafine size. Precirol, Myverol, and Pluronic used for incorporation of lutein in NLC via ultra-sonication method resulted in a stable formulation (Ucisik et al., 2015). The ultrasound homogenization technique was employed to formulate NLCs loaded with different types of terpens. Lipid blend consisted of modified beeswax as solid lipid and caprylic triglyceride as liquid lipid. Zeta potential studies revealed improved physical stability for different terpene loaded NLC (Huang et al., 2019; Lasoń et al., 2016). Currently, three different NLC structures imperfect, amorphous and multiple exists. Imperfect type provides holes due to chemical difference in solid and liquid lipid allowing nutraceutical molecule to be accommodated in these imperfections. Amorphous NLCs are so called due to solidification of lipid blend in the amorphous state. Multiple type NLC is produced by multiple oil in fat in water (O/F/W) emulsification and induce liquid nanocompartment in solid lipid matrix, thereby enhancing encapsulation efficiency (Liu et al., 2010).

Comprehensively, lipid nanoparticles offer promising alternative to nanodispersions and liposomes, demonstrating the increased encapsulation efficiency due to lipid nature, augmented bioavailability with better penetration capacity and improved stability of the formulation.

6.3.4 LIPOSOME

Liposomes derived from two Greek words lipo means fat and soma means structure. It is a closed vesicular system made by lipid by layer in which hydrophilic portion faces aqueous solution from both sides, whereas hydrophobic portion forms inner side and helps in entrapment of active pharmaceutical ingredients. It is biocompatible and mostly used because it could be administered into the body via different routes. This technology is needed for increasing stability, loading of Nutraceuticals in combination with drug delivery or for controlling the release of the compound or for maintaining its degradation. There are various methods for preparation of liposome such as mechanical dispersion method, solvent dispersion method, detergent removal method, French pressure cell: extrusion, freeze-thawed liposomes, lipid film hydration, micro-emulsification, membrane extrusion, dried reconstituted vesicles, ether or ethanol injection, reverse phase evaporation method and sonication. Liposome can be characterized by different techniques such as Particle size and zeta potential, ultraviolet spectra, Surface morphology by scanning electron microscopy, differential scanning calorimetry, Fourier transform infrared spectroscopy, X-ray diffraction, and nuclear magnetic resonance.

Biagi et al. prepared a combination of nutraceuticals by using fermented red rice, liposomal berberine, and curcumin. They used it once a day for 8 weeks and the result showed an improved lipid profile and reduce markers of inflammation in low-risk dyslipidemia patient (Biagi et al., 2019). Zhang et al. prepared Gallic acid liposome decorated with lactoferrin by thin-layer dispersion method. This showed anti-bacterial activity against *E. coli and S. aureus* (Zhang et al., 2019). Claudia Bonechi et al. prepared zwitterionic liposomes by Thin-film hydration for improving bioavailability of three natural antioxidants, which is Tyrosol, hydroxytyrosol and oleuropein. Result showed significant increase in bioavailability and also proved that this method could be used for further preparation of antioxidant (Bonechi et al., 2019). Jin et al. prepared liposome of curcumin by thin hydration method with the help of milk fat globule membrane phospholipids and soybean lecithin. Milk fat globules phospholipids also have potential as excipient for preparation of liposome. This compound is having more advantage over others because it is prepared by a phospholipid which is obtained from milk (Jin et al., 2016). Agiba et al. prepared liposome by the thin film hydration method by using glucosamine sulfate (GluS) and chondroitin sulfate for increasing its oral permeation for the treatment of osteoarthritis. Result showed significant increase in its bioavailability 60% for glucosamine sulfate (GluS) and 64% for hondroitin sulfate (Agiba et al., 2018).

6.3.5 MISCELLANEOUS NANOFORMULATIONS

6.3.5.1 PHYTOSOME

Phytosomes are made up of two words 'phyto' and 'some' which means plant with cell like structure. It has a plant extract with a layer of lipid. Some phytoconstituents are poorly absorbed in the body due to their poor water solubility generally they are large in size so they cannot diffuse due to poor lipid solubility that leads to loss of active compounds. Some of active compounds of plants are destroyed by the stomach environment so, making a complex with lipid enhance its bioavailability and absorption of active compounds. Phytosomes are miscible in water and lipid. The following methods are used for the preparation of Phytosome Solvent evaporation, Freeze-drying, Anti-solvent precipitation, Anhydrous co-solvent lyophilisation. Among all solvent evaporation is the most commonly used method. In the presence of aprotic solvent appropriate molar proportion ranging from 0.5–2.0 mixture of phospholipids and extracts were taken as depicted in Figure 6.2. The rotary evaporator is used for the solvent evaporation thin layer of phytoactive constituents obtained. For isolation of Phytosome different procedure can be performed by adding non solvent or by freeze drying or spray drying. Phytosome were obtained by using precipitation, lyophilisation or spray drying. Natural or synthetic phospholipids can be used for preparation of Phytosome. It's widely found in Soyabean oil, egg yolk, sunflower seed and different other plant seed. Some widely used phospholipids, which used for the preparation of phytosomes are enlisted here glycerophospholipids include phosphatidylcholine (PC), phosphatidylethanolamine (PE), phosphatidylserine (PS), phosphatidic acid (PA), phosphatidylinositol (PI), and phosphatidylglycerol (PG).

Flavonoids are phenolic compound extracted from plants. Flavonoids maintain organoleptic properties in fruit and also used as therapeutic agents. Now days it is used for various pharmacological actions so it is termed as "Nutraceuticals." Franceschi and Giori (2006) patented preparation technique of phospholipid complexes of olive fruits or leaves. They used phosphatidylcholine and drug in ethanol refluxed it for 1 hour and concentrated this under reduced pressure. The prepared suspension increased extract solubility (Mazumder et al., 2016). Silibinin and sinigrin Phytoconstituents are specially used for treatment of cancer because of their cytotoxic effects. Silibinin is used for different variety of cancers such as for breast, prostate and skin cancer, whereas sinigrin is used for skin cancer. Silibin and singrin

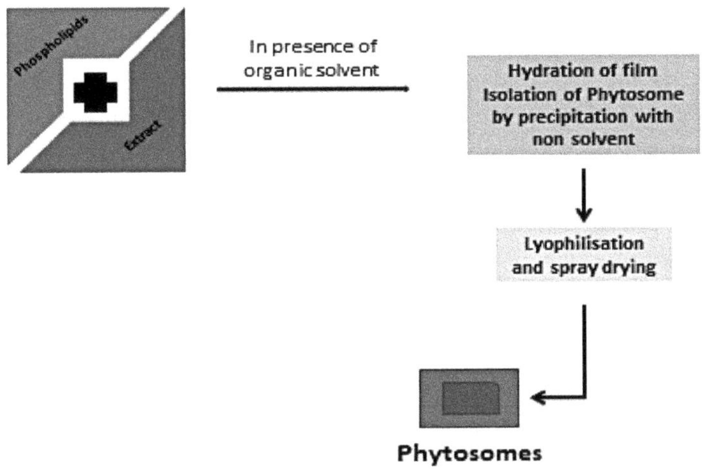

FIGURE 6.2 Preparation technique of liposome.

Phytosome are prepared by phospholipid complexation result suggested that bioavaibility of drug increase 6 times than uncomplexed drug. It could be used for skin disease and wound healing (Arora et al., 2015). Curcumin is a polyphenol extracted from turmeric plant. Major problem of curcumin is its absorption in blood and tissue. Giori et al. (2007) prepared curcumin complex by using phospholipid. The phospholipids complexes of curcumin showed higher systemic levels of the parent agent than uncomplexed curcumin. The improved bioavailability of phospholipid complex of curcumin increases the potential scope of medical applications for curcumin as a chemopreventive agent (Giori et al., 2007).

6.3.5.2 NIOSOME

Niosome is a vesicular system made up of non-ionic surfactant bi layer. It is biodegradable, non-toxic, nonimmunogenic and non-carcinogenic. The main disadvantage of niosome is its vesicular instability which is improved in proniosome is a modified form of niosome which have vesicular stability that was lacking in niosome. Ghoyal et al. prepared gugulipid-loaded proniosomal gel and evaluated its anti-inflammatory activity and compared to commercial formulations of NSAIDs. Thus, in conclusion, proniosomal formulation of gugulipid holds an immense potential for development of topical herbal anti-inflammatory formulation comparable to topical NSAIDs (Ghoyal et al., 2011).

6.3.5.3 ETHOSOME

Ethosome is made up of two words "Etho" means ethanol and "some" means structure. Ethosomes are non-invasive vesicle which helps in deep penetration of the drug into layers of the skin. The size ranges from 30 nm to few microns. Kalita et al. prepared ethosomal gel for enhancing the penetration power of epigallocatechingallate (EGCG) by using lipid based vesicle delivery that is Ethosome. Ethosome were prepared by using mechanical dispersion method, result showed significant improvement in penetration power and stability of epigallocatechingallat (Kalita et al., 2013). **Ramadon** et al. prepared ethosomal gel of Quercetin. Quercetin ethosomes were prepared by thin-film hydration method this further incorporated in carbopol gel. By choosing a right administration route they increase the bioavailability of Quercetin compared to oral route (Ramadon et al., 2017). The difference between liposome, niosome, phytosome and ethosome are shown in Figure 6.3.

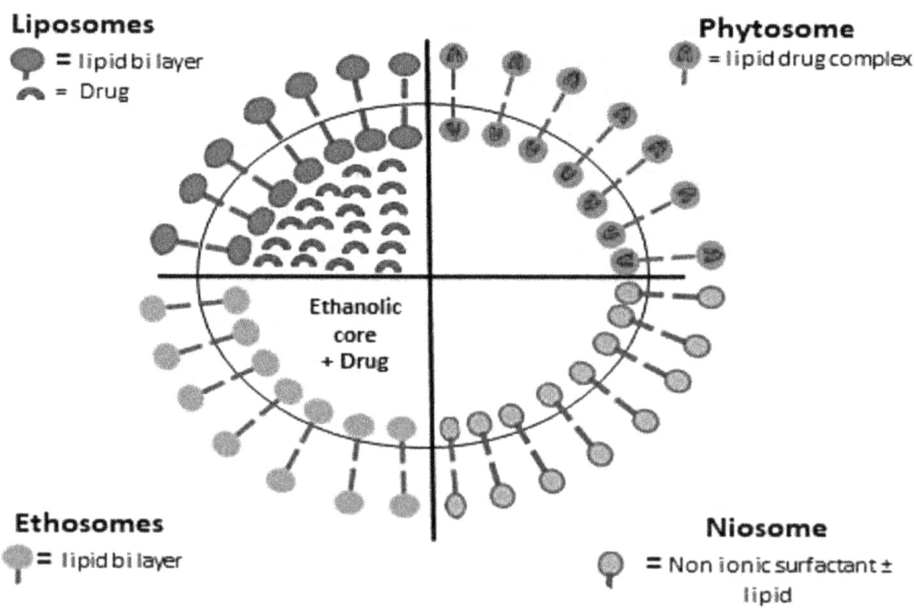

FIGURE 6.3 Difference between liposome, niosome, phytosome, and ethosome.

6.3.6 *MISCELLANEOUS NANOFORMULATIONS*

Dendrimers is derived from a Greek work "Dendron" which means tree. It is a branching polymer dividing from a central core. The size of Dendrimers is less than 15 nm but have higher molecular weight. They subdivide into three part initiator core, interior layers and exterior layers. Dendrimers can be synthesized by different methods which are divergent method, convergent method, double exponential mixed growth, hypercores and branched growth (Dasgupta et al., 2018). Curcumin, its utility is limited due to its hydrophobic structure. Debnath et al. (2013) prepared curcumin derivatives (tetrahydro-curcumin, curcumin mono-carboxylic acid, curcumin mono-galactose, curcumin mono-alkyne and Dendrimers-curcumin conjugate). They checked all the complexes for its cytotoxicity and water solubility. Among all complexes only dendrimer-curcumin conjugate showed water solubility and cytotoxicity against SKBr3 and BT549 breast cancer cells (Debnath et al., 2013; Mollazade et al., 2013). Carbon nanotubes are nanometric diameter material made up of carbon. It helps in delivery of drug/phytoconstituents because it forms a very strong bond with drugs/phytoconstituents. It is a versatile carrier for drug delivery, tissue engineering, gene therapy and in bio-sensing. Fu et al. (2017) used milk protein and prepared α-lactalbumin nano-tubes by using self-assembly technique. The result showed that it increases the stability of α-lactalbumin in various conditions such as pasteurization and freeze drying at extreme temperatures. Nanofiber is a normal fiber, but having size range of 2–600 nm. The mechanical strength of the fiber is much more than steel fiber. There are three methods of preparation for nanofibers which are electro spinning, melt processing and interfacial polymerization. Among all electro spinning is a commonly used method for its preparation. Nanocomposites is made up of two words nano means nanosized material composite means matrix so it's like nanosized particles are embedded in matrix having a size range up to 100nm. Ceramic, glass and polymers are the preferable choice for matrix because they enhanced thermal and mechanical stability (Fu et al., 2017).

6.4 CHALLENGES

Nutraceuticals receiving enormous attention being a simple approach towards prevention and treatment of various lifestyle diseases. However, the potency of nutraceuticals depends on its bioavailability. Nanoscale delivery systems are designed to combat bioavailability problems, enhancing absorption characteristics, modulate pharmacokinetic profile of bioactive in the GI tract, thereby boosting bioaccessibility and their health benefits. Experiencing rapid evolution, nanonutraceutical segment is implementing the use of diverse nanocarriers or absorption enhancers or food adjuvants that are known to improve solubility, permeability as well as the stability of bioactives (Gupta et al., 2010). Nevertheless, there are few obstacles that still have to overcome to establish this wellness approach to its full potential.

Exclusive use of food-grade compounds for nutraceutical formulation poses the basic challenge in development of nanonutraceuticals (Cerquerira et al., 2013). Secondly, achieving stability of nanocarrier system in a food matrix throughout the processing and shelf-life is still being worked upon. Thirdly, designing delivery systems with augmented permeability through biomembrane and controlling the release of bioactive at the desired site without any physico-chemical modification in GIT passage still remains a challenge for scientists. Absorption enhancers create passage for bioactive molecule by stressing epithelial barrier. The potential toxicity with regular and prolonged use of nutraceutical formulations employing absorption enhancers has not been addressed (Dias-Souza et al., 2018). The most relevant concern is the risk assessment of the use of nanocarriers in food owing to the alteration of bioactive loaded nano delivery systems during digestion and its intact delivery into systemic circulation. The robust data regarding risk of its long term ingestion is not available in the literature (Goncalves et al., 2018). So, more pragmatic in vitro digestion models must necessarily be designed to understand the pharmacokinetics of delivery system. Additionally, newer analytical methods must be devised for quantitative evaluation of nutraceuticals. These technical tools may help in safety assessment of nutraceuticals. With the arrival of novel nanomaterials in the food industry, assurance of human and environmental safety through extensive research is expected for safer future. In this context, there is an earnest need of balanced regulatory guidelines for nanonutraceutical manufacturing, claims and safety (Shishir et al., 2018).

6.5 CONCLUSIONS

There is a sweeping wave of consciousness advocating healthier lifestyle throughout the world, resulting in enhanced relevance and impact of nutraceutical for the contemporary world. A strong connect between balanced nutrition and wellness has been deciphered and achieved through the ingestion of natural concentrated bioactives. The aware global population has conferred nutraceutical, a decisive portrayal, in their daily health regimen. However, biopharmaceutical and stability concerns and large potions requisites, necessitated emergence of nanotechnologically engineered advanced nutraceuticals. Revolutionized nutraceuticals with the nanoscale carrier system will uphold to be a majestic wand for forthcoming engenderment. In spite of propitious fabrication of nanonutraceuticals, nanoscale-based toxicity and possible bioaccumulation leading to potential human health hazards need to be preferentially addressed. Preclinical and clinical efficacy as well as toxicity investigations is recommended for safe use of nanonutraceuticals. Furthermore, the amicable attempt by regulatory agencies to enforce the relevant regulations and guidelines to strengthen the manufacturer-consumer relationship and attain viable eco-friendly nanotechnological perspective will positively contribute to the betterment of humanity.

ETHICS DECLARATION

This manuscript does not include any studies with human participants or animals performed by any of the authors.

CONFLICT OF INTEREST

The authors declare having no conflict of interest.

KEYWORDS

- **casein phosphopeptides (CPP)**
- **chitosan-tripoly phosphatepentasodium (CS-TPP)**
- **epigallocatechin gallate (EGCG)**
- **ferulic acid (FA)**
- **gallic acid (GA)**
- **nanostructured lipid carriers (NLC)**

REFERENCES

Aboalnaja, K. O., Yaghmoor, S., Kumosani, T. A., & McClements, D. J., (2016). Utilization of nanoemulsions to enhance bioactivity of pharmaceuticals, supplements, and nutraceuticals: Nanoemulsion delivery systems and nanoemulsion excipient systems. *Expert Opin. Drug Deliv. 13,* 1327–1336.

Aditya, N. P., Espinosa, Y. G., & Norton, I. T., (2017). Encapsulation systems for the delivery of hydrophilic nutraceuticals: Food application. *Biotechnol. Adv. 35,* 450–4457.

Aghbashlo, M., Mobli, H., Madadlou, A., & Rafiee, S., (2012). Integrated optimization of fish oil microencapsulation process by spray drying. *J. Microencapsul. 29,* 790–804.

Agiba, A. M., Nasr, M., Abdel-Hamid, S., Eldin, A. B., & Geneidi, A. S., (2018). Enhancing the Intestinal Permeation of the Chondroprotective Nutraceuticals Glucosamine Sulfate and Chondroitin Sulfate Using Conventional and Modified Liposomes. *Curr. Drug Deliv. 15,* 907–916.

Ajazuddin, Saraf, S. (2010). Applications of novel drug delivery system for herbal formulations. *Fitoterapia, 81,* 680–689.

Alvarez-Román, R., Barré, G., Guy, R., & Fessi, H., (2001). Biodegradable polymer nanocapsules containing a sunscreen agent: preparation and photoprotection. *Eur. J. Pharm. Biopharm. 52,* 191–195.

Anand, P., Kunnumakkara, A. B., Newman, R. A., & Aggarwal, B. B., (2007). Bioavailability of Curcumin: Problems and Promises. *Mole. Pharm. 4,* 807–818.

Andlauer, W., & Furst, P. F., (2002). Nutraceuticals: a piece of history, present status and outlook, Food Research International, *35,* 171–176.

Arora, S., Sharma, A., & Kaur, P., (2015). Preparation and Characterization of Phytosomal-Phospholipid Complex of *P. amarus* and its Tablet Formulation. *J. Pharm. Technol. Res. Manag. 1,* 1–18.

Peñalva, R., Esparza, I., González-Navarro, C. J., Quincoces, G., Peñuelas, I., & Irache, J. M., (2015). Zein nanoparticles for oral folic acid delivery, *Journal of Drug Delivery Science and Technology, 30,* 450–457.

Augustin, M. A., & Hemar, Y., (2009). Nano- and micro-structured assemblies for encapsulation of food ingredients. *Chem. Rev. 38,* 902–912.

Axelos, M. A. V., & Voorde, M. V., (2016). *Nanotechnology in Agriculture and Food Science.* Wiley-VCH; 2017.

Bagchi, D., & Nair, S. *Developing New Functional Food and Nutraceutical Products*, Elsevier Academic Press Inc, London, U.K., pp. 481–494.

Bennet, D., Marimuthu, M., Kim, S., & An, J., (2012). Dual drug-loaded nanoparticles on self-integrated scaffold for controlled delivery. *Int. J. Nanomedicine. 7,* 3399–3419.

Berge, S. M., Bighley, L. D., & Monkhouse, D. C., (1977). Pharmaceutical Salts. *J. Pharma. Sci. 66,* 1–19.

Bharimalla, A. K., Deshmukh, S. P., Vigneshwaran, N., Patil, P. G., & Prasad, V., (2017). Nanocellulose-Polymer Composites for Applications in Food Packaging: Current Status, Future Prospects and Challenges. *Polym. Plast. Technol. Eng. 56,* 805–823.

Bhatt, R., Singh, D., Prakash, A., & Mishra, N., (2015). Development, characterization and nasal delivery of rosmarinic acid-loaded solid lipid nanoparticles for the effective management of Huntington's disease. *Drug Deliv. 22,* 931–939.

Bhusnure, O. G., Gholve, S. B., Bhange, M. M., & Pentewar, R. S., (2015). Formulation of nanosuspension and nanoemulsion as a new approach for the delivery of poorly soluble drugs. *Indo American J. Pharma. Res. 11,* 5.
Biagi, M., Minoretti, P., Bertona, M., & Emanuele, E., (2019). Effects of a nutraceutical combination of fermented red rice, liposomal berberine, and curcumin on lipid and inflammatory parameters in patients with mild-to-moderate hypercholesterolemia: an 8-week, open-label, single-arm pilot study. *Arch. Med. Sci. Atheroscler. 3,* 137–41.
Biesalski, H. K., (2002). Nutraceuticals: the link between nutrition and medicine. *J Toxicol. Cutaneous Ocul. Toxicol. 21,* 9–30.
Bikram, M., Gobin, A. M., Whitmire, R. E., & West, J. L., (2007). Temperature-sensitive hydrogels with SiO2-Au nanoshells for controlled drug delivery. *J. Control Release. 123,* 219–227.
Blasiak, P., Horzela, A., Penson, K. A., & Solomon, A. I., (2006). Dobiński-type relations: some properties and physical applications. *J. Phys. A Math. Theor. 39,* 4999–5006.
Bonechi, C., Donati, A., Tamasi, G., Pardini, A., Rostom, H., & Leone, G., (2019). Chemical characterization of liposomes containing nutraceutical compounds: Tyrosol, hydroxytyrosol and oleuropein. *Biophys. Chem. 246,* 25–34.
Boon, C. S., McClements, D. J., Weiss, J., & Decker, E. A., (2010). Factors Influencing the Chemical Stability of Carotenoids in Foods. *Crit Rev. Food Sci. Nutr. 50,* 515–532.
Bouwmeester, H., Brandhoff, P., Marvin, H. J. P., Weigel, S., & Peters, R. J. B., (2014). State of the safety assessment and current use of nanomaterials in food and food production. *Trends Food Sci. Technol. 40,* 200–210.
Brower, V. (1998). Nutraceuticals: Poised for a healthy slice of the healthcare market? *Nat Biotechnol. 16,* 728–731.
Cerqueira, M. A., Pinheiro, A. C., Ramos, O. L., Silva, H., Bourbon, A. I., & Vicente, A. A. (2017). Advances in Food Nanotechnology, In: Busquets, R. (Eds.), *Emerging Nanotechnologies in Food Science,* Elsevier, USA, pp. 11–38.
Cerqueira, M. A., Pinheiro, A. C., Silva, H. D., Ramos, P. E., Azevedo, M. A., & Flores-López, M. L., (2013). Design of bio-nanosystems for oral delivery of functional compounds. *Food Eng. Rev. 6,* 1–19
Chau, C. F., Wu, S. H., & Yen, G. C., (2007). The development of regulations for food nanotechnology. *Trends Food Sci. Technol. 18,* 269–280.
Chowdhury, M. A., (2018). Design, Development and Applications of the Silica-Based Contrast Agents in Magnetic Resonance Imaging. *Curr. Nanomed. 8,* 3–27.
Cook, M. T., Tzortzis, G., Charalampopoulos, D., & Khutoryanskiy, V. V., (2012). Microencapsulation of probiotics for gastrointestinal delivery. *J. Control. Release. 162,* 56–67.
Couvreur, P., Barratt, G., Fattal, E., & Vauthier, C., (2002). Nanocapsule Technology: A Review. *Crit. Rev. Ther. Drug Carrier Sys. 19,* 99–134.
Cruz, L., Schaffazick, S. R., Costa, T. D., Soares, L. U., Mezzalira, G., & Da Silveira N. P. (2006a). Physico-Chemical Characterization and In Vivo Evaluation of Indomethacin Ethyl Ester-Loaded Nanocapsules by PCS, TEM, SAXS, Interfacial Alkaline Hydrolysis and Antiedematogenic Activity. *J. Nanosci. Nanotechnol. 6,* 3154–3162.
Cruz, L., Soares, L. U., Costa, T. D., Mezzalira, G., Da Silveira, N. P., & Guterres, S. S., (2006b). Diffusion and mathematical modeling of release profiles from nanocarriers. *Int. J. Pharm. 313,* 198–205.
Dang, R. (2017). Nutraceuticals for Healthy Life. *Indian J. Pharm. Edu. Res. 51,* 148–151.

Agboola, O., Popoola, P., Sadiku, R., Sanni, S. E., Fayomi, S. O., & Fatoba, O. S. (2020). Nanotechnology in Wastewater and the Capacity of Nanotechnology for Sustainability, In: Dasgupta, N., Ranjan, S., & Lichtfouse, E. (eds.). *Environmental Nanotechnology*, Springer Nature Switzerland AG, pp. 1–46.

Davidov-Pardo, G., & McClements, D. J., (2015). Nutraceutical delivery systems: Resveratrol encapsulation in grape seed oil nanoemulsions formed by spontaneous emulsification. *Food Chem. 167*, 205–212.

De Vos, P., Faas, M. M., Spasojevic, M., & Sikkema, J., (2010). Encapsulation for preservation of functionality and targeted delivery of bioactive food components. *Int Dairy J. 20,* 292–302.

Debnath, S., Saloum, D., Dolai, S., Sun, C., Averick, S., & Raja, K. (2013). Dendrimer-Curcumin Conjugate: A Water Soluble and Effective Cytotoxic Agent Against Breast Cancer Cell Lines. *Anti-Cancer Agents in Med. Chem. 13,* 1531–1539.

Dillard, C. J., & German, J. B., (2000). Phytochemicals: nutraceuticals and human health. *J Sci. Food Agri. 80,* 1744–1756.

Dixon, K., & Phifer, M. (2008). *Hydraulic and Physical Properties of MCU Saltstone.* Savannah River Site (S. C.), 10. 2172/928559.

Dizaj, S. M., Vazifehasl, Z., Salatin, S., Adibkia, K., & Javadzadeh, Y. (2015). Nanosizing of drugs: Effect on dissolution rate. *Res. Pharma. Sci. 10,* 95–108.

Dureja, H., Kaushik, D., & Kumar, V. (2003). *Indian Journal of Pharmacology, 35,* 363–372. Educational Forum Developments in Nutraceuticals [Internet]. [cited 2019 Jun 1]. Available from: http://medind.nic.in/ibi/t03/i6/ibit03i6p363.pdf.

Doublier, J-L., Garnier, C., & Cuvelier, G. (2017). Gums and Hydrocolloids: Functional Aspects, In Eliasson, A.-C, (Ed.) *Carbohydrates in Food*, Third edition. CRC Press/Taylor & Francis, Boca Raton, pp. 285–332.

Ernest, U., Chen, H. Y., Xu, M. J., Taghipour, Y. D., Asad, M. H. H., & Bin, Rahimi, R., (2018). Anti-Cancerous Potential of Polyphenol-Loaded Polymeric Nanotherapeutics. *Molecules*, *23,* 2787.

Esfanjani, A. F., & Jafari, S. M. (2016). Biopolymer nano-particles and natural nano-carriers for nano-encapsulation of phenolic compounds. *Colloids Surf. 146*, 532–543.

Espín, J. C., García-Conesa, M. T., & Tomás-Barberán, F. A., (2007). Nutraceuticals: Facts and fiction. *Phytochem. 68,* 2986–3008.

Waraho, T., McClements, D. J., & Decker, E. A., (2011). Mechanisms of lipid oxidation in food dispersions. *Trends in Food Science & Technology, 22*, 3–13.

Ezhilarasi, P. N., Karthik, P., Chhanwal, N., & Anandharamakrishnan, C., (2013). Nanoencapsulation Techniques for Food Bioactive Components: A Review. *Food Bioproc Tech. 6,* 628–47.

Fathi, M., Martín, Á., & McClements, D. J. (2014). Nanoencapsulation of food ingredients using carbohydrate based delivery systems. *Trends Food Sci. Technol. 39,* 18–39.

Fathi, M., Mozafari, M. R., & Mohebbi, M., (2012). Nanoencapsulation of food ingredients using lipid based delivery systems. *Trends Food Sci. Technol. 23,* 13–27.

Franceschi, F., & Giori A., (2006). Phospholipid complexes of olive fruits or leaves extracts having improved bioavailability, WO2007118631A1.

Fu, W. C., Opazo, M. A., Acuña, S. M., & Toledo, P. G. (2017). New route for self-assembly of α-lactalbumin nanotubes and their use as templates to grow silver nanotubes. In: Permyakov, E. A. (eds). Plos One, *12,* e0175680.

Gangwar, R. K., Tomar, G. B., Dhumale, V. A., Zinjarde, S., Sharma, R. B., & Datar, S., (2013). Curcumin conjugated silica nanoparticles for improving bioavailability and its anticancer applications. *J. Agric. Food Chem. 61,* 130926133947000.

Garti, N., Amar-Yuli, I., Spernath, A., Hoffman, R. E., (2004). Transitions and loci of solubilization of nutraceuticals in U-type nonionic microemulsions studied by self-diffusion NMR. *Chem. Phys. 6,* 2968.

Giori A., & Federico, F. (2007). *Phospholipid Complexes of Curcumin Having Improved Bioavailability*. PAT: EP1991244.

Gonçalves, R. F. S., Martins, J. T., Duarte, C. M. M., Vicente, A. A., & Pinheiro, A. C. (2018). Advances in nutraceutical delivery systems: From formulation design for bioavailability enhancement to efficacy and safety evaluation. *Trends Food Sci. Technol. 78,* 270–291.

Ingale, A. G., & Chaudhari, A. N., (2018). Nanotechnology in the Food Industry, In: Gothandam, K. M., Ranjan, S., Dasgupta, N., Ramalingam, C., & Lichtfouse, E. (Eds.). *Nanotechnology, Food Security and Water Treatment,* pp. 87–128.

Goyal, C., Ahuja, M., & Sharma, S. K., (2011). Preparation and evaluation of anti-inflammatory activity of gugulipid-loaded proniosomal gel. *Acta. Pol. Pharm. 68,* 147–150.

Gramdorf, S., Hermann, S., Hentschel, A., Schrader, K., Müller, R. H., & Kumpugdee-Vollrath, M., (2008). Crystallized miniemulsions: Influence of operating parameters during high-pressure homogenization on size and shape of particles. *Colloids Surf. A Physicochem. Eng. Asp. 331,* 108–13.

Grishkewich, N., Mohammed, N., Tang, J., & Tam, K. C., (2017). Recent advances in the application of cellulose nanocrystals. *Curr. Opin. Colloid Interface Sci. 29,* 32–45.

Gupta, S., Chauhan, D., Mehla, K., Sood, P., & Nair, A., (2010). An overview of nutraceuticals: current scenario. *J Basic Clin. Pharm. 1,* 55–62.

Gupta, S., & Ghosh, M., (2013). In Vitro Antioxidative Evaluation of α- and β-Carotene, Isolated from Crude Palm Oil. *J. Anal. Methods Chem.* 351671.

Guterres, S. S., Beck, R. C. R., & Pohlmann, A. R., (2009). Spray-drying technique to prepare innovative nanoparticulated formulations for drug administration: a brief overview. *Brazilian J. Phys. 39,* 205–209.

Ha, T. V. A., Kim, S., Choi, Y., Kwak, H. S., Lee, S. J., & Wen, J., (2015). Antioxidant activity and bioaccessibility of size-different nanoemulsions for lycopene-enriched tomato extract. *Food Chem. 178,* 115–121.

Hafez, M. M., Maghrabi, I. A., & Zaki, N. M., (2013). Toward an Alternative Therapeutic Approach for Skin Infections: Antagonistic Activity of Lactobacilli Against Antibiotic-Resistant Staphylococcus aureus and Pseudomonas aeruginosa. Probiotics Antimicrob Proteins. *5,* 216–226.

Huang, J., She, J., Yang, X., Liu, J., Zhou, X., & Yang, B., (2019). A new macrodiolide and two new polycyclic chromones from the fungus *Penicillium* sp. SCSIO041218. *Molecules, 24,* 1686.

Izquierdo, P., Esquena, J., Tadros T. F., Dederen C., Garcia M. J., & Azemar N., (2002). Formation and Stability of Nano-Emulsions Prepared Using the Phase Inversion Temperature Method. *Langmuir. 18*(1), 26–30.

Jampilek, J., Kos, J., Kralova, K., Jampilek, J., Kos, J., & Kralova, K., (2019). Potential of Nanomaterial Applications in Dietary Supplements and Foods for Special Medical Purposes. *Nanomater., 9,* 296.

Jin, H. H., Lu, Q., & Jiang, J. G., (2016). Curcumin liposomes prepared with milk fat globule membrane phospholipids and soybean lecithin. *J Dairy Sci. 99,* 1780–1790.

Jose, S., Anju, S. S., Cinu, T. A., Aleykutty, N. A., Thomas, S., & Souto, E. B., (2014). In vivo pharmacokinetics and biodistribution of resveratrol-loaded solid lipid nanoparticles for brain delivery. *Inter. J. Pharma. 474*, 6–13.

Joye, I. J., Davidov-Pardo, G., & McClements, D. J., (2014). Nanotechnology for increased micronutrient bioavailability. *Trends Food Sci. Technol. 40,* 168–182.

Kalita, B., Das, M. K., & Sharma, A. K., (2013). Novel Phytosome formulations in making herbal extracts more effective. *Res. J. Pharmacy Technol. 6,* 1295–301.

Kalra, E. K. (2003). Nutraceutical-definition and introduction. *AAPS PharmSci. 5,* E25.

Kannappan, R., Gupta, S. C., Kim, J. H., Reuter, S., & Aggarwal, B. B., (2011). Neuroprotection by Spice-Derived Nutraceuticals: You Are What You Eat! *Mole. Neurobiol. 44,* 142–59.

Katouzian, I., Esfanjani, A. F., Jafari, S. M., & Akhavan, S., (2017). Formulation and application of a new generation of lipid nano-carriers for the food bioactive ingredients. *Trends Food Sci. Technol. 68,* 14–25.

Keservani, R. K., Kesharwani, R. K., Vyas, N., Jain, S., Raghuvanshi, R., & Sharma, A. K., (2010). Nutraceutical and Functional Food as Future Food: A Review, *Der Pharmacia Lettre, 2,* 106–116.

Keservani, R. K., Kesharwani, R. K., Sharma, A. K., Gautam, S. P., & Verma, S. K., (2017). Nutraceutical Formulations and Challenges. Developing New Functional Food and Nutraceutical Products, Bagchi, D., Nair, S (Eds). 161–177.

Dias-Souza, M. V., Nascimento, A. Á., & Vasconcelos, T. Y. L., (2018). In: Keservani, R. K., Sharma, A. K. (Eds.) Nanodispersions for drug delivery, Apple academic press, New York, pp. 91–106.

Khalil, N. M., Nascimento, T. C. F., Casa, D. M., Dalmolin, L. F., De Mattos, A. C., & Hoss, I., (2013). Pharmacokinetics of curcumin-loaded PLGA and PLGA-PEG blend nanoparticles after oral administration in rats. *Colloids Surf. 101*, 353–60.

Kumar, P. (2006). Special Issue in Honor of Dr. K. L. Mittal, *Adv. Colloid Interface Sci. 123–126*, 1–486.

Kumari, A., Yadav, S. K., & Yadav, S. C., (2010). Biodegradable polymeric nanoparticles based drug delivery systems. *Colloids Surf. 75,* 1–18.

Gunasekaran, S., & Ko, S., (2014). Rationales of Nano- and Microencapsulation for Food Ingredients, In: Kwak, H.-S. Nano- and microencapsulation for foods. Wiley-Blackwell, New Jersey, pp. 43–64.

Lasoń, E., Sikora, E., Ogonowski, J., Tabaszewska, M., & Skoczylas, Ł., (2016). Release study of selected terpenes from nanostructured lipid carriers. *Colloids Surf. 510*, 87–92.

Lemes, A. C., Sala, L., Ores, J. D. C., Braga, A. R. C., Egea, M. B., & Fernandes, K. F., (2016). A Review of the Latest Advances in Encrypted Bioactive Peptides from Protein-Rich Waste. *Int J Mole. Sci., 17,* 950.

Liu, C. H., & Wu, C. T., (2010). Optimization of nanostructured lipid carriers for lutein delivery. *Colloids Surf. 353*, 149–56.

Livney, Y. D. Beta-lactoglobulin-polysaccharide nanoparticles for hydrophobic bioactive compounds – US8791064B2.

Lopes, E., Pohlmann, A. R., Bassani, V., & Guterres, S. S., (2000). Polymeric colloidal systems containing ethionamide: preparation and physico-chemical characterization. *Die Pharmazie*, *55,* 527–530.

Luo, P., Morrison, I., Dudkiewicz, A., Tiede, K., Boyes, E., & O'toole, P., (2013). Visualization and characterization of engineered nanoparticles in complex environmental and food matrices using atmospheric scanning electron microscopy. *J. Microsc. 250*, 32–41.

Madadlou, A., Emam-Djomeh, Z., Mousavi, M. E., Mohamadifar, M., & Ehsani, M., (2010). Acid-induced gelation behavior of sonicated casein solutions. *Ultrason. Sonochem. 17,* 153–158.

Manjunath, K., Reddy, J. S., & Venkateswarlu, V., (2005). Solid lipid nanoparticles as drug delivery systems. *Method Find. Exp. Clin. 27,* 127.

Martins, J. T., Bourbon, A. I., Pinheiro, A. C., Fasolin, L. H., & Vicente, A. A. (2018). Protein-Based Structures for Food Applications: From Macro to Nanoscale. *Front. Sustain. Food syst. 2,* 77.

Mazumder, A., Dwivedi, A., Fox, L. T., Brümmer, A., Du Preez, J. L., & Gerber, M., (2016). In vitro skin permeation of sinigrin from its phytosome complex. *J. Pharmacy Pharmacol. 68,* 1577–1583.

McClements, D. J. (2015). Selection of Delivery Systems: Case Studies. In: Nanoparticle- and microparticle-based delivery systems: encapsulation, protection and release of active compounds, CRC Press, London, pp. 401–452.

McClements, D. J. (2005). Theoretical Analysis of Factors Affecting the Formation and Stability of Multilayered Colloidal Dispersions. *Langmuir. 21,* 9777–9785.

Mcclements, D. J., & Decker, E. A. (2000). Lipid Oxidation in Oil-in-Water Emulsions: Impact of Molecular Environment on Chemical Reactions in Heterogeneous Food Systems, *J Food Sci. 65,* 1270–1280

McClements, D. J., Decker, E. A., Park, Y., & Weiss, J., (2009). Structural Design Principles for Delivery of Bioactive Components in Nutraceuticals and Functional Foods. *Critical Rev. Food Sci. Nutr. 49,* 577–606.

Mehnert, W., & Mäder, K., (2001). Solid lipid nanoparticles: production, characterization and applications. *Adv. drug deliv. rev. 47,* 165–96.

Misra, R., & Sahoo, S. K., (2011). Coformulation of Doxorubicin and Curcumin in Poly(d, l-lactide-co-glycolide) Nanoparticles Suppresses the Development of Multidrug Resistance in K562 Cells. *Mol. Pharm. 8,* 852–866.

Moinard-Checot, D., Chevalier, Y., Briançon, S., Fessi, H., & Guinebretière, S., (2006). Nanoparticles for Drug Delivery: Review of the Formulation and Process Difficulties Illustrated by the Emulsion-Diffusion Process. *J. Nanosci. Nanotechnol. 6,* 2664–2681.

Mollazade, M., Nejati-Koshki, K., Akbarzadeh, A., Zarghami, N., Nasiri, M., & Jahanban-Esfahlan, R., (2013). PAMAM dendrimers augment inhibitory effects of curcumin on cancer cell proliferation: possible inhibition of telomerase. *Asian Pac. J. Cancer Prev. 14,* 6925–6928.

Muller, R., Petersen, R., Hommoss, A., & Pardeike, J., (2007). Nanostructured lipid carriers (NLC) in cosmetic dermal products. *Adv. Drug Deliv. Rev. 59,* 522–530.

Müller, R. H., Mäder, K., & Gohla, S., (2000). Solid lipid nanoparticles (SLN) for controlled drug delivery – a review of the state of the art. *Eur. J. Pharm. Biopharm. 50,* 161–177.

Nagavarma, B. V. N., Yadav, H. K. S., Ayaz, A., Vasudha, L. S., & Shivkumar, H. G., (2012). Different techniques for preparation of polymeric nanoparticles – a review, *Asian J. Pharm. Clin. Res. 5,* 16–23.

Garti, N., Aserin, A., Spernath, A., & Amar, I., (2007). Nano-sized self-assembled liquid dilutable vehicles, United States Patent No.: US 7,182,950 B2.

Nasr, M., (2016). Development of an optimized hyaluronic acid-based lipidic nanoemulsion co-encapsulating two polyphenols for nose to brain delivery. *Drug Deliv. 23,* 1444–1452.

Nedovic, V., Kalusevic, A., Manojlovic, V., Levic, S., & Bugarski, B., (2011). An overview of encapsulation technologies for food applications. *Procedia Food Sci. 1,* 1806–1815.

Neethirajan, S., & Jayas, D. S., (2011). Nanotechnology for the Food and Bioprocessing Industries. *Food Bioproc. Tech. 4,* 39–47.

Hilton, J. (2017). Chapter 1 Growth patterns and emerging opportunities in nutraceutical and functional food categories: market overview in Bagchi, D., Nair, S. Developing new functional food and nutraceutical products.

Panwar, R., Sharma, A. K., Dutt, D., Pruthi, V., Panwar, R., & Sharma, A. K., (2015). Phenolic Acids from Parthenium hysterophorus: Evaluation of Bioconversion Potential as Free Radical Scavengers and Anticancer Agents. *Adv. Biosci. Biotechnol.*, *6,* Article ID: 53115.

Panwar, R., Sharma, A. K., Kaloti, M., Dutt, D., & Pruthi, V., (2016). Characterization and anticancer potential of ferulic acid-loaded chitosan nanoparticles against ME-180 human cervical cancer cell lines. *Appl. Nanosci. 6,* 803–13.

Parthasarathi, S., Muthukumar, S. P., & Anandharamakrishnan, C., (2016). The influence of droplet size on the stability, in vivo digestion, and oral bioavailability of vitamin E emulsions. *Food Funct. 7,* 2294–2302.

Patra, J. K., Das, G., Fraceto, L. F., Campos, E. V. R., Rodriguez-Torres, M. D. P., Acosta-Torres, L. S., Diaz-Torres, L. A., Grillo, R., Swamy, M. K., Sharma, S., Habtemariam, S., & Shin, H. S., (2018). Nano based drug delivery systems: recent developments and future prospects. *J. Nanobiotechnology 16,* 71.

Paul, W., & Sharma, C. P., (2020). Inorganic nanoparticles for targeted drug delivery. Biointegration of Medical Implant Materials, In: C. P. Sharma (Eds), Woodhead Publishing, Cambridge, 333–373.

Pinto Reis P. C., Neufeld R. J., Ribeiro, A. J., & Veiga, F., (2006). Nanoencapsulation I. Methods for preparation of drug-loaded polymeric nanoparticles. *Nanomed. 2,* 8–21.

Radhika, P. R., Brij, R., Singh, M., Sivakumar, T., Radhika, P. R., Pharm, M., Radhika, P. R., Singh, R. B. M., & Sivkumar, T., (2011). Nutraceuticals: An area of tremendous scope. *Int. J. Res. Ayurveda Pharm. 2,* 410–405.

Raja, M. A., Shah, Z., Arif, M., & Liu, C., (2016). Self-assembled nanoparticles based on amphiphilic chitosan derivative and arginine for oral curcumin delivery. *Int. J. Nanomed. 11,* 4397–4412.

Rajam, R., & Anandharamakrishnan, C., (2015). Microencapsulation of Lactobacillus plantarum (MTCC 5422) with fructooligosaccharide as wall material by spray drying. *LWT – Food Sci. Technol. 60,* 773–80.

Ramadon, D., Anwar, E., & Harahap, Y., (2017). In vitro Penetration and Bioavailability of Novel Transdermal Quercetin-loaded Ethosomal Gel. *Indian J Pharma. Sci. 79,* 948–56.

Rao, J., Decker, E. A., Xiao, H., & McClements, D. J., (2013). Nutraceutical nanoemulsions: influence of carrier oil composition (digestible versus indigestible oil) on β-carotene bioavailability. *J Sci. Food Agric. 93,* 3175–3183.

Rao, J. P., & Geckeler, K. E., (2011). Polymer nanoparticles: Preparation techniques and size-control parameters. *Prog. Polym. Sci. 36,* 887–913.

Reis, S., Neves, M., Lúcio, S., Martins, J. L. C., & Lima, S., (2013). Novel resveratrol nanodelivery systems based on lipid nanoparticles to enhance its oral bioavailability. *Int. J. Nanomed. 8,* 177.

Rosano, H. L., Cavallo, J. L., Chang, D. L., & Whittam, J. H., (1987). Microemulsions: A commentary on their preparation, *J. Soc. Cosmet. Chem. 39,* 201–209

Ruiz, J. C., & Campos, M. R. S. (2016). New Polymers for Encapsulation of Nutraceutical Compounds. John Wiley & Sons, Incorporated; 352 p.

Sanguansri, P., & Augustin, M. A. (2006). Nanoscale materials development – a food industry perspective. Trends *Food Sci. Technol. 17,* 547–56.
Singh, P. K., & Singh, H., (2017). Challenges in the Development of Functional Foods: Role of Nanotechnology, In Sen, S., & Pathak, Y. (Eds.). *Nanotechnology in Nutraceuticals: Production to Consumption,* CRC Press, pp. 233–264.
Shima, F., Akagi, T., Uto, T., & Akashi, M., (2013). Manipulating the antigen-specific immune response by the hydrophobicity of amphiphilic poly(γ-glutamic acid) nanoparticles. *Biomaterials, 34,* 9709–9716.
Shishir, M. R. I., Xie, L., Sun, C., Zheng, X., & Chen, W., (2018). Advances in micro and nano-encapsulation of bioactive compounds using biopolymer and lipid-based transporters. *Trends Food Sci. Technol. 78,* 34–60.
Singh, A. P., Saraf, S. K., & Saraf, S. A., (2012). SLN approach for nose-to-brain delivery of alprazolam. *Drug Deliv. Translational Res. 2,* 498–507.
Singh, T., Shukla, S., Kumar, P., Wahla, V., Bajpai, V. K., & Rather, I. A., (2017). Application of Nanotechnology in Food Science: Perception and Overview. *Front. Microbiol. 8,* 1501.
Sohaimy, S. A. (2012). El Functional Foods and Nutraceuticals-Modern Approach to Food Science. *World Appl. Sci. J. 20,* 691–708.
Soppimath, K. S., Aminabhavi, T. M., Kulkarni, A. R., & Rudzinski, W. E. (2001). Biodegradable polymeric nanoparticles as drug delivery devices. *J. Control. Release. 70,* 1–20.
Spernath, A., & Aserin, A. (2006). Microemulsions as carriers for drugs and nutraceuticals. *Adv. Colloid Interface Sci. 128–130,* 47–64.
Sun, F., Ju, C., Chen, J., Liu, S., Liu, N., & Wang, K., (2012). Nanoparticles based on hydrophobic alginate derivative as nutraceutical delivery vehicle: Vitamin D 3 loading. Artif. *Cells Blood Substit. Immobil. Biotechnol. 40,* 113–119.
Tamjidi, F., Shahedi, M., Varshosaz, J., & Nasirpour, A., (2013). Nanostructured lipid carriers (NLC): A potential delivery system for bioactive food molecules. *Innov. Food Sci. Emerg. Technol. 19,* 29–43.
Tiyaboonchai, W., Tungpradit, W., & Plianbangchang, P., (2007). Formulation and characterization of curcuminoids loaded solid lipid nanoparticles. *Int. J Pharma. 337,* 299–306.
Tsai, Y. M., Chien, C. F., Lin, L. C., & Tsai, T. H., (2011). Curcumin and its nano-formulation: The kinetics of tissue distribution and blood-brain barrier penetration. *Int. J. Pharma. 416,* 331–338.
Ucisik, M. H., Sleytr, U. B., & Schuster, B., (2015). Emulsomes meet S-layer proteins: an emerging targeted drug delivery system. *Curr. Pharma. Biotechnol. 16,* 392–405.
Weir, A., Westerhoff, P., Fabricius, L., Hristovski, K., & Von Goetz, N., (2012). Titanium Dioxide Nanoparticles in Food and Personal Care Products. *Environ. Sci. Technol. 46,* 2242–2250.
Weiss, J., Decker, E. A., McClements, D. J., Kristbergsson, K., Helgason, T., & Awad, T., (2008). Solid lipid nanoparticles as delivery systems for bioactive food components. *Food Biophys. 3,* 146–154.
Yao, M., McClements, D. J., & Xiao, H., (2015). Improving oral bioavailability of nutraceuticals by engineered nanoparticle-based delivery systems. *Curr. Opin. Food Sci. 2,* 14–9.

Yi, J., Lam, T. I., Yokoyama, W., Cheng, L. W., & Zhong, F., (2014). Controlled Release of β-Carotene in β-Lactoglobulin-Dextran-Conjugated Nanoparticles' in Vitro Digestion and Transport with Caco-2 Monolayers. *J. Agric. Food Chem. 62,* 8900–8907.

Zaki, N. (2014). Progress and Problems in Nutraceuticals Delivery: An Expert Review. *J. Bioequivalence & Bioavailability*, *6,* 75–77.

Zamarioli, C. M., Martins, R. M., Carvalho, E. C., Freitas, L. A. P., Zamarioli, C. M., & Martins, R. M. (2015). Nanoparticles containing curcuminoids (*Curcuma longa*): development of topical delivery formulation. Rev. Bras. *Farmacogn. 25,* 53–60.

Zhang, Y., Pu, C., Tang, W., Wang, S., & Sun, Q., (2019). Gallic acid liposomes decorated with lactoferrin: Characterization, in vitro digestion and antibacterial activity. *Food Chem. 293*, 315–322.

CHAPTER 7

Comprehensive Potency of Nanomaterials as Medicinal Applicant of Nanomaterials

M. PANDEESWARAN, S. DHARANEEDHARAN, N. RENUGA DEVI, and S. VIGNESH

G.T.N. Arts College (Autonomous), Dindigul, Tamil Nadu, India

ABSTRACT

Zero-dimensional nanomaterials (NMs) are another name for nanoparticles. Nanoparticles have several advantages in modern medicine. This will serve as a broad introduction to the role of nanoparticles in medicine rather than an in-depth examination in the future. Furthermore, the focus of this study was on technologies that have either already proceeded to clinical application or have undergone *in vivo* testing. This study has examined medical uses of nanoparticles under the broad areas of medical imaging and drug/gene delivery. Readers have been directed to the numerous wide-ranging assessments currently accessible within each application area wherever feasible. Finally, the influence of nanoparticles in contemporary medicine on the environment and society has been examined.

7.1 INTRODUCTION

Nanomaterials (NMs) are particles having size scaled in nanometers (nm), i.e., less than 100 nm. Currently, these materials have emerged as important participants in current drug discovery, with uses ranging from contrasting tools to acting as vectors for DNA drug delivery into specific cells. NMs

Advanced Nanomaterials for Biological, Nutraceutical, and Medicinal Applications.
Sreeraj Gopi, Augustine Amalraj, and T. R. Sreeraj (Eds.)

differ from bulk materials in a variety of ways, including their virtual size, chemical reactivity, biochemical energy absorption, and biomolecule movement (Murthy, 2007). Furthermore, zero-dimensional NMs, such as nanowires and nanotubes, and multi-dimensional NMs, such as self-assembled monolayer films, are referred to as zero-dimensional NMs since all of their dimensions are in the nanoscale.

Due to the vast diversity of their structural features, the rapid incremental stride in nanomaterial uses in commercial products and biomedical sectors provides a major challenge for effective and timely screening of potentially harmful effects in biological environments. Size, form, and topology are inherent features of NMs, as are chemical composition, surface morphology, solubility, porosity, density, band gap, aggregation state, surface area, and polarity, with the latter qualities arising as a result of ecological interactions (Chang et al., 2015). Furthermore, these physicochemical parameters have a significant impact on nanoparticle bioactivity.

The advantages and applications of nanoparticles in modern medicine are numerous, and they make diagnosis and treatment much more reliable and straightforward. Nanoparticles, on the other hand, exhibit toxicity due to their unique environmental and biological characteristics.

As a result, rather than being a thorough analysis, the current chapter is meant to serve as a general introduction to the function of nanoparticles in medicine.

7.2 NANOPARTICLES IN MEDICAL IMAGING

Traditional biological imaging of cells and tissues using fluorescence microscopy, as well as magnetic resonance imaging (MRI) of diverse areas of the body and their distinct chemical composition utilizing nanoparticles, might yield vastly enhanced medical pictures. Table 7.1 summarizes the many NMs that have been described and their medical imaging applications.

7.2.1 OPTICAL IMAGING

Organic dyes like fluorescein isocyanate (FITC) and rhodamine are commonly used to image cells and tissue sections because they may bind to biomolecules that are preferentially attached to cells or cell components via ligand/receptor interactions. Inadequate fluorescence intensity and photobleaching are common difficulties with this sort of imaging.

TABLE 7.1 Summary of Current Nanoparticle Technologies in Medicine

Area	Nanoparticle Type	Major In Vivo Applications	Significant Characteristics	References
Optical imaging	Quantum dots	Site-specific imaging *in-vivo*	• Imaging of lymph nodes, lung blood vessels, and tumors. • Greater intensity and resistance to photobleaching compared with conventional methods. • Site-specific targeting via surface functionalization. • Subcutaneous imaging without surgical incisions.	Akerman et al. (2002); Gao et al. (2004); Kim et al. (2004)
MRI	Superparamagnetic iron oxide nanoparticles	Cancer detection	• Enhanced contrast for imaging of liver, lymph nodes, and bone marrow. • Paramagnetic properties that can alter magnetic resonance relaxation times of selected regions or fluids *in vivo*.	Harisinghani et al. (2003); Huh et al. (2005)
Drug and gene delivery	Polymer- and liposome-based nanoparticles	Cancer therapy	• Targeted delivery by surface functionalization. • Strategies for prolonging residence times *in vivo* (e.g., PEG attachment). • Strategies for solubilizing water-insoluble drugs (e.g., paclitaxel). • Multi-layer and multi-functional (e.g., chemotherapeutic and anti-angiogenic).	Duncan (2003); Allen & Cullis (2004); Micha et al. (2006); Sengupta et al. (2005)
		Neurodegenerative disease therapy	• Transport across blood brain barrier (e.g., by PEG incorporation). • Superior to direct drug administration. • Therapies for diseases unresponsive to small molecule drugs (gene therapy).	Schlachetzki et al. (2004); Garcia-Garcia et al. (2005); Popovic & Brundin (2006)
		HIV/AIDS therapy	• Solubilizing water-insoluble drugs by emulsification. • Ability to transfect cells by DNA incorporation in nanoparticle.	De Jaeghere et al. (2000); Olbrich et al. (2001); Tabatt et al. (2004)
		Ocular disease therapy	• Ability to prolong drug residence times within ocular mucus layer or retina. • Alternative to frequent application of high-drug concentration drops.	Pignatello et al. (2002, 2002b); Ludwig (2005)
		Respiratory disease therapy	• Mitigation of inflammatory responses in respiratory tract.	John et al. (2003); Kumar et al. (2003)

Quantum dots (QDs) are nanoparticles made up of inorganic semiconductor molecules that generate bright fluorescent light when exposed to UV radiation. The wavelength (color) emitted is sensitive to particle size, which is a unique property resulting from these materials' band gap. The band gap is the difference in energy between the valence and conduction bands caused by the introduction of energy of a given wavelength (excitation), generally in the form of a photon. A hole is left behind when an electron passes from the valence band to the conduction band (this is a term given to an energy level lacking an electron and is not a physical feature). The electrons travel back to the valence band after the excitation stops, releasing their extra energy. This energy is totally discharged as light in the case of QDs. Larger QDs contain more electron-hole pairs and may collect and release more energy as a result. Because energy is inversely proportional to wavelength ($E = hc/\lambda$), because energy and wavelength are inversely linked ($E = hc/\lambda$), the wavelength of produced light drops as QD size grows. When compared to typical organic dyes, QDs may generate light that is substantially more powerful and resistant to photobleaching, which could be a considerable benefit in 3-D tissue imaging, particularly during the collection of consecutive sections in the z-direction.

Inorganic QD materials are insoluble in aqueous solutions, which is an important attribute for biological and medicinal applications. As a result, they are coated with a thin layer of a water-soluble substance that selectively attaches to a certain cell or cell component. Each QD contains a vast number of spots on its surface that can be used to anchor soluble and/or bioactive compounds. Furthermore, each QD can be connected to several types of molecules, giving it many functions. Michalet et al. (2005) highlighted numerous surface modification tactics such as targeting and prolonged retention in the circulation in a review of the application of QDs for live cell and *in vivo* imaging.

In mice and pigs, Kim et al., (2004) used oligomeric phosphine coated QDs to map lymph nodes. These QDs are constructed of CdTe capped with CdSe, a combination that can emit light when excited in the near infrared. The capacity to map lymph nodes up to 1 cm below the skin surface without the requirement for surgical incisions is significant. The authors extrapolated that the quantities employed were below recognized harmful thresholds because the toxicity of the injected QDs was not evaluated in this investigation.

By covering the QD surface with appropriate chemicals, QDs can be targeted to specific organs within the body. By covering the QD surface with a peptide sequence, CGFECVRQCPERC, which is known to bind to endothelial cells in lung blood arteries, Akerman et al. (2002) demonstrated that ZnS-capped CdSe QDs may be guided to the lungs of mice. QDs were directed to blood or lymphatic arteries within tumors in rats using the same manner. The QDs were ingested by the targeted cells via endocytosis in both cases, but not by cells in the surrounding tissue.

Gao & colleagues (2004) encased semiconductor QDs in an amphiphilic copolymer and added targeting ligands to the polymer surface (Figure 7.1). The QDs were manufactured of CdSe with a ZnS cap and a coordinating ligand to prevent aggregation in solution (tri-n-octylphosphine oxide [TOPO]). The copolymer was made up of three segments: butylacrylate, ethylacrylate, and methacrylic acid; the first two are more hydrophobic than the latter. The hydrophobic segments of the copolymer are attracted to the TOPO in solution, resulting in the configuration depicted in Figure 7.1, with the hydrophilic segment's carboxylic acid groups jutting out. These acid groups can be used to link molecules like polyethylene glycol (PEG) or affinity ligands to them. The composite particles have a diameter of 2,030 nm. The tumor was targeted by attaching an antibody against a prostate-specific membrane antigen (PSMA) to the particle surface (PSMA). Following intravenous injection, these particles did indeed selectively bind to tumors in mice, as predicted (Figure 7.2). There was no buildup in the brain, kidneys, or lungs.

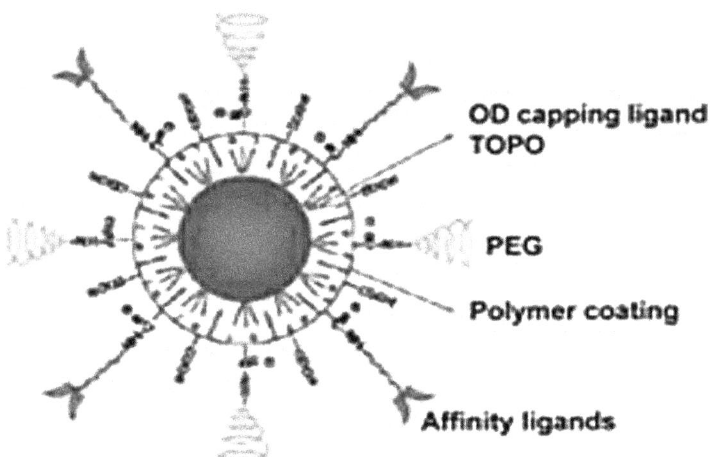

FIGURE 7.1 Quantum dots (QDs) used in tumor imaging. Surface modification of the CdSe/ZnS QD with a capping ligand TOPO which keeps QDs from aggregating in solution; this assembly is enclosed by an amphiphilic polymer whose hydrophobic segments bind to TOPO and whose hydrophilic carboxylic acid groups can bind to affinity ligands (such as a tumor-specific antibody) or PEG.

Source: Reprinted with permission from authors. Copyright © 2007 Dove Medical Press Limited.

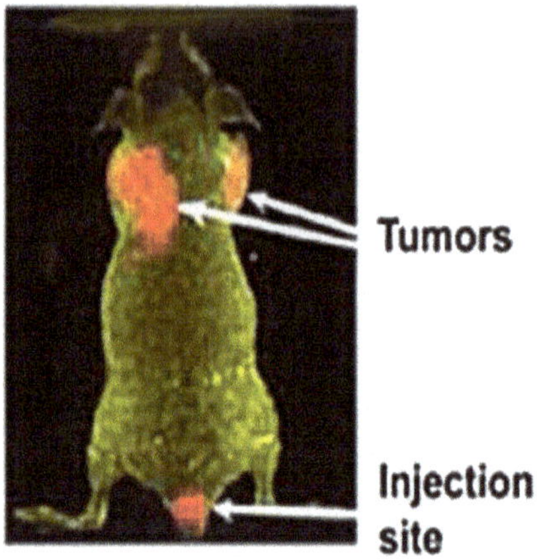

FIGURE 7.2 Quantum dots (QDs) used in fluorescence image of a live mouse showing targeted delivery of QDs to a tumor.

Source: Adapted from Gao et al. (2004) with permission from Macmillian Publishers Ltd. Nature Biotechnology.

7.2.2 MAGNETIC RESONANCE IMAGING (MRI)

Magnetic resonance imaging (MRI) is a non-invasive, three-dimensional examination of the body used to diagnose and treat most illnesses of the brain, spine, and musculoskeletal system. Magnetic resonance spectroscopy is used in MRI to examine hydrogen atoms found naturally in tissue. A transverse radio frequency (RF) pulse is used to activate the magnetic dipoles within hydrogen nuclei in a sample that is put in a strong stationary magnetic field. The spinning nuclei are aligned with the magnetic field prior to the RF pulse. The additional energy provided by the RF pulse causes these nuclei to spin at a different frequency and in a different (transverse) direction. The hydrogen nuclei revert to a condition of equilibrium in alignment with the stationary magnetic field after the RF pulse. T1 and T2 are two parameters that are commonly used to describe the relaxation process. T1 is the time it takes for nuclear spins to return to their original alignment with the stationary field; T2 is the time it takes for the nuclei's transverse magnetization to vanish. Different T1 and T2 relaxation periods can be used to distinguish hydrogen nuclei and different types of tissue. MRI scans need the acquisition of several images based on spatial location as well as T1 or T2 weighting. In a T1-weighted picture, a sample with a low T1 appears bright.

The inherent differences in relaxation periods across regions of interest (such as normal versus scar tissue) are minimal in most clinical applications, necessitating the employment of contrast management. Contrast managers are paramagnetic compounds that can change the relaxation periods of certain areas or types of tissue/fluid in the body. Gadolinium compounds have been successfully utilized as contrast managers for many years, having the capacity to choose regions like as the kidney and brain (Mornet et al., 2004).

SPIOnps (superparamagnetic iron oxide nanoparticles) have recently been discovered to be good contrast managers for T2-weighting scans, hence complementing gadolinium-based scans. When imaging the liver, lymph nodes, and bone marrow, T2 weighting is crucial. The relaxation durations of SPIOnps are observed to be longer than gadolinium-based drugs (Mornet et al., 2004).

Using a mouse xenograft model, Huh & colleagues (2005) demonstrated how SPIOnps may be utilized to detect cancer *in vivo*. The nanoparticles were linked to herceptin, a cancer-targeting antibody, in this study. Before conjugation with herceptin, SPIOnps were produced by thermal breakdown of iron acetylacetonate and made water-soluble by binding with 2,3-dimercaptosuccinic acid. When given intravenously to mice, a fast shift in the T2-weighted MRI signal from the tumor in the animals' thigh was seen. The specificity of antibody binding was confirmed in a control experiment, in which the same iron oxide nanoparticles were linked to a nonspecific antibody and no change in the MRI signal was observed.

SPIOnps could also be used to predict features that would otherwise be undetectable by traditional MRI. Harisinghani & colleagues (2003) used dextran-coated SPIOnps to detect tiny lymph node metastases in human prostate cancer patients, increasing their retention in the bloodstream and progressive uptake into the lymph nodes, where they are absorbed by macrophages. SPIO nanoparticles, Harisinghani et al. (2003), enable the imaging of metastases that standard MRI can only hazily detect. The relevance of this research is that individuals with localized cancer can now receive early surgical treatment rather than being limited to radiation therapy.

7.3 NANOPARTICLES IN DRUG AND GENE DELIVERY

One of the most important medical benefits of nanoparticles is their ability to deliver drugs safely. This is owing in great part to the success of polymer and liposome-based drug delivery methods (Brigger et al., 2002), several of which are currently in clinical use.

Polymeric drugs, polymer-protein conjugates, polymer-drug conjugates, and polymeric micelles are all examples of polymer-based drug delivery methods (Duncan, 2003). Polymers could be emulsified into nano-scaled particles with drug ingredients trapped inside. Natural polymers with antiviral or anticancer activities are known as polymeric medications. PEG is a well-known polymer protein conjugate with high water solubility, great biocompatibility, and enhanced solubility due to drug attachment. PEG attachment also lowers drug clearance in the kidneys and improves cell receptor-mediated absorption. These features could extend a drug's half-life and minimize dose frequency. Low molecular weight medications' solubility and selectivity are improved through polymer-drug conjugation. Finally, polymeric micelles are often made with amphiphilic polymers that form micelles in solution and contain a medication.

Liposomes are vesicles made up of phospholipid molecules with hydrophobic and hydrophilic components that can form bilayers and entrap fluid. When two layers of orientated lipid molecules come together in such a way that their hydrophobic sides are in contact with one another, a bilayer is created. Lipid molecules create vesicles (size ranging from 10 to 1,000 nm) when conditions are favorable, in which a volume of fluid is encapsulated by lipid bilayers (Torchilin & Weissig, 2003). Drug compounds can be introduced into the fluid contained within vesicles or lipid bilayers. The structure of these biocompatible and biodegradable synthetic bilayers is comparable to that of biological membranes in the body. Liposomes are distinct from other particles because they lack a solid core that establishes their identity. However, colloidal entities, including nanoscale polymers, make up a large percentage of nanoscale drug delivery systems. Another type of nanoparticle is solid lipid nanoparticles (SLNs), which are comprised of lipids and solidify at normal temperature (Muller et al., 2000). These nanoparticles are made by emulsifying a molten lipid with a medicine and a surfactant, then cooling it.

With a few examples classified by disease category, the use of polymer and liposome-based nanoparticles in medication delivery is demonstrated. Table 7.1 contains a summary of notable examples.

7.3.1 NANOMEDICINE FOR CANCER

As seen by the numerous nanoparticle-based medications and delivery systems in clinical use, nanoparticles have had a tremendous influence in the treatment of several forms of cancer. In previous reviews, various liposome and polymer-based medicines or therapeutic agents were mentioned (Duncan, 2003; Allen & Cullis, 2004).

Cancer is one of the most serious causes of death worldwide, and despite efforts to reduce risk factors in recent decades, the prevalence of cancer has continued to rise. Current cancer treatment protocols include accurate cancer staging, chemotherapy, radiation, and/or surgical resection. Because they are non-specific in targeting cells, radiotherapy and chemotherapy are recognized for their significant side effects. Furthermore, anticancer medicines with poor pharmacokinetic qualities due to poor solubility, stability, and metabolism face toxicity, inefficacy, and limited biodistribution. As a result, it's critical to produce effective formulations that may solve the aforementioned obstacles while also providing selective tumor targeting without causing considerable damage to healthy tissues' viability (Lina et al., 2019).

Paclitaxel is a well-known anticancer medicine that is used to treat a variety of cancers, including ovarian, cutaneous, esophageal, and lung cancers (Kikuchi et al., 2005; Abratt et al., 2006; Chao et al., 2006; De Giorgi et al., 2006; Roof et al., 2006; Worden et al., 2006). This medicine interferes with cancer cells' functioning by stabilizing microtubules, eventually leading to apoptosis (Koziara et al., 2006). This fat-soluble medicine is most commonly administered as an ethanol solution (Taxol®), which is combined with a solvent, polyoxyethylated castor oil (Cremophor® EL). The side effects of Cremophor®, particularly hypersensitivity reactions, have been a key drawback of this method, necessitating the use of steroids and antihistamines as premedications (Zhang et al., 2005; Micha et al., 2006). Abraxane®, a new type of paclitaxel, was approved for clinical usage in early 2005. Paclitaxel is injected into nanoparticles of albumin, a natural polymer, using a high-pressure emulsification method in this form. This soluble version of paclitaxel has been found to not only eradicate the negative effects of Cremophor® (Micha et al., 2006), but also to provide some additional advantages. When compared to Taxol®, the albumin carrier improves drug delivery from the bloodstream to the tumor site and allows for higher drug doses (Ibrahim et al., 2002).

However, paclitaxel nanoparticle loading did not address multidrug resistance, an issue that occurs when cancer cells adjust to stimuli by producing efflux transporters or other proteins on the surface (Gottesman et al., 1996; Tomonaga et al., 1996). By putting paclitaxel into emulsifying wax nanoparticles, Koziara & colleagues (2006) attempted to overcome this difficulty. The wax is polyoxyethylene 20-sorbitan mono-oleate, a commercial substance (Tween 80®). The nanoparticles were made by emulsifying a mixture of medicine, wax, and a surfactant that had been heated. The efficacy of these drug-loaded nanoparticles was tested in a mouse xenograft model (HCT-15), which contains tumor cells that exhibit the efflux transporter p-glycoprotein. The ensuing stop of tumor growth was discovered as the result of a combination of overcoming resistance (via nonspecific cytoskeletal disruption) and the antiangiogenic activity of paclitaxel utilizing a control experiment with Taxol®. These examples of various paclitaxel formulations show how various nanoparticle-based drug delivery techniques can be used to adjust and improve a medication's performance.

The interaction of chemotherapeutic and antiangiogenic drugs is a significant factor in tumor therapy. According to Sengupta & colleagues (2005), disruption of tumor blood vessels can disrupt chemotherapeutic drug delivery while also increasing the expression of factors linked to drug resistance. They also synthesized two layered nanoparticle drug delivery system, namely, a core of poly(lactic-co-glycolic) acid (PLGA) conjugated with doxorubicin enclosed within a liposome composed of phospholipids conjugated with PEG and combretastatin. The doxorubicin is the chemotherapeutic agent and combretastatin is the antiangiogenic agent. The size of these bilayered particles ranged from 80 to 120 nanometers. The basic concept was to get the particles to the tumor location and then slowly release the medicine when the PLGA core degraded. The particles were rapidly taken up by the tumor when supplied intravenously to mice with tumors produced by carcinoma or melanoma cells, consistent with the extended residence time resulting from the PEG conjugation (Harris & Chess, 2003) and the recognized leakiness' of tumors (Yuan et al., 1995). The nanoparticles significantly slowed tumor development and extended the animals' lives.

7.3.2 NANO-DRUG FOR NEURODEGENERATIVE DISEASES

Drug delivery to the central nervous system remains a difficulty in the development of effective neurodegenerative disease therapies (Garcia-Garcia et al., 2005; Popovic & Brundin, 2006). Overcoming the inherent propensity of the blood brain barrier (BBB) to hinder drug transport is a significant aspect of this difficulty. The purpose of this barrier is to protect the brain from foreign chemicals and blood-borne illnesses; however, it is unable to identify many therapeutic agents. As a result, substantial dosages must be given, increasing the likelihood of negative side effects.

Nanoparticle-based systems, ranging from polymer particles to liposomes, have been studied for years as a way to circumvent this constraint. Garcia-Garcia and colleagues provided a comprehensive overview of studies in this field (2005).

Drug transport across the BBB has been demonstrated to be aided by nanoparticles made from poly(hexadecyl cyanoacrylate) and similar compounds. Kreuter & colleagues (2003) coated poly(butyl cyanoacrylate) (PBCA) nanoparticles with dalargin (an analgesic) and exhibited penetration through the BBB in rats. In a mouse model of Alzheimer's disease, Siegemund & colleagues (2006) demonstrated that PBCA nanoparticles injected with thioflavins may target-fibrillar amyloid cells. Calvo & colleagues (2002, 2001) developed a nanoparticle system made of a PEG/poly(hexadecyl cyanoacrylate) copolymer (PHDCA). Because PEG is hydrophilic and PHDCA is hydrophobic, the copolymer molecules organize themselves into particles with an insoluble PHDCA core and a PEG surface layer in an aqueous environment. Because PEG is not detected as a foreign substance by macrophages in the blood, it is commonly used in many drug delivery systems. This improves the half-life of drug carriers in the blood (Harris & Chess, 2003). Indeed, combining PEG with PHDCA improves PHDCA's capacity to penetrate the BBB. Copolymers of PEG and compounds comparable to PEG, such as PEG, might potentially be used to make polymeric micelles (propylene oxide). Pluronic® P-85 polymer, for example, is a commercial product, and P-85 micelles have been used to transport analgesics over the BBB in mice (Witt et al., 2002).

Drug delivery to the central nervous system using liposome-based drug delivery devices has also been widely researched (Garcia-Garcia et al., 2005). Surface covering with PEG works well in these systems as well. Schmidt & colleagues (2003) created liposomes with sizes ranging from 90 to 100 nanometers to encapsulate the medication Prednisolone, which is used to treat multiple sclerosis. Within 2 hours after intravenous administration, mice with experimental autoimmune encephalomyelitis had an elevated amount of liposomes in the central nervous system. In a spinal cord segment, Figure 7.3 shows gold-(black) labeled liposomes amid astrocytes and microglia, demonstrating BBB penetration. The drug-loaded liposomes caused BBB integrity to be restored, as well as a decrease in inflammation and macrophage infiltration. This medication was shown to be more effective than giving free glucocorticosteroids, which is a common treatment for multiple sclerosis.

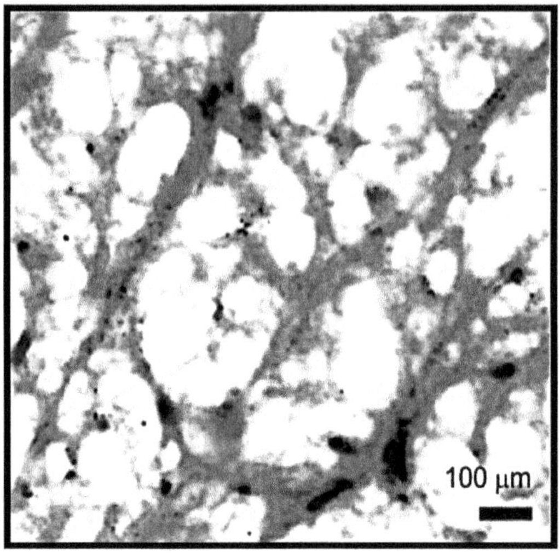

FIGURE 7.3 Liposome-based drug delivery to the nervous system. Gold-labeled liposomes (colored black in image) among astrocytes and microglia in rat spinal cord sections indicating penetration of the blood-brain barrier (astrocytes and microglia stained red); scale bar = 100 μm.

Source: Adapted from Schmidt et al. (2003) by permission of Oxford University Press.

Figure 7.3 shows drug transport to the nervous system through liposomes. In rat spinal cord slices, gold-labeled liposomes (colored black in picture) amid astrocytes and microglia indicate blood-brain barrier (BBB) penetration (astrocytes and microglia dyed red); scale bar = 100 m. By permission of Oxford University Press, this article was adapted from Schmidt et al. (2003).

Gene delivery across the BBB is another use for PEG-conjugated liposomes. This method is being used to discover treatments for chronic neurological illnesses that don't respond to small-molecule medications (such as Huntington's disease, Rett syndrome, and Fragile-X syndrome) (Schlachetzki et al., 2004). Shi & colleagues (2001) crossed the BBB in rats with plasmid DNA expressing-galactosidase. A targeting monoclonal antibody anti-TFR, which targets the brain, liver, and spleen, was linked to some of the PEG molecules on the liposome surfaces. The liposomes were delivered to specified locations using antibodies, and plasmid-induced gene expression in the brain was evaluated for at least 6 days after liposome administration. The capacity to transport genes that would ordinarily be blocked *in vivo* by endonucleases by loading them into liposomes with targeting capabilities is crucial.

7.3.3 NANO-DRUG FOR HIV/AIDS

De Jaeghere & colleagues (2000) used pH-sensitive nanoparticles produced from a copolymer of methacrylic acid and ethyl acrylate to study the transport of CGP 70726, an HIV-1 protease inhibitor. Eudragit® L10055 is the brand name for this copolymer on the market. Because of its pH-dependent solubility, the copolymer was chosen. CGP 70726 and other antiviral medications have been shown to stop HIV-1 from replicating (Robins & Plattner, 1993). Poor water solubility is a major issue when administering drugs like CGP 70726. De Jaeghere and colleagues made nanoparticles by emulsifying a copolymer solution with a CGP 70726 and benzyl alcohol combination. Oral delivery of the nanoparticles in dogs resulted in effective drug release, as evidenced by blood sample analysis.

The HIV-1 Tat protein has emerged as a possible candidate for an HIV-1/AIDS preventive or therapeutic vaccination (Cafaro et al., 1999; Caputo et al., 2004). An SLN-based system incorporating compressed DNA and a Tat protein was disclosed by Rudolph & colleagues (2004). This was part of a larger research by the same group, which demonstrated that SLNs loaded with DNA could transfect mammalian cells *in vitro* (Olbrich et al., 2001; Tabatt et al., 2004). Electrostatic forces were used to adsorb DNA and the Tat peptide onto the nanoparticle surface. When delivered to the lungs of mice through intratracheal instillation or aerosol application, increased gene expression was detected, suggesting effective transfection of the SLNs.

While there are less reports of *in vivo* nanoparticle-based medication delivery in the domain of HIV/AIDS than in areas like cancer and neurological illnesses, multiple recent *in vitro* studies may surely predict activity in the area (Berton et al., 1999, 2001; Nam et al., 2002; Cui & Mumper, 2003; Becker et al., 2004; Sawant et al., 2006).

7.3.4 NANO-DRUG FOR OCULAR DISEASES

The ability to lengthen medication residence periods by trapping the drug in the ocular mucus layer is the primary rationale for employing nanoparticle-based drug delivery methods in ophthalmic applications (Ludwig, 2005). This layer, which acts as a macromolecule diffusion barrier, is released by goblet cells in the conjunctiva and protects the cornea's epithelial layer. The majority of ocular illnesses are treated using eye drops. Because of the fast precorneal loss induced by mucus flow during blinking, these solutions are generally highly concentrated and require frequent administration.

According to Ludwig's complete assessment, nanoparticles have established an efficient approach to overcome this obstacle (2005).

Pignatello & colleagues (2002a, b) delivered non-steroidal and anti-inflammatory medications (flurbiprofen and ibuprofen) to rabbit eyes using commercially available Eudragit® polymers. These medications are commonly used to reduce the inflammatory response that occurs after ocular surgery. Eudragit® RS and RL polymers were copolymers of poly(ethyl acrylate), poly(methyl methacrylate), and poly(methyl methacrylate) used in these studies (chlorotrimethyl-aminoethyl-methacrylate). These polymers are insoluble and expand when exposed to physiological conditions, making them ideal for preventing release. Ethanol-polymer mixtures were emulsified (with the assistance of water and a surfactant) to generate drug-embedded nanoparticles with a diameter of 100 nm. The rabbit eyes' conjunctive sacs were injected with saline solutions of these nanoparticles. Both medicines (flurbiprofen and ibuprofen) were infused into nanoparticles, which significantly suppressed inflammatory responses following surgical stress and were comparable to traditional eye-drop controls. The significance of this finding is that the nanoparticle system was built with a lower drug concentration than the eye-drop control, and the nanoparticle system was also able to create greater drug availability in the vitreous humor due to the longer drug residence time in the polymer matrix.

Drugs must be administered to the retinal part of the eye in some illness states, such as cytomegalovirus (CMV) retinitis. The retina, choroid (the area beneath the retina), iris, and surrounding tissue would all be permanently damaged if infected with CMV. The use of bovine serum albumin (BSA) nanoparticles to transport ganciclovir, a medicine used to treat CMV infection, was explored by Merodio & colleagues (2002). The drug was incubated in an aqueous solution with BSA, and droplets were formed by adding ethanol in an emulsification procedure; the resultant nanoparticles had a diameter of around 280 nm. These nanoparticles were resuspended in saline before being delivered into the eye. The presence of nanoparticles in a thin layer on the retina was found for up to two weeks after injection, and histological investigation revealed no inflammatory reactions or alterations in tissue shape when compared to normal eye controls.

7.3.5 NANOMEDICINE FOR RESPIRATORY DISEASES

Only a few research on the use of nanoparticle-based medication delivery methods in respiratory illnesses have been discovered. There are various instances of treatments that have been shown to be successful in the treatment of allergy, hereditary, and infectious illnesses of the respiratory system in the literature (Pison et al., 2006).

In a mouse model of allergic asthma, John & colleagues (2003) used a liposome-based nanoparticle technology to reduce inflammation. P-selectin receptors on active endothelial cells in circulation were inhibited, which reduced interactions between endothelial cells and leukocytes. As a result, peribronchial inflammation is less likely to occur. By adding fucose and sulfate ester groups on the liposome surface, the 73 nm nanoparticles were intended to imitate the physiological P-selectin super ligand (PSGL-1). LPS and cockroach antigen caused lung inflammation and airway hyperreactivity in mice. The liposomal nanoparticles were shown to bind preferentially to selectins on activated endothelial cells in both cases (Figure 7.4). When comparing mice treated with nanoparticles to controls, histological examination revealed a substantial decrease in peribronchial inflammation and airway hyper-reactivity.

Figure 7.4 shows selective binding of fucose and sulfate ester liposomes to active endothelial cells in mouse lungs after allergen exposure. Negative control (A) (liposomes without fucose and sulfate ester groups); and (B) fucose and sulfate ester groups in liposomes. In both pictures, the scale bars are 30 meters. John et al.'s work has been adapted with their consent (2003).

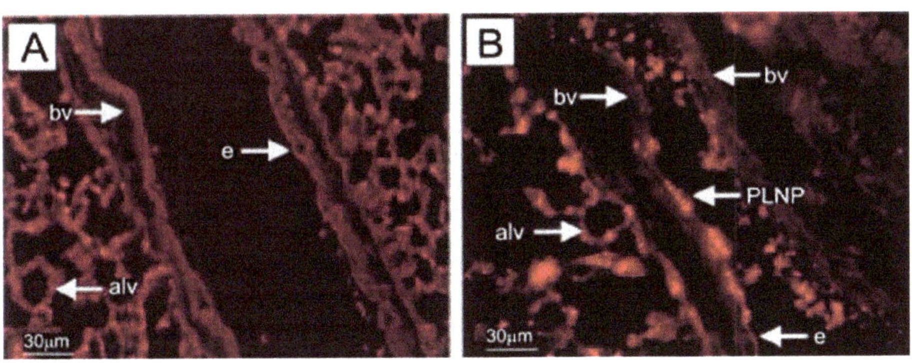

FIGURE 7.4 Selective binding of liposomes presenting fucose and sulfate ester groups to activated endothelial cells in mouse lungs following allergen challenge. (A) Negative control (liposomes without fucose and sulfate ester groups). And (B) liposomes with fucose and sulfate ester groups. Scale bars in both images = 30 μm.

Source: Reprinted with permission from John et al., 2003

Kumar & colleagues (2003) demonstrated that chitosan (CS)/interferon-pDNA nanoparticles, a polymer-drug conjugate, can attenuate allergen-induced airway inflammation. It is well understood that allergic illnesses (such as asthma) produce a decrease in interferon (IFN) production in individuals, making them vulnerable to airway inflammation and hyper-responsiveness. Kumar & colleagues (2003) propose to treat IFN-insufficiency by administering it intravenously as a polymer-drug combination. Nanoparticle treatment boosted IFN-expression by epithelial cells in allergen-challenged mice, allowing for a decrease in inflammation and restoration of lung shape within 3–6 hours.

7.4 ENVIRONMENTAL AND SOCIETAL CONSIDERATIONS

In recent years, the impact of NMs on the environment and public health has gotten a lot of attention. However, as nanomedicine and the larger area of nanotechnology advance, much more work need to be done, particularly because various NMs pose different types of dangers. This section will concentrate on the nanoparticle kinds mentioned in the previous applications. *In vivo* investigations will be highlighted, as they have been in prior sections, although major *in vitro* work will also be discussed.

7.5 TOXICITY OF QUANTUM DOTS

QDs are inorganic nanoparticles with an organic coating that renders them biocompatible or bioactive, as explained in the section on medical imaging. The greatest toxicological concern associated with the *in vivo* usage of QDs is the exposure of the inorganic core due to organic layer breakdown. CdSe, ZnS, CdTe, InP, InAs, and GaAs are just a few examples of inorganic-metal complexes that may be used to make QDs. Each of these compounds has its own chemical features, which can have a significant impact on its toxicological. Although the literature on the *in vivo* toxicity of such chemicals is limited, there are reports that raise substantial concerns and point to the need for more research. Hardman just released a comprehensive study on quantum dot toxicity (2006). Table 7.2 provides a summary of current toxicological studies on QDs.

TABLE 7.2 Toxicological Effects of Nanoparticles Associated with Medical Applications

Nanoparticle Type	Toxicological Effects	References
Quantum dots	Potential for exposure to inorganic core (e.g., cadmium) and resulting cytotoxic effects (e.g., liver damage).	Derfus et al. (2004)
	Toxicity risk greatly reduced by coating with ZnS and soluble polymers (such as PEG).	Ballou et al. (2004)
	Risks associated with production, handling, and storage of QDs need to be evaluated.	Oberdorster et al. (2005); Hardman (2006)
Metallic	Iron oxide and gold nanoparticles are not toxic.	Weissleder et al. (1989); Connor et al. (2005); Muldoon et al. (2005); Hainfeld et al. (2006)
	Surface functionalization may influence toxicity.	Goodman et al. (2004)
Polymeric/ liposomal	Not toxic since these nanoparticles have natural or highly biocompatible components (e.g., chitosan, PEG).	Alonso (2004); de Campos et al. (2004)

Abbreviations: PEG: poly(ethylene glycol); QD: quantum dots.

Derfus & colleagues (2004) used an *in vitro* model of primary rat hepatocytes to investigate the toxicity of a variety of cadmium-based QDs. The liver is the major target of Cd exposure; hence this cell type was chosen for this study. At a concentration of 62.5 g/mL, QDs with CdSe cores capped with mercaptoacetic acid (MAA) and TOPO were shown to be acutely cytotoxic. The release of Cd^{2+} ions during oxidation of the CdSe lattice by air and UV radiation was linked to this phenomenon. Coating the CdSe particles with ZnS practically eliminated surface oxidation and cytotoxicity, and polymer or protein coating on top of the ZnS improved things much further. While this was a promising outcome, commercial QDs produced with CdSe/ZnS with a polyacrylate and streptavidin coating also released some Cd.

In mice, Ballou & colleagues (2004) investigated the *in vivo* toxicity of CdSe/ZnS QDs coated with amphiphilic poly(acrylic acid) or PEG. The animals were given QDs at a dosage of 20 pmol/g animal weight. There was no necrosis in the liver, spleen, or bone marrow, where the QDs were found to deposit, and the animals remained alive for 133 days after tissue examination.

The preceding two instances demonstrate how difficult it is to assess the toxicity of QDs. Because of intrinsic changes in experimental design (QD concentration measurement being only one of many components of this), and differences in the organic coating, the *in vitro* and *in vivo* experiments cannot be directly compared. Nonetheless, they emphasize the importance of long-term animal research before QDs may be authorized for commercial usage, as well as the probable necessity for regulated storage and handling settings. Questions about the safety of QD production procedures, such as the danger of QD exposure by inhalation or skin contact, must also be decisively resolved (Oberdorster et al., 2005; Hardman, 2006).

7.6 TOXICITY OF METALLIC NANOPARTICLES (MNPS)

In rats, Muldoon and colleagues evaluated the toxicity of SPIOnps employed as MRI contrast agents. The nanoparticles were injected either intracerebrally or intraarterially into the brain. Although the MRI signal strength decreased with time (weeks to months), normal rats' brain tissue showed no pathological abnormalities. Weissleder and colleagues conducted a toxicity investigation of iron oxide nanoparticles in mice and dogs about two decades ago, and their findings were consistent (1989). The nanoparticles were given intravenously in this study, and there were no acute or subacute toxicity reactions in the histology of the targeted tissues or in blood tests. The safety of several iron oxide-based nanoparticles employed as contrast agents in clinical practice is well-known (Lubbe et al., 1999, 2001; Neuberger et al., 2005).

The toxicity of gold nanoparticles (AuNPs) was investigated by Hainfeld & colleagues (2006) in the context of their application as X-ray contrast agents. When mice were given the substance intravenously, it accumulated in their kidneys and tumors (retention was low in the liver and spleen). Up to 30 days after injection, organ histology and blood tests revealed no signs of harm. These findings are in accord with Connor & colleagues' (2005) findings from an *in vitro* investigation of gold nanoparticle toxicity using a human leukemia cell line.

The nanoparticles were taken up by the cells, but no cytotoxicity was seen. The toxicity of AuNPs can, however, be modified by chemical surface modification. Goodman & colleagues (2004) recently shown that coating AuNPs with a cationic polymer monolayer (alkyl thiol with a quaternary ammonium group) can make them cytotoxic. Cytotoxic behavior was not seen when an anionic monolayer (alkyl chain with carboxylate end group) was attached.

7.7 TOXICITY OF POLYMERIC AND LIPOSOMAL NANOPARTICLES

Because the particles are frequently made from or coated with natural or very biocompatible polymers, this group of nanoparticles is arguably the least harmful (such as PEG). These particles are commonly used in drug delivery applications to transport medications that are cytotoxic by design (to kill cancer cells) but are prohibited from targeting other parts of the body.

Natural polymers like chitosan (CS) and natural lipids are advantageous in the production of polymer- or liposome-based nanoparticles because they are not recognized as alien by the body and are easily digested (Alonso, 2004; de Campos et al., 2004). Depending on the polymer type and content, nanoparticles generated from synthetic polymers can have a broad range of clearance rates from the bloodstream and accumulation in mononuclear phagocytic system (MPS) organs (such as the liver and spleen) (Moghimi et al., 2001; Owens & Peppas, 2006). As previously mentioned, the presence of PEG in the nanoparticle structure might delay the elimination of nanoparticles from the bloodstream. Because they are less likely to saturate the MPS, PEG-coated particles are thought to be less hazardous than uncoated particles (Peracchia et al., 1999; Plard & Bazile, 1999).

7.8 CONCLUSIONS

NMs are important in the pharmaceutical industry because they help to visualize medical imaging and deliver therapeutic materials to specific cellular targets. The bioactivity of NMs is greatly influenced by their inherent characteristics. As a result, this chapter focuses mostly on recent clinical research technologies. Furthermore, the influence of NMs on medical diagnostics and therapeutic applications, as well as their social and environmental impact, was explored.

ACKNOWLEDGMENT

The authors are very thankful to acknowledge the author Murthy S, for providing permission to use the images from the article "Nanoparticles in modern medicine: State of the art and future challenges." We also acknowledge the editors in the Dove Press for granting permission to reuse the images in our book chapters.

KEYWORDS

- **drug delivery**
- **medical imaging**
- **nanomaterials**
- **nanoparticles**
- **quantum dots**
- **toxicity**

REFERENCES

Abratt, R. P., Lee, J. S., Han, J. Y., Tsai, C. M., Boyer, M., Mok, T., & Lehnert, M., (2006). Phase II trial of gemcitabine-carboplatin-paclitaxel as neoadjuvant chemotherapy for operable non-small cell lung cancer. *J. Thorac. Oncol., 1*, 135–140.

Akerman, M. E., Chan, W. C. W., Laakkonen, P., Laakkonen, P., Bhatia, S. N., & Ruoslahti, E., (2002). Nanocrystal targeting *in vivo*. *Proc. Nat. Acad. Sci. U S A, 99*, 12617–12621.

Allen, T. M., & Cullis, P. R., (2004). Drug delivery systems: Entering the mainstream. *Science, 303*, 1818–1822.

Alonso, M. J., (2004). Nanomedicines for overcoming biological barriers. *Biomed. Pharmacother., 58*, 168–172.

Ballou, B., Lagerholm, B. C., Ernst, L. A., Bruchez, M. P., & Waggoner, A. S., (2004). Noninvasive imaging of quantum dots in mice. *Bioconjugate Chem., 15*, 79–86.

Becker, M. L., Bailey, L. O., & Wooley, K. L., (2004). Peptide-derivatized shell-cross-linked nanoparticles. 2. Biocompatibility evaluation. *Bioconjugate Chem., 15*, 710–717.

Berton, M., Allemann, E., Stein, C. A., & Gurny, R., (1999). Highly loaded nanoparticulate carrier using an hydrophobic antisense oligonucleotide complex. *Eur. J. Pharm. Sci., 9*, 163–170.

Berton, M., Turelli, P., Trono, D., Allémann, E., & Gurny, R., (2001). Inhibition of HIV-1 in cell culture by oligonucleotide-loaded nanoparticles. *Pharm. Res., 18*, 1096–1101.

Brigger, I., Dubernet, C., & Couvreur, P., (2002). Nanoparticles in cancer therapy and diagnosis. *Adv. Drug Deliv. Rev., 54*, 631–651.

Cafaro, A., Caputo, A., Fracasso, C., Maggiorella, M. T., Goletti, D., Baroncelli, S., & Borsetti, A., (1999). Control of SHIV-89.6P-infection of cynomolgus monkeys by HIV-1 tat protein vaccine. *Nat. Med., 5*, 643–650.

Calvo, P., Gouritin, B., Chacun, H., Desmaële, D., D'Angelo, J., Noel, J. P., & Couvreur, P., (2001). Long-circulating pegylated polycyanoacrylate nanoparticles as new drug carrier for brain delivery. *Pharm. Res., 18*, 1157–1166.

Calvo, P., Gouritin, B., Villarroya, H., Eclancher, F., Giannavola, C., Klein, C., & Couvreur, P., (2002). Quantification and localization of pegylated polycyanoacrylate nanoparticles in brain and spinal cord during experimental allergic encephalomyelitis in the rat. *Eur. J. Neurosci., 15*, 1317–1326.

Caputo, A., Gavioli, R., & Ensoli, B., (2004). Recent advances in the development of HIV-1 tat-based vaccines. *Curr. HIV Res., 2*, 357–376.

Chang, E. H., Harford, J. B., Eaton, M. A., Boisseau, P. M., Dube, A., Hayeshi, R., Swai, H., & Lee, D. S., (2015). Nanomedicine: Past, present and future – a global perspective. *Biochem. Biophys. Res. Commun., 468*, 511–517.

Chao, Y., Li, C. P., Chao, T. Y., Su, W. C., Hsieh, R. K., Wu, M. F., & Cheng, A. L., (2006). An open, multi-center, phase II clinical trial to evaluate the efficacy and safety of paclitaxel, UFT, and leucovorin in patients with advanced gastric cancer. *Br. J. Cancer, 95*, 159–163.

Connor, E. E., Mwamuka, J., Gole, A., Murphy, C. J., & Wyatt, M. D., (2005). Gold nanoparticles are taken up by human cells but do not cause acute cytotoxicity. *Small, 1*, 325–327.

Cui, Z. R., & Mumper, R. J., (2003). Microparticles and nanoparticles as delivery systems for DNA vaccines. *Crit. Rev. Ther. Drug Carr. Syst., 20*, 103–137.

De Campos, A. M., Diebold, Y., Carvalho, E. L. S., Carvalho, E. L., Sánchez, A., & Alonso, M. J., (2004). Chitosan nanoparticles as new ocular drug delivery systems: *In vitro* stability, *in vivo* fate, and cellular toxicity. *Pharm Res., 21*, 803–810.

De Giorgi, U., Giannini, M., Frassineti, L., Kopf, B., Palazzi, S., Giovannini, N., & Marangolo, M., (2006). Feasibility of radiotherapy after high-dose dense chemotherapy with epirubicin, preceded by dexrazoxane, and paclitaxel for patients with high-risk stage II-III breast cancer. *Int. J. Radiat. Oncol. Biol. Phys., 65*, 1165–1169.

De Jaeghere, F., Allemann, E., Kubel, F., Galli, B., Cozens, R., Doelker, E., & Gurny, R., (2000). Oral bioavailability of a poorly water-soluble HIV-1 protease inhibitor incorporated into pH-sensitive particles: Effect of the particle size and nutritional state. *J. Control Release, 68*, 291–298.

Derfus, A. M., Chan, W. C. W., & Bhatia, S. N., (2004). Probing the cytotoxicity of semiconductor quantum dots. *Nano Lett., 4*, 11–18.

Duncan, R., (2003). The dawning era of polymer therapeutics. *Nat. Rev. Drug Discov., 2*, 347–360.

Gao, X. H., Cui, Y. Y., Levenson, R. M., Chung, L. W., & Nie, S., (2004). *In vivo* cancer targeting and imaging with semiconductor quantum dots. *Nat. Biotechnol., 22*, 969–976.

Garcia-Garcia, E., Andrieux, K., Gil, S., & Couvreur, P., (2005). Colloidal carriers and blood-brain barrier (BBB) translocation: A way to deliver drugs to the brain? *Int. J. Pharm., 298*, 274–292.

Goodman, C. M., Mccusker, C. D., Yilmaz, T., & Rotello, V. M., (2004). Toxicity of gold nanoparticles functionalized with cationic and anionic side chains. *Bioconjugate Chem., 15*, 897–900.

Gottesman, M. M., Pastan, I., & Ambudkar, S. V., (1996). P-glycoprotein and multi-drug resistance. *Curr. Opin. Genet. Dev., 6*, 610–617.

Hainfeld, J. F., Slatkin, D. N., Focella, T. M., & Smilowitz, H. M., (2006). Gold nanoparticles: A new x-ray contrast agent. *Br. J. Radiol., 79*, 248–253.

Hardman, R. A., (2006). Toxicologic review of quantum dots: Toxicity depends on physicochemical and environmental factors. *Environ. Health Perspect., 114*, 165–172.

Harisinghani, M. G., Barentsz, J., Hahn, P. F., Deserno, W. M., Tabatabaei, S., Van De, K. C. H., & Weissleder, R., (2003). Noninvasive detection of clinically occult lymphnode metastases in prostate cancer. *N. Engl. J. Med., 348*, 2491–2499.

Harris, J. M., & Chess, R. B., (2003). Effect of pegylation on pharmaceuticals. *Nat. Rev. Drug Discov., 2*, 214–221.

Huh, Y. M., Jun, Y. W., Song, H. T., Kim, S., Choi, J. S., Lee, J. H., & Cheon, J., (2005). *In vivo* magnetic resonance detection of cancer by using multifunctional magnetic nanocrystals. *J. Am. Chem. Soc., 127*, 12387–12391.

Ibrahim, N. K., Desai, N., Legha, S., Soon-Shiong, P., Theriault, R. L., Rivera, E., & Ellerhorst, J. A., (2002). Phase I and pharmacokinetic study of ABI-007, a cremophor-free, protein-stabilized, nanoparticle formulation of paclitaxel. *Clin. Cancer Res., 8*, 1038–1044.

John, A. E., Lukacs, N. W., Berlin, A. A., Palecanda, A., Bargatze, R. F., Stoolman, L. M., & Nagy, J. O., (2003). Discovery of a potent nanoparticle p-selectin antagonist with anti-inflammatory effects in allergic airway disease. *FASEB J.*, *17*, 2296–228.

Kikuchi, A., Sakamoto, H., & Yamamoto, T., (2005). Weekly carboplatin and paclitaxel are safe, active, and well tolerated in recurrent ovarian cancer cases of Japanese women previously treated with cisplatin containing multidrug chemotherapy. *Int. J. Gynecol. Cancer, 15*, 45–49.

Kim, S., Lim, Y. T., Soltesz, E. G., De Grand, A. M., Lee, J., Nakayama, A., & Cohn, L. H., (2004). Near-infrared fluorescent type II quantum dots for sentinel lymph node mapping. *Nat. Biotechnol., 22*, 93–97.

Koziara, J. M., Whisman, T. R., Tseng, M. T., & Mumper, R. J., (2006). *In-vivo* efficacy of novel paclitaxel nanoparticles in paclitaxel-resistant human colorectal tumors. *J. Control Release, 112*, 312–319.

Kreuter, J., Ramge, P., Petrov, V., Hamm, S., Gelperina, S. E., Engelhardt, B., & Begley, D. J., (2003). Direct evidence that polysorbate-80-coated poly(butylcyanoacrylate) nanoparticles deliver drugs to the CNS via specific mechanisms requiring prior binding of drug to the nanoparticles. *Pharm. Res., 20*, 409–416.

Kumar, M., Kong, X., Behera, A. K., Hellermann, G. R., Lockey, R. F., & Mohapatra, S. S., (2003). Chitosan IFN-γ-pDNA nanoparticle (CIN) therapy for allergic asthma. *Genet. Vaccines Ther., 1*, 3.

Lubbe, A. S., Alexiou, C., & Bergemann, C., (2001). Clinical applications of magnetic drug targeting. *J. Surg. Res., 95*, 200–206.

Lübbe, A. S., Bergemann, C., Brock, J., & McClure, D. G., (1999). Physiological aspects in magnetic drug-targeting. *J. Magn. Magn. Mater., 194*, 149–155.

Ludwig, A., (2005). The use of mucoadhesive polymers in ocular drug delivery. *Adv. Drug Deliv. Rev., 57*, 1595–1639.

Merodio, M., Irache, J. M., Valamanesh, F., & Mirshahi, M., (2002). Ocular disposition and tolerance of ganciclovir-loaded albumin nanoparticles after intravitreal injection in rats. *Biomaterials, 23*, 1587–1594.

Micha, J. P., Goldstein, B. H., Birk, C. L., Rettenmaier, M. A., & Brown, III. J. V., (2006). Abraxane in the treatment of ovarian cancer: The absence of hypersensitivity reactions. *Gynecol. Oncol., 100*, 437, 438.

Michalet, X., Pinaud, F. F., Bentolila, L. A., Tsay, J. M., Doose, S. J. J. L., Li, J. J., & Weiss, S., (2005). Quantum dots for live cells, *in vivo* imaging, and diagnostics. *Science, 307*, 538–544.

Moghimi, S. M., Hunter, A. C., & Murray, J. C., (2001). Long-circulating and target-specific nanoparticles: Theory to practice. *Pharmacol. Rev., 3*, 283–318.

Mornet, S., Vasseur, S., Grasset, F., & Duguet, E., (2004). Magnetic nanoparticle design for medical diagnosis and therapy. *J. Mater. Chem.*, *14*, 2161–2175.

Muldoon, L. L., Sàndor, M., Pinkston, K. E., & Neuwelt, E. A., (2005). Imaging, distribution, and toxicity of superparamagnetic iron oxide magnetic resonance nanoparticles in the rat brain and intracerebral tumor. *Neurosurgery, 57*, 785–976.

Muller, R. H., Mader, K., & Gohla, S., (2000). Solid lipid nanoparticles (SLN) for controlled drug delivery – a review of the state of the art. *Eur. J. Pharm. Biopharm., 50*, 161–177.

Murthy, S. K., (2007). Nanoparticles in modern medicine: State of the art and future challenges. *Int. J. Nanomedicine, 2*, 129–141.

Nam, Y. S., Park, J. Y., Han, S. H., & Chang, I. S., (2002). Intracellular drug delivery using poly(D,L-lactide-co-glycolide) nanoparticles derivatized with a peptide from a transcriptional activator protein of HIV-1. *Biotechnol. Lett., 24*, 2093–208.

Neuberger, T., Schöpf, B., Hofmann, H., Hofmann, M., & Von, R. B., (2005). Superparamagnetic nanoparticles for biomedical applications: Possibilities and limitations of a new drug delivery system. *J. Magn. Magn. Mater., 293*, 483–496.

Oberdorster, G., Oberdorster, E., & Oberdorster, J., (2005). Nanotoxicology: An emerging discipline evolving from studies of ultrafine particles. *Environ. Health. Perspect., 113*, 823–839.

Olbrich, C., Bakowsky, U., Lehr, C. M., Müller, R. H., & Kneuer, C., (2001). Cationic solid-lipid nanoparticles can efficiently bind and transfect plasmid DNA. *J. Control Release, 77*, 345–355.

Owens, D. E., & Peppas, N. A., (2006). Opsonization, biodistribution, and pharma-cokinetics of polymeric nanoparticles. *Int. J. Pharm., 307*, 93–102.

Peracchia, M. T., Fattal, E., Desmaele, D., Besnard, M., Noel, J. P., Gomis, J. M., & Couvreur, P., (1999). Stealth (R) pegylated polycyanoacrylate nanoparticles for intravenous administration and splenic targeting. *J. Control Release, 60*, 121–128.

Pignatello, R., Bucolo, C., Ferrara, P., Maltese, A., Puleo, A., & Puglisi, G., (2002a). Eudragit Rs100 (R) nano-suspensions for the ophthalmic controlled delivery of ibuprofen. *Eur. J. Pharm. Sci., 16*, 53–61.

Pignatello, R., Bucolo, C., Spedalieri, G., Maltese, A., & Puglisi, G., (2002b). Flurbiprofen-loaded acrylate polymer nanosuspensions for ophthalmic application. *Biomaterials, 23*, 3247–3255.

Pison, U., Welte, T., Giersig, M., & Groneberg, D. A., (2006). Nanomedicine for respiratory diseases. *Eur. J. Pharmacol., 533*, 341–350.

Plard, J. P., & Bazile, D., (1999). Comparison of the safety profiles of PLA(50) and Me. PEG-PLA(50) nanoparticles after single dose intravenous administration to rat. *Colloid Surf. B-Biointerfaces, 16*, 173–183.

Popovic, N., & Brundin, P., (2006). Therapeutic potential of controlled drug delivery systems in neurodegenerative diseases. *Int. J. Pharm., 314*, 120–126.

Robins, T., & Plattner, J., (1993). HIV protease inhibitors – their anti-HIV activity and potential role in treatment. *J. Acquir. Immune Defic. Syndr. Hum. Retrovirol., 6*, 162–170.

Roof, K. S., Coen, J., Lynch, T. J., Wright, C., Fidias, P., Willett, C. G., & Choi, N. C., (2006). Concurrent cisplatin, 5-FU, pacli-taxel, and radiation therapy in patients with locally advanced esophageal cancer. *Int. J. Radiat. Oncol. Biol. Phys., 65*, 1120–1128.

Rudolph, C., Schillinger, U., Ortiz, A., Tabatt, K., Plank, C., Müller, R. H., & Rosenecker, J., (2004). Application of novel solid lipid nanoparticle (SLN)-gene vector formulations based on a dimeric HIV-1 TAT-peptide *in vitro* and *in vivo*. *Pharm. Res., 21*, 1662–1669.

Sawant, R. M., Hurley, J. P., Salmaso, S., Kale, A., Tolcheva, E., Levchenko, T. S., & Torchilin, V. P., (2006). "SMART" drug delivery systems: Double-targeted pH-responsive pharmaceutical nanocarriers. *Bioconjugate Chem., 17*, 943–949.

Schlachetzki, F., Zhang, Y., Boado, R. J., & Pardridge, W. M., (2004). Gene therapy of the brain – the trans-vascular approach. *Neurology., 62*, 1275–1281.

Schmidt, J., Metselaar, J. M., Wauben, M. H., Toyka, K. V., Storm, G., & Gold, R., (2003). Drug targeting by long-circulating liposomal glucocorticosteroids increases therapeutic efficacy in a model of multiple sclerosis. *Brain, 126*, 1895–1904.

Sengupta, S., Eavarone, D., Capila, I., Zhao, G., Watson, N., Kiziltepe, T., & Sasisekharan, R., (2005). Temporal targeting of tumor cells and neovasculature with a nanoscale delivery system. *Nature, 436*, 568–572.

Shi, N. Y., Boado, R. J., & Pardridge, W. M., (2001). Receptor-mediated gene targeting to tissues *in vivo* following intravenous administration of pegylated immunoliposomes. *Pharm Res., 18*, 1091–1095.

Siegemund, T., Paulke, B. R., Schmiedel, H., Bordag, N., Hoffmann, A., Harkany, T., & Härtig, W., (2006). Thioflavins released from nanoparticles target fibrillar amyloid beta in the hippocampus of APP/PS1 transgenic mice. *Int. J. Dev. Neurosci., 24*, 195–201.

Tabatt, K., Sameti, M., Olbrich, C., Müller, R. H., & Lehr, C. M., (2004). Effect of cationic lipid and matrix lipid composition on solid lipid nanoparticle-mediated gene transfer. *Eur. J. Pharm. Biopharm., 57*, 155–162.

Tomonaga, M., Oka, M., Narasaki, F., Fukuda, M., Nakano, R., Takatani, H., & Cowan, K. H., (1996). The multidrug resistance-associated protein gene confers drug resistance in human gastric and colon cancers. *Jpn. J. Cancer Res., 87*, 1263–1270.

Torchilin, P. V., Torchilin, V. P., Torchilin, V., & Weissig, V., (2003). *Liposomes: A Practical Approach* (No. 264). Oxford University Press.

Weissleder, R. A., Stark, D. D., Engelstad, B. L., Bacon, B. R., Compton, C. C., White, D. L., & Lewis, J., (1989). Superparamagnetic iron-oxide – pharmacokinetics and toxicity. *Am. J. Roentgenol., 152*, 167–173.

Witt, K. A., Huber, J. D., Egleton, R. D., & Davis, T. P., (2002). Pluronic P85 block copolymer enhances opioid peptide analgesia. *J. Pharmacol. Exp. Ther., 303,* 760–767.

Worden, F. P., Moon, J., Samlowski, W., Clark, J. I., Dakhil, S. R., & Williamson, S., (2006). Southwest oncology group, the head and neck working group. A phase II evaluation of a 3-hour infusion of paclitaxel, cisplatin, and 5-fluorouracil in patients with advanced or recurrent squamous cell carcinoma of the head and neck – southwest oncology group study 0007. *Cancer, 107*, 319–327.

Yuan, F., Dellian, M., Fukumura, D., Leunig, M., Berk, D. A., Torchilin, V. P., & Jain, R. K., (1995). Vascular-permeability in a human tumor xenograft – molecular-size dependence and cutoff size. *Cancer Res., 55*, 3752–3756.

Zhang, J. A., Anyarambhatla, G., Ma, L., Ugwu, S., Xuan, T., Sardone, T., & Ahmad, I., (2005). Development and characterization of a novel Cremophor (R) EL free liposome-based paclitaxel (LEP-ETU) formulation. *Eur. J. Pharm. Biopharm., 59*, 177–187.

CHAPTER 8

Who's Superior: Comparison of Market Analysis Between Biological and Synthetic Nanomaterials

KARTHIK VARMA,[1] SREERAJ GOPI,[1] and JÓZEF T. HAPONIUK[2]

[1]R&D Center, Aurea Biolabs (P) Ltd., Kolenchery, Cochin, Ernakulam, Kerala, India

[2]Chemical Faculty, Gdańsk University of Technology, Gdańsk, Poland

ABSTRACT

The nanotechnology and nanomaterial industry has gained a lot of accolades in the research and consumer field on account of their global market potential and diverse range of applications. Both biologically and synthetically synthesized nanomaterials (NMs) have their own scientific superiorities and limitations. Irrespective of the origin, both of the materials have laid their mark in the global industry. Commercialization of nanotechnology products possesses some constraints and setbacks. The global demand of nanomaterial industry lies in the field of drug delivery, electronics, and material science, aeronautical, and aerospace, construction, food, and agriculture industry, etc. The compound annual growth rate (CAGR) for the nanomaterial is always on the increasing side. The chapter presents an overview on the different application of the NMs and the dominating factors and market constraints.

8.1 INTRODUCTION

Nanotechnology and nanomaterials (NMs) have paved the way and has gained a lot of research interest in recent years on account of their performance. The diverse range of application of the NMs ranges from biomedical

Advanced Nanomaterials for Biological, Nutraceutical, and Medicinal Applications.
Sreeraj Gopi, Augustine Amalraj, and T. R. Sreeraj (Eds.)

to environmental applications (Pelaz et al., 2017). The word 'nano' is derived from the Greek, meaning dwarf. The general classification of the NMs is based on their origin as biological and synthetic. The seven general classifications of the NMs include carbon based NMs, nanocomposites, metals, and alloys, biological NMs, nanoglasses, nanopolymers, and nanoceramics (Anbazhagan et al., 2016). Biologically synthesized NMs have the superiority of low toxicity compared to the synthetic counterparts, hence extending the umbrella of applications. There is a wide variety of biomaterials or biologically synthesized material which can be used for a wide variety of applications. Bionanomaterials poses excellent biodegradability, biocompatibility, superior encapsulation, etc. (Mahapatro et al., 2011). These advantages make biological NMs to be used in multiple biomedical and pharmaceutical applications. Nanoparticles can be formulated in different forms like Hydrogels (Gopi et al., 2016), nanocomposites, etc. (Gopi et al., 2019a, b). Synthetic NMs in comparison to the natural NMs have comparatively higher levels of toxicity. The growing demand of the market has led researchers to focus more on the use of application of different NMs and nanotechnology. There has been a considerable increase in the market potential of the NMs on the account of their versatility of use and extend of application. The chapter presents an overview on the different market potential of the different NMs, their different applications and factors affecting the nanomaterial industry.

8.2 MARKET POTENTIAL OF BIOLOGICAL AND SYNTHETIC NANOMATERIALS (NMS)

The increase in the demand of the products in the market has led researchers to develop and deliver innovative technologies in the field of NMs irrespective of their origin. The wide array of market potential of NMs range from biomedical applications to industrial applications and even in the automobile industries (Figure 8.1). The NMs used can be a single compound or as a composite.

8.2.1 BIOMEDICAL AND PHARMACEUTICAL APPLICATIONS

Recent few decades have witnessed the growing interest of research for the use of different NMs. Biologically synthesized NMs are widely used in biomedical and pharmaceutical applications on account of their low toxicity and biodegradability (Su et al., 2020). Metal particle and metal particle oxides can be biologically synthesized to be used in a wide variety of applications

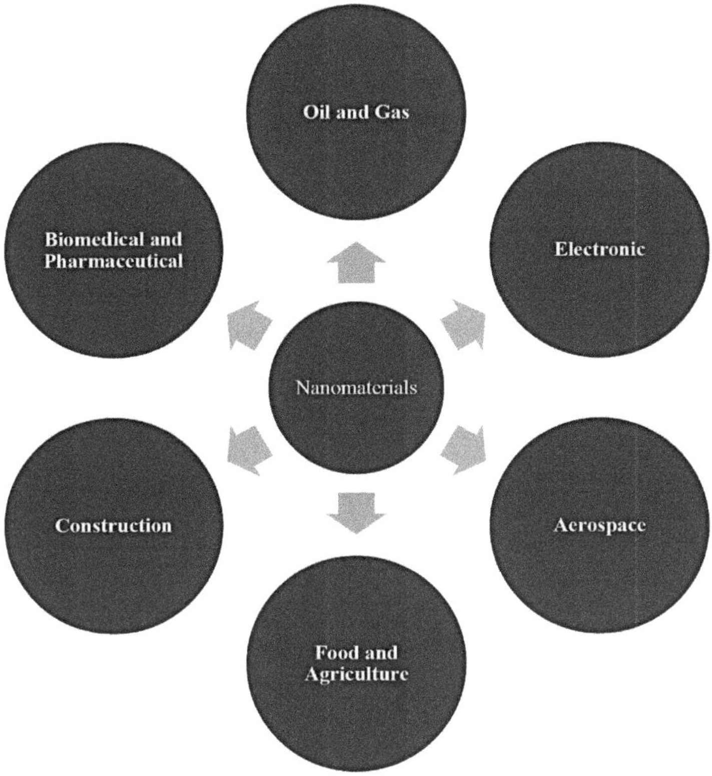

FIGURE 8.1 Applications of nanomaterials.

(Kulkarni et al., 2014). Due to the small size, biologically synthesized NMs can be used in different biomedical applications such as the detection of diseases, surgical fields, molecular imaging and detection, tissue engineering (TE), etc. Recent years have witnessed the use of biological nanoparticles and NMs in different pharmaceutical applications like targeted delivery systems (Karlsson et al., 2018), antioxidant delivery (Kim et al., 2019), gene therapy (Rai et al., 2019) and oral drug delivery (Diaz et al., 2010), and implantable device with biodegradable materials. Proteins and nucleic acids fall under the most common biological NMs, which are used in medical and surgical implants (Sahoo et al., 2015). Other important applications of NMs in medicinal and biological applications include biodetection of pathogens (Edelstein et al., 2000), protein detection (Tirrell et al., 2002), TE (Ma et al., 2003), hyperthermia (Yoshida et al., 1999), separation of biological molecules (Molday et al., 1982) phagokinetic studies, etc. (Brcuhez–Parak) (Figure 8.2). Industrial residues of natural products are explored to utilize the material to deliver value-added NMs. NMs are often employed as a sustainable source of natural carrier to encapsulate bioactive materials for various applications and whose efficacy is established by means of preclinical studies. Several bioactives can be incorporated into nanocomposites to open the umbrella of application of biological/natural NMs (Gopi et al., 2017, 2018; Amalraj et al., 2020). These natural NMs have the scientific superiorities like enhanced efficacy, low toxicity and better encapsulation efficiency.

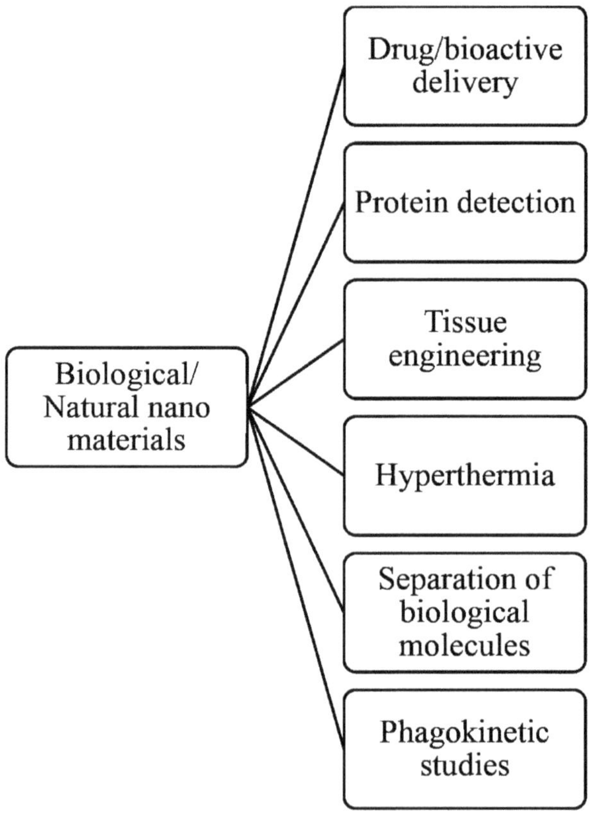

FIGURE 8.2 Applications of biological/natural nanomaterials.

8.2.2 CONSTRUCTION INDUSTRY

NMs and nanocomposites are widely used in construction industries also. The major application of the NMs includes flooring materials (Bi et al., 2017), insulating materials (Irie et al., 2004), self-cleaning windows (Somasundaram et al., 2019), etc. The NMs when used in construction industries finds application in enhancing vital parameters like strength (Becher et al., 1991), durability (Ge et al., 2008), lightness (Sobolev et al., 2005), heat insulation, anti-fogging (Kontos et al., 2007), construction safety (Zhang et al., 2006), structural health, etc. Carbon nanotubes (CNTs) are used in various areas of the construction field like concrete, ceramics, solar cell, etc., to increase the mechanical durability, crack prevention, enhancing the mechanical and thermal properties, electron mediation, etc. (Makar et al., 2003). SiO_2 and TiO_2 nanoparticle are used in areas like ceramic, windows, and solar cells in construction fields to enhance many physicochemical properties like anti-fogging, flame proofing, anti-reflection coatings, fire resistance and many other (Firoozi et al., 2016). Another important nanoparticle, which find use in various applications of the construction industry, include Cu nanoparticles, Ag nanoparticles, Fe_2O_3 nanoparticles, etc. (Mohajerani et al., 2019).

8.2.3 FOOD AND AGRICULTURAL INDUSTRY

Nanoparticles/NMs are also used in food and packaging industries because of the versatile properties. The early 2000 have witnessed around 860-million-dollar business in sales. Nanostructured food ingredients have the advantage of enhanced taste, structure, and stability. NMs have the advantage of increased shelf life and decreased wastage of food due to microbial infestation (Pradhan et al., 2015). Nanostructured materials have better encapsulation and releasing efficiency than the normal products (Nakagawa et al., 2014). It can also control/mask odor and taste of food products, restrict interaction of active constituents with the food matrix, sustained release of active agents, targeted delivery, higher penetration rate, etc. (Ubbink et al., 2006). NMs are also used in various food packaging industry on account of their multiple scientific superiorities. NMs can improve the food packaging with the encapsulation of bioactives like essential oils and other natural products thereby providing antimicrobial activity (Galvez et al., 2007). Nanoparticles like Titanium oxide/zinc oxide, etc., have reported antibacterial activity (Ottaway et al., 2010). NMs are also used as a barrier in food packaging to prevent thermal and mechanical shock (Sorrentino et al., 2007). Nanoscale fillers such as clay, silica, chitin/chitosan (CS), etc., offers lighter, stronger, fire resistance and excellent thermal stability, etc. (Duncan et al., 2011; Othman et al., 2014) in food industry in a wide variety of applications there are some reported toxicity studies also on their usage.

8.2.4 AEROSPACE AND AUTOMOBILE INDUSTRY

The growing market demand for natural NMs in the automobile and aerospace industry is also gaining lots of research interest nowadays. Natural or biologically synthesized and synthetic NMs are used mainly in a wide range of automobile industry and aeronautical industry. NMs can be used in the automobile industry to enhance the strength, durability, performance, etc. (Mathew et al., 2018). The use of NMs whether it is synthetic or biological is not limited to any particular segment in the automobile industry. NMs are used in various segments of automobile industries like used to manufacture body parts, light weight body parts, scratch resistant materials, automobile engines, tires, ultra-reflecting mirrors and glasses, interior parts, fuel cells, nanofilters for air cleaning, etc. (Shafique et al., 2019). NMs are also used in the field of aeronautical and military in the areas of weaponry, protection, and communication.

8.2.5 ELECTRONIC DEVICES

Innovative ideas at the nanotechnology can play a significant role in future developments of electronic devices, which use massive electronic circuit boards for communication and other signaling models. The application of NMs is not limited to simple electronic circuit devices alone. Wireless devices are also being developed to give promising information like temperature, weight, speed, and acoustics and to record the data constantly. The use of NMs has led to the development of a wide variety of sophisticated devices and applications (Dwivedi et al., 2015).

8.2.6 OIL AND GAS INDUSTRY

The multi-varied application of NMs has paved the way to the application of nanotechnology in the oil and gas industry. The recent trend of nanomaterial technology in the oil field is in the area of the oil recovery process. Mostly polymer coated nanoparticles have been applied in the oil recovery processes (Diasty et al., 2015). Many industrial giants like TOTAL, SHELL, British Petroleum, etc., have laid research collaboration with many institutes for developing different technologies and strategies for oil recovery process (Wachter et al., 2009). Compared to the traditional techniques, nanofluids technique has the potential to increase the oil recovery (Li et al., 2015). Researches in the oil industry have concentrated on the production and recovery of heavy oil, bitumen, which possess high density and molecular weight, etc. For the high-density liquids, nanofluid flooding technology is incorporated to increase the oil recovery. TiO_2 nanoparticles are most commonly used for this purpose (Ehtesabi et al., 2014).

8.3 FACTORS AFFECTING THE NANOMATERIAL INDUSTRY

Even though the marketing industry of NMs is growing day by day, there are many factors affecting the nanomaterial industry. The major factors affecting the marketing potential of the nanomaterial industry include market fluctuation in the prize of the materials, increase in market demand, increase in public and private expenditure towards nanotechnology research, qualified manpower and lack of proper instrumentation, proper production facilities, toxicity of the NMs and their by-products discharged to the environment, etc. The US department of commerce has listed a number of 22 factors that can influence the nanomaterial market, which includes the cost of infrastructure for production, regulatory, and patent bodies, etc., and the same has also been depicted by many other studies. In 2009, Hobson et al. reported the safety testing of nanoproducts as one of the barriers in the nanomarket. Absorptive capacity problems and difficulties in identifying the application of areas also can have a threat on the commercialization of NMs. The versatile applications of the nanomaterial have laid the foundation to overcome the hurdles of the many conventional extraction process, but limited on account of their higher productivity cost and low finished product cost.

8.4 WORLD NANOMATERIAL MARKET

Throughout the globe, the market potential of nanoindustry is growing at a considerable rate irrespective of their origin. An estimate projection by Mordor Intelligence have witnessed a global demand of nanomaterial industry and predicted the expected compound annual growth rate (CAGR) of 22%, accounting USD 4.1 billion in 2015 and expected to reach around USD 11.3 billion by 2020 (Mordor, 2017). Deltoite Touche Tohmatsu limited has reported a lower CAGR of 15.5% for the global NMs market for the time period 2012–2020 (Dickson, 2017). Taking on account of the material wise trend analysis, a clearer picture on the global nanomarket analysis will obtain. A study report by the Allied market group states that the predominant material that dominates the nanomaterial industry are the metal and non-metal oxides. The CAGR of the metal and non-metal market has an expected growth rate of 20.7% for the duration of 2017–2022 (Nanomaterials Market – Global Opportunity Analysis and Industry Forecast, 2014–2022,

Allied Market Research September 2016). Titanium dioxide (TiO_2) and silicon are the major NMs, which come under the dominant metal and non-metal oxide NMs. Global leaders in the nanomaterial industry predict the increasing demand of silver nanoparticles (AgNPs), which has a high degree of commercialization. The market potential of the AgNP finds the application in numerous industries like electronics, health care, beauty, and textile industry. The world market analysis has witnessed a global consumption of 50% of the total nanomaterial industry with a CAGR of almost 13% (Silver Nanoparticles Market Size by Application, Industry Analysis Report, Regional Outlook, Growth Potential, Price Trends, Competitive Market Share and Forecast, 2016–2024, Global Market Insights Inc., January 2017). The nanoclays and nanocellulose (NC) also have numerous applications as predicted by Transparency Market Research (TMR) with a forecast of CAGR of 24.9% for the nanoclay and 19% for the NC. Nanocomposites are also used in the separation of gases (Nanoclay market – Global Industry analysis, Size, Share, Growth, Trends, and Forecast 2015–2023, Transparency Market Research, 2016).

8.5 CONCLUSION

The Nanomaterial industry is taking steady and vast interest in the research field irrespective of their origin. The use of nanomaterial has unlocked many hurdles and opened the way to many research ideas in various fields. Irrespective of the origin, natural, and synthetic materials are used in a wide variety of applications. Biological/natural NMs have the scientific superiority of low toxicity. Moreover, more studies need to be done to evaluate the bio-efficacy and side effects of nano-based materials.

KEYWORDS

- **industrial applications**
- **market analysis**
- **nanomaterials**
- **nanomaterials industry**
- **nanotechnology**

REFERENCES

Adam, H., & Youssef, A., (2019). Economic impacts of nanotechnology industry: Case study on Egypt. In: *2019 Proceedings of 10th International Conference on Digital Strategies for Organizational Success*.

Amalraj, A., Jacob, J., Varma, K., & Gopi, S., (2020). Preparation and characterization of liposomal β-caryophyllene (rephyll) by nanofiber weaving technology and its effects on delayed onset muscle soreness (DOMS) in humans: A randomized, double-blinded, crossover-designed, and placebo-controlled study. *ACS. Omega.* doi.org/10.1021/acsomega.0c03456.

Anbazhagan, M., Ramachandran, S., Subramanian, P., Nachimuthu, R., & Gothandam, K. M., (2016). *Nanomaterials: Classification, Biological Synthesis and Characterization, 3*, 31–71.

Becher, P. F., (1991). Microstructural design of toughened ceramics. *J. Am. Ceram. Soc., 74*, 255–269.

Bi, Y., Zaikova, T., Schoepf, J., Herckes, P., Hutchison, J. E., & Westerhoff, P., (2017). The efficacy and environmental implications of engineered TiO_2 nanoparticles in a commercial floor coating. *Environmental Science: Nano, 4*, 2030–2042.

Bruchez, M., Moronne, M., Gin, P., Weiss, S., & Alivisatos, A. P., (1998). Semiconductor nanocrystals as fluorescent biological labels. *Science, 281*, 2013–2016.

Chan, W. C. W., & Nie, S. M., (1998). Quantum dot bio conjugates for ultrasensitive nonisotopic detection. *Science, 28*, 2016–2018.

Chen, J., Guo, Z., Tian, H., & Chen, X., (2016). Production and clinical development of nanoparticles for gene delivery. *Mol. Ther. Methods Clin. Dev., 3*, 16023.

Colton, R. J., (2000). The BARC biosensor applied to the detection of biological warfare agents. *Biosensors Bioelectron., 14*, 805–813.

David, H., (2009). Commercialization of nanotechnology. Wiley interdisciplinary reviews. *Nanomedicine and Nanobiotechnology, 1*, 189–202.

De La Isla, A., Brostow, W., Bujard, B., Estevez, M., Rodriguez, J. R., Vargas, S., & Castano, V. M., (2003). Nanohybrid scratch resistant coating for teeth and bone viscoelasticity manifested in tribology. *Mat. Resr. Innovat., 7*, 110–114.

Díaz, A., David, A., Pérez, R., González, M. L., Báez, A., Wark, S. E., Zhang, P., et al., (2010). Nanoencapsulation of insulin into zirconium phosphate for oral delivery applications. *Biomacromolecules, 11*, 2465–2470.

Dickson, D., (2015). *Advanced Manufacturing in a Highly Connected World.* Deloitte Touche Tohmatsu Limited.

Duncan, T. V., (2011). Applications of nanotechnology in food packaging and food safety: Barrier materials, antimicrobials and sensors. *J. Colloid. Interface. Sci., 363*, 1–24.

Dwivedi, N., Kumar, S., Carey, J. D., & Dhand, D., (2015). Functional nanomaterials for electronics, optoelectronics, and bioelectronics. *J. Nanomater*. Article ID 136465.

Edelstein, R. L., Tamanaha, C. R., Sheehan, P. E., Miller, M. M., Baselt, D. R., Whitman, L. J., & Colton, R. J., (2000). The BARC biosensor applied to the detection of biological warfare agents. *Biosensors Bioelectron., 14*, 805–813.

Ehtesabi, H., Ahadian, M. M., Taghikhani, V., & Ghazanfari, M. H., (2014). Enhanced heavy oil recovery in sandstone cores using TiO_2 nanofluids. *Energ. Fuel, 28,* 423–430.

El-Diasty, A. I., & Aly, A. M., (2015). *Understanding the Mechanism of Nanoparticles Applications in Enhanced Oil Recovery*. SPE North Africa Technical Conference and Exhibition, Cairo.

Firoozi, A., Olgun, G., & Mobasser, S., (2016). Carbon nanotube and civil engineering. *Saudi J. Eng. Technol., 1*, 1–14.

Gálvez, A., Abriouel, H., López, R. L., & Omar, N. B., (2007). Bacteriocin based strategies for food biopreservation. *Int. J. Food. Microbiol., 120,* 51–70.

Ge, Z., & Gao, Z., (2008). Applications of nanotechnology and nanomaterials in construction. *First International Conference on Construction in Developing Countries*, 235–240.

Gopi, S., Amalraj, A., & Thomas, S., (2016). Effective drug delivery system of biopolymers based on nanomaterials and hydrogels—A review. *Drug Des., 129,* 2169–0138.100012.

Gopi, S., Amalraj, A., Jude, S., Thomas, S., & Guo, Q., (2019a). Bionanocomposite films based on potato, tapioca starch and chitosan reinforced with cellulose nanofiber isolated from turmeric spent. *J. Taiwan Inst. Chem. Eng., 96*, 664–671.

Gopi, S., Amalraj, A., Jude, S., Varma, K., Sreeraj, T. R., Haponiuk, J. T., & Thomas, S., (2017). Preparation, characterization and anti-colitis activity of curcumin-asafoetida complex encapsulated in turmeric nanofiber. *Mater. Sci. Eng. C., 81*, 20–31.

Gopi, S., Amalraj, A., Kalarikkal, N., Zhang, J., Thomas, S., & Guo, Q., (2019b). Preparation and characterization of nanocomposite films based on gum Arabic, maltodextrin and polyethylene glycol reinforced with turmeric nanofiber isolated from turmeric spent. *Mater. Sci. Eng. C., 97*, 723–729.

Gopi, S., Amalraj, A., Varma, K., Jude, S., Reddy, P. B., Divya, C., Haponiuk, J. T., & Thomas, S., (2018). Turmeric nanofiber-encapsulated natural product formulation act as a phytogenic feed additive – a study in broilers on growth performance, biochemical indices of blood, and *E. coli* in cecum. *Int. J. Polym. Mater. Polym. Biomater., 67*, 581–588.

Irie, H., Sunada, K., & Hashimoto, K., (2004). Recent developments in TiO_2 photocatalysis: Novel applications to interior ecology materials and energy saving systems. *Electrochemistry, 72*, 807–812.

Karlsson, J., Vaughan, H. J., & Green, J. J., (2018). Biodegradable polymeric nanoparticles for therapeutic cancer treatments. *Annu. Rev. Chem. Biomol. Eng., 9*, 105–127.

Kim, K. S., Song, C. G., & Kang, P. M., (2019). Targeting oxidative stress using nanoparticles as a theranostic strategy for cardiovascular diseases. *Antioxid. Redox. Signal., 30*, 733–746.

Kontos, A. I., Kontos, A. G., Tsoukleris, D. S., Vlachos, G. D., & Falaras, P., (2007). Superhydrophilicity and photocatalytic property of nanocrystalline titania sol gel films. *Thin Solid Films, 515*, 7370–7375.

Li, M. C., Wu, Q., Song, K., Qing, Y., & Wu, Y., (2015). Cellulose nanoparticles as modifiers for rheology and fluid loss in bentonite water-based fluids. *ACS Appl. Mater. Interfaces, 7*, 5006–5016.

Ma, J., Wong, H., Kong, L. B., & Peng, K. W., (2003). Biomimetic processing of nanocrystallite bioactive apatite coating on titanium. *Nanotechnology, 14*, 619–623.

Mah, C., Zolotukhin, I., Fraites, T. J., Dobson, J., Batich, C., & Byrne, B. J., (2000). Microsphere-mediated delivery of recombinant AAV vectors *in vitro* and *in vivo*. *Mol. Therapy, 1*, S239.

Mahapatro, A., & Singh, D. K., (2011). Biodegradable nanoparticles are excellent vehicle for site directed in-vivo delivery of drugs and vaccines. *J. Nanobiotechnology*, 9–55.

Mahtab, R., Rogers, J. P., & Murphy, C. J., (1995). Protein-sized quantum dot luminescence can distinguish between "straight", "bent", and "kinked" oligonucleotides. *J. Am. Chem. Soc., 117*, 9099, 9100.

Makar, J., & Beaudoin, J., (2003). *Carbon Nanotubes and Their Application in the Construction Industry*, 292.
Mathew, J., Joy, J., & George, S. C., (2019). Potential applications of nanotechnology in transportation: A review. *J. King. Saud. Univ. Sci., 4,* 586–594.
Mohajerani, A., Burnett, L., Smith, J. V., Kurmus, H., Milas, L., Arulrajah, A., Horpibulsuk, S., & Kadir, A. A., (2019). Nanoparticles in construction materials and other applications, and implications of nanoparticle use. *Materials (Basel)., 12*, 3052.
Molday, R. S., & MacKenzie, D., (1982). Immunospecific ferromagnetic iron dextran reagents for the labelling and magnetic separation of cells. *J. Immunol. Methods, 52,* 353–367.
Nam, J. M., Thaxton, C. C., & Mirkin, C. A., (2003). Nanoparticles-based bio-bar codes for the ultrasensitive detection of proteins. *Science, 301*, 1884–1886.
Nanoclay Market – Global Industry analysis, Size, Share, Growth, Trends and Forecast 2015–2023, (2016). Transparency Market Research.
Nanomaterials Market – Global Opportunity Analysis and Industry Forecast, (2014–2022). Allied Market Research.
Othman, S. H., (2014). Bio-nanocomposite materials for food packaging applications: Types of biopolymers and nano-sized filler. *Agric. Sci. Proc., 2*, 296–303.
Panatarotto, D., Prtidos, C. D., Hoebeke, J., Brown, F., Kramer, E., Briand, J. P., Muller, S., et al., (2003). Immunization with peptide-functionalized carbon nanotubes enhances virus-specific neutralizing antibody responses. *Chemistry& Biology, 10,* 961–966.
Parak, W. J., Boudreau, R., Gros, M. L., Gerion, D., Zanchet, D., Micheel, C. M., Williams, S. C., et al., (2002). Cell motility and metastatic potential studies based on quantum dot imaging of phagokinetic tracks. *Adv. Mater., 14*, 882–885.
Pelaz, B., Alexiou, C., Alvarez-Puebla, R. A., Alves, F., Andrews, A. M., Ashraf, S., Balogh, L. P., et al., (2017). Diverse applications of nanomedicine. *ACS Nano, 11*, 2313–2381.
Pradhan, N., Singh, S., Ojha, N., Srivastava, A., Barla, A., Rai, V., & Bose, S., (2015). Facets of nanotechnology as seen in food processing, packaging, and preservation industry. *Bio. Med Res. Int.*, 365672.
Rai, R., Alwani, S., & Badea, I., (2019). Polymeric nanoparticles in gene therapy: new avenues of design and optimization for delivery applications. *Polymers (Basel), 11*, 745.
Sahoo, N., Sahoo, R. K., Biswas, N., Guha, A., & Kuotsu, K., (2015). Recent advancement of gelatin nanoparticles in drug and vaccine delivery. *Int. J. Biol. Macromol., 81*, 317–331.
Salinas, B. J., Xu, Z. Y., Agrawal, G., & Richard, B., (2012). Controlled electrolytic metallics – an intervention less nanostructured platform. *SPE International Oil Field Nanotechnology Conference and Exhibition*. Noordwijk.
Shafique, M., & Luo, X., (2019). Nanotechnology in transportation vehicles: An overview of its applications, environmental, health and safety concerns. *Materials, 12*, 2493.
Silver Nanoparticles Market Size by Application, (2017). Industry Analysis Report, Regional Outlook, Growth Potential, Price Trends, Competitive Market Share & Forecast, 2016–2024 (Global Market Insights Inc.
Singh, T., Shukla, S., Kumar, P., Wahla, V., Bajpai, V. K., & Rather, I. A. (2017). Application of Nanotechnology in Food Science: Perception and Overview. *Front. Microbiol. 8*, 1501. doi: 10.3389/fmicb.2017.01501 (accessed on 23 August 2023).
Sobolev, K., & Gutierrez, M. F., (2005). How nanotechnology can change the concrete world. *Am. Ceram. Soc. Bull., 84*, 16–20.
Sorrentino, A., Gorrasi, G., & Vittoria, V., (2007). Potential perspectives of bio nanocomposites for food packaging applications. *Trends. Food. Sci. Technol., 18*, 84–95.

Su, S., & Kang, P. M., (2020). Systemic review of biodegradable nanomaterials in nanomedicine. *Nanomaterials, 10,* 656.

Tahmassebi, N., Moradian, S., Ramezanzadeh, B., Khosravi, A., & Behdad, S., (2010). Effect of addition of hydrophobic nano silica on viscoelastic properties and scratch resistance of an acrylic/melamine automotive clear coat. *Tribol. Int., 43*, 685–693.

Tirrell, M., Kokkoli, E., & Biesalski, M., (2002). The role of surface science in bioengineered materials. *Surf. Sci., 500*, 61–83.

Wachter, W. A., McCarthy, S. J., Beck, J. S., & Stern, D. L., (2009). *Exxonmobil Research and Engineering Company*, INVS. US 7504021.

Wang, S., Mamedova, N., Kotov, N. A., Chen, W., & Studer, J., (2002). Antigen/antibody immuno complex from CdTe nanoparticle bioconjugates. *Nano Letters, 2*, 817–822.

Weissleder, R., Elizondo, G., Wittenburg, J., Rabito, C. A., Bengele, H. H., & Josephson, L., (1990). Ultrasmall superparamagnetic iron oxide: Characterization of a new class of contrast agents for MR imaging. *Radiology, 175*, 489–493.

Zhang, W., Suhr, J., & Koratkar, N., (2006). Carbon nanotube/polycarbonate composites as multifunctional strain sensors. *J. Nanosci. Nanotechnol., 6*, 960–964.

CHAPTER 9

Nanomaterials: A Backbone to Various Biological Applications

HEMANT BORASE,[1] ABHISHEK SHARMA,[1] PRITTESH PATEL,[1] and SATISH PATIL[2]

[1]C. G. Bhakta Institute of Biotechnology, Uka Tarsadia University, Bardoli, Gujarat, India

[2]School of Life Sciences, Kavayitri Bahinabai Chaudhari North Maharashtra University, Jalgaon, Maharashtra, India

ABSTRACT

Nanomaterials (NMs) have established themselves as a potent and promising alternative to glitches, limitations, and grand challenges existing in different areas of biological science. Currently, antimicrobial resistance, low agriculture productivity, malnutrition, stress management, lifestyle associated diseases are some of the key issues in bioscience. Medicine is most influenced sectors by nanotechnology interventions; different NMs are likely candidates for answering alarming issue of microbial drug resistance, targeted delivery of therapeutic agents, cancer management, diagnostic imaging as well as in the development of artificial implants with more longevity and acceptance inside the body. In food science, NMs drastically improved some important areas such as detection of food adulteration, improvement in bioavailability of active ingredients, increasing shelf life, and ecofriendly food packaging. Agriculture is the backbone of developing countries, NMs are coming up with exciting products such as nanofertilizers, nanopesticides, precision farming, detection, and inhibition of phytopathogens. Apart from above well-established application of NMs, there are several areas of biology in which these tiny entities can play determining role, such as wastewater

Advanced Nanomaterials for Biological, Nutraceutical, and Medicinal Applications.
Sreeraj Gopi, Augustine Amalraj, and T. R. Sreeraj (Eds.)

purification, development of smart sensors, recombinant DNA technology, and gene therapy to name a few. Nowadays, many reports have suggested toxic effects exhibited by NMs, and this aspect is more important when we are intended to apply NMs in biological applications. Therefore, during fabrication of NMs, safe by design approach should be used so as NMs will be safe not only for end users but also for non-target organisms present in soil and water reservoirs.

9.1 INTRODUCTION

The existence of living systems on earth is largely sustained by the processes operating at the nanoscale, such as the transmission of nerve impulses by neurotransmitters, muscle contraction using actin-myosin, transport of oxygen by hemoglobin, pathways belonging to the central dogma of molecular biology. Nanomaterials (NMs) are the materials with nanoscale (1 to 100 nm) appearance and are conveniently divided into zero-dimensional (nanoparticles, quantum dots), one-dimensional (nanofilms, nanofibers), two-dimensional (thin films, nanocoatings) and three dimensional such as nanocomposites, multiple nanolayers (Anandharamakrishnan, 2020). The physical, chemical, and biological properties of the NMs are different and are not observed in their bulk counterparts. High surface-to-volume ratio, size-dependent effects, high catalytic potential, electrical conductivity, optics, magnetism, and other characteristics make NMs highly beneficial in industrial processes and in diverse consumer products (Rasmussen et al., 2018).

Currently, NMs are mostly fabricated by chemical and physical methods, such as chemical reduction, gas phase deposition, nanolithography, laser ablation, ball milling, aerosol, electrochemical, sonochemical, and heat decomposition (Hasan et al., 2018). NMs synthesized by these methods contain traces of potentially toxic chemicals hazardous to living organisms and the environment. These toxic chemicals present in nanoproducts restricted their applications associated with biology, such as in health, food, and in agriculture. There is an emerging trend in NMs synthesis that is researchers are more interested in using biologically synthesized NMs (using bacteria, fungi, and plant extracts) over the NMs fabricated by chemical and physical methods. Since biofabricated NMs are inherently safer at the interface of biological systems and NMs (Hasan et al., 2018). Moreover, biological synthesis is low cost, rapid method operated at normal environmental parameters of temperature and pH (Patil et al., 2016). Despite of this, industrial nanosynthesis by biological methods is still in infancy, may be due to variability in biological agents (as the synthesis is performed by a consortium of different metabolites from microbes and plant extracts) (Borase et al., 2014; Salukhe et al., 2011).

Nanotechnology is a disruptive technology with great potential to alter the currently available product and technology with more superior options. The global market value of nanotechnology was about 49 billion U.S. dollars (USD) in 2017 and it is expected to reach up to nearly 76 billion USD by 2020 with an estimated growth rate of 30% per year (Statista, 2020). The world-wide production of NMs is around 10,000 tons/year (Piccinno et al., 2012). Several commercially important NMs are titanium dioxide (TiO_2NPs), zinc oxide nanoparticles (ZnONPs), silicon dioxide nanoparticles (Nano-SiO_2), carbon nanotubes (CNTs), silver nanoparticles (AgNPs), gold nanoparticles (AuNPs), fullerenes, and quantum dots (Piccinno et al., 2012; Borase et al., 2014; Anjum et al., 2020). NMs are now very commonly used in different scientific disciplines such as in physics (solar panels, data storage, lab on chip, display, nanosensors), chemistry (nanocatalyst, nanocomposites, pollutant removal, coating, textiles), automobile and aerospace engineering (Bondarenko et al., 2013; Fytianos et al., 2020).

Moreover, the field of biology is revolutionized by applying principles and products of nanotechnology (Chaudhry et al., 2018). The term bio-nanotechnology and nano-biotechnology is often used interchangeably with the general assumption of exploration of NMs in different areas of biology. However, bio-nanotechnology has specifically involved the study of structure, understanding the working of biomolecules present in the living system and are having nanostructures such as hemoglobin, histone, chromosomes, glucose, and many more (Wooley, 2000). On the other hand, nano-biotechnology typically associated with the application of nanotechnology to the life science. It is primarily expanding the knowledge of natural nanomachine to fabricate similar nano-assemblies in the lab such as respirocytes, microbivores, nanorobots, and others (Surendran et al., 2009).

Development of rapid, accurate diagnostic assays, point of care biosensors for health and environmental remediation are also major objectives of nano-biotechnology (Surendran et al., 2009; Nagamune et al., 2017). Considering the potential of NMs in biology, researchers working in diverse disciplines of biology are utilizing and connecting nanotechnology with their core research areas, and it is reflected with a rapid surge in the number of research publications on applications of NMs in biological sciences (Figure 9.1).

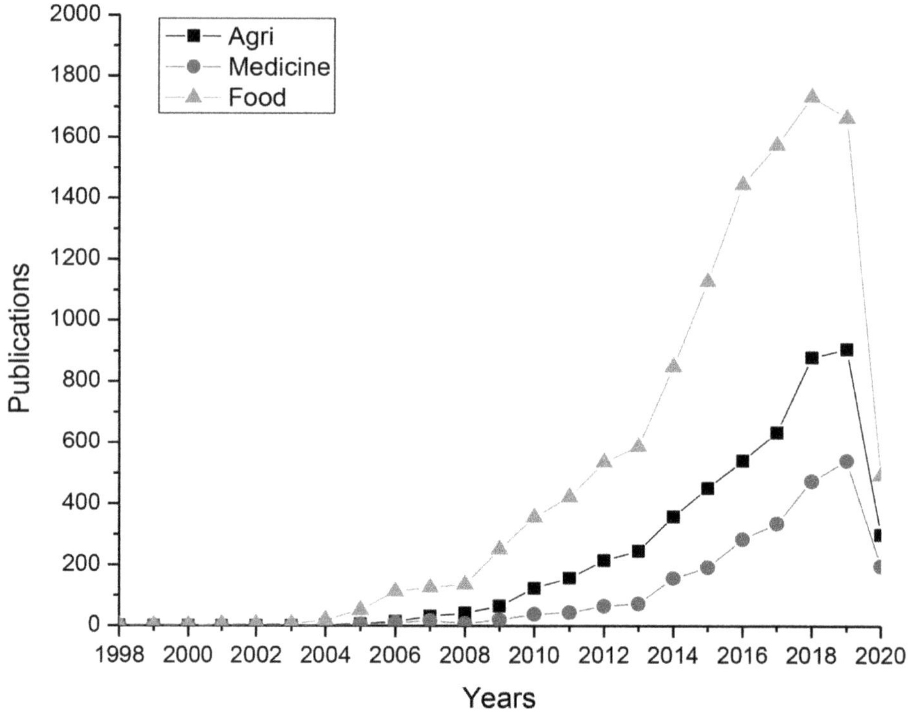

FIGURE 9.1 Number of publications on applications of nanomaterials in agriculture, medicine, and food since 1998 to 2020.

Source: Adapted from http://www.ncbi.nlm.nih.gov/pubmed.

In this chapter, we have explored and discussed the challenges faced by medicine, agriculture, food industries and correlated nanotechnology with present and future applications in the above areas. Introductory and applications-based information on nanotechnology will enable readers about hot areas for nanotechnology research. Apart from this light is thrown on the properties of NMs that are vital before applying NMs in biological science and the current status on the regulatory system governing the use of NMs in biology.

9.2 APPLICATIONS OF NANOMATERIALS (NMS) IN MEDICINE

Nanomedicine deals with the use of processes and products working on the nanoscale for diagnosis, prevention, and treatment of diseases. Nanomedicine is currently utilizing nanoparticles, nanocomposites, nano-emulsions and other NMs in the healthcare sector (Wu et al., 2020). Since 1995 to 2017, around 50 nano-pharmaceuticals have received an approval from the FDA and are in clinical use (Ventola et al., 2017). NMs have proved themselves as a promising candidate in different aspects of medicine due to its unique three-dimensional construction with multiple therapeutic entities at nanoscale. This complex nature demands for careful and controlled synthesis at large scale, stringent characterization to achieve desired pharmacological effects.

9.2.1 NANOPARTICLES AS NEW-GENERATION ANTIMICROBIAL AGENTS

Antibiotic is an indispensable tool to treat infections, to human, plants, and animals. The blind and unnecessary use of antimicrobials in the prevention and treatment of diseases has led to the development of antimicrobial resistance in pathogens having current public health concern (Wang et al., 2014). Antimicrobial resistance threatens the efficient prevention and treatment of an ever increasing range of infections caused by bacteria, parasites, viruses, and fungi. Antibiotic resistance results in survival and growth of bacteria in the presence of one or more antibiotics. Pathogenic microorganisms have developed resistance to commercial antibiotics by several ways, such as selective pressure, rapid removal, sequestration, metabolic resistance, overproduction of detoxification enzymes, gene transfer and mutations in antibiotics target site (Despres et al., 2007). Metal nanoparticles (MNPs) are very effective against the pathogens, parasites, and vectors of different diseases due to multiple modes of action. MNPs can inhibit multiple microbial enzymes, change membrane permeability, generation of free radicals, cytoplasmic leakage and hence making it very difficult for pathogens to devise resistance strategy (Patil et al., 2012). There is an enormous room at the nano-surface awaiting exploration for effective surface coupling chemistry to prepare bioactive NMs. The interaction of a variety of molecules such as antibiotics, vaccines with different nanoparticles in a controlled manner has emerged a new alternative to tackle drug resistance (Nayak et al., 2010; Puttappa et al., 2019).

The antibacterial effects of norvancomycin-capped AgNP, suggests the use of functionalized nanoparticles as novel bactericidal agents (Wei et al., 2007). Antibiotics-capped nanoparticles could be assembled through hydrogen bonding between amino and carboxylic acid functional groups and with different approaches like electrostatic interaction, covalent binding by conjugation chemistry, and non-covalent affinity-based receptor-ligand binding (Wei et al., 2007; Allahverdiyev et al., 2011). In such cases, the molecules are held together to the surface of the particle by van der Waals forces, hydrogen bridges, Ag-thiol bonds (from cysteine residues) or by hydrophobic interaction. Therefore, these antibiotics – NMs conjugate will exert multiple actions on pathogens as described above with lower chances of resistance emergence and will also arise as the better approach for minimizing side effects due to higher doses of antibiotics. Curcuminoid-loaded lipid nanoparticles showed higher bioavailability (because such formulation bypasses gastrointestinal (GI) degradation) and increased *in vitro* antimalarial activity as compared with free curcuminoid (Nayak et al., 2010). Heparin functionalized solid lipid nanoparticles (SLNs) encapsulated with chloroquine was highly effective against chloroquine sensitive (D6) and chloroquine resistant W2 strain when compared to the free chloroquine under *in vitro* conditions. Primaquine is a unique drug targeting late liver stage of *Plasmodium vivax* and *Plasmodium ovale*, but it is suffering from toxicity and side effects. Lipid nanoemulsion encapsulated with primaquine reported to reach in liver at high concentration of 45%, thereby decreasing side effects and increases antiplasmodial activity (Puttappa et al., 2019). *Acanthamoeba castellanii* is an opportunistic pathogen of human. *A. castellanii* have a growing trophozoites stage and dormant cyst stage. Antiamoebic drugs such as fluconazole, neomycin, amphotericin is inactive against the cyst stage with drug resistance, toxicity, and recurrence after stopping therapy. AgNP fabricated by leaf extract of *Jatropha curcas*, *Jatropha gossypifolia* and *Euphorbia milii* were able to destroy 74% trophozoites at a concentration of 25 μg/ml. Interestingly the AgNP was also safe on peripheral blood mononuclear cells (Borase et al., 2013).

9.2.2 NANOTECHNOLOGY IN CANCER MANAGEMENT

Cancer is ranked second in global deaths (around 9.6 million deaths in 2018) after heart diseases. One in every five deaths is because of cancer with high frequency in low- and middle-income countries (WHO, 2018). In 2010, the annual economic cost of cancer was about 1.16 trillion USD. Lung, colorectal, stomach, liver, and breast cancer are the most common cancer types responsible for the deaths. Three factors are very much important for the survival of cancer patients: (i) Early and accurate detection; (ii) targeted therapies; and (iii) personalized medicine (Ye et al., 2018). Traditional chemotherapy and radiotherapy are suffering with numerous side effects on the patient, such as non-selectivity, loss of weight and hairs, decrease in immunity. Some strategies, including radio immunotherapy, immunotoxins, drug immunoconjugates increases efficiency, but their targeted delivery is still a major hurdle (Mishra et al., 2010).

The success of cancer treatment depends upon delivering the required amount of anticancer drug to the tumor site so as to destroy cancer cells and minimal damage to healthy cells. NMs have distinct properties to revolutionize cancer treatment. Nanomedicine is itself have a diagnostic and therapeutic effects and can carry large quantity of anticancer drug to target site. Multiple cancer cell targeting molecules can be attached to nanosystem to increase specificity and affinity. Use of NMs in cancer treatment decrease drug resistance and toxicity (Acharya et al., 2009).

Liposomes are lipid bilayer containing spherical vesicle used as delivery cargo for drugs and nutrients. Liposomal delivery of polyethylene glycol (PEG)-ylated liposomal doxorubicin achieved a prolonged circulation required for an effective cure. Folate-receptor-targeted liposomes having doxorubicin have been circumvented drug resistance in *in vitro* cultured tumor cells (Immordino et al., 2006). Both organic and inorganic nanoparticles are also used in cancer treatment. Acharya et al. (2009) synthesized rapamycin-loaded polymeric poly(lactide-co-glycolide) (PLGA) nanoparticles conjugated with antibodies to the epidermal growth factor receptor (EGFR-Rapa-NPs). EGFR is expressed in breast cancer cells. It was observed that EGFR-Rapa-NPs have high antiproliferative activity as compared to unconjugated nanoparticles and native rapamycin due to greater cellular uptake in malignant breast cancer cells. Bismuth NPs were used in radiation therapy with an increase in radiation dose to tumor by 90%. Anticancer gene therapy involves the incorporation of external genes into the tumor cell genome for tumoricidal properties. Viral vectors used for gene delivery to give rise to toxicity, immune responses and targeting issues. Liposomes and nanoparticles are currently emerging as a superior non-viral vectors. Akt1 is an oncogene. It was found that biodegradable nano-polymeric carrier has effectively delivered Akt1 small interference-RNA and thereby masking production of Akt1 protein as well as cancer cell proliferation and metastasis (Jere et al., 2009).

9.2.3 NANOTECHNOLOGY IN DIAGNOSTIC BIOMARKERS DETECTION

A biomarker is an indicator of the severity or presence of any disease. The old definition of the biomarker was limited to biomolecules from specimen sample, but nowadays radiological characters are considered as biomarkers that is both specimen biomarkers and imaging biomarkers (Ye et al., 2018). The currently available cancer screening methods (pap smear, mammography, prostate-specific antigen (PSA) blood test, colonoscopy)) are inadequately specific/sensitive and chances of over diagnosis are also high (Cottet et al., 2006; Ye et al., 2018). NMs must be decorated with tumor specific ligands (antibody, peptide, aptamer) having high specificity and affinity for cancer biomarkers. The ligand attached to NMs will react specifically with tumor marker and undergoes spontaneous change that be amplified into measurable response. The ligand should maintain its selectivity after attachment with NPs without steric hindrance. It was reported that when as antibody-coated magnetic NPs were added directly to blood plasma, the PSA present in prostate cancer reacts specifically and induced a binding and cluster formation of NPs. The limit of detection (LOD) for PSA is as low as 400 fmol/L within a short period of 14 min. (Ranzoni et al., 2012). Peptide-based targeting is another approach in cancer detection, since peptides are smaller in size, easy to manufacture at low cost and has less immunogenicity. RGD peptide consisting of arginine/glycine/asparagine recognize and bind to cancer angiogenesis and metastasis surface marker integrin $\alpha v\beta 3$. Nanorods of zinc oxide coated with RGD peptide was found to selectively bind with integrin $\alpha v\beta 3$ positive human glioblastoma over integrin $\alpha v\beta 3$ negative human breast cancer cells, thus acting as differentiating assay for integrin positive and negative cancer cells (Hong et al., 2011). Apart from direct involvement in cancer detection, NMs are increasing the sensitivity of instrument-based conventional tools such as mass spectroscopy, optical detection, fluorescence, luminescence, and electrochemical detection. Gold nanostructure, single wall CNTs, and quantum dots are good imaging contrast agents for *in vivo* molecular imaging techniques like positron emission tomography (PET), magnetic resonance imaging (MRI), and computed tomography (CT) (Ye et al., 2014).

Cardiovascular diseases (CVDs) represent 30% of total deaths around the globe. It is projected that by 2030 almost 23.6 million people will die from CVD (WHO). Like that of cancer, early and accurate identification of acute myocardial infarction (AMI) is of great clinical importance as mortality is increasing at an alarming rate. Electrocardiograms (ECG), echocardiograms, CT angiograms, and coronary angiography are commonly used to get information about the onset of CVD. These tools require an expert operator, and a sophisticated laboratory setup. Cardiac biomarkers such as creatine kinase (CK), CK-MB (creatine kinase myocardial band), troponin I, myoglobin, lactate dehydrogenase (LDH), aspartate transaminase are vital for sensitive detection of cardiac attack onset. AuNPs functionalized with iminodiacetic acid moiety (IDA) undergoes aggregation and color change after interaction with myoglobin, similar approach was used for detection of another cardiac biomarkers such as troponin I CK-MB, and brain natriuretic peptide (BNP) using polyaniline (PANI) nanowire and single-walled carbon nanotube (Singh & Tripathi, 2017).

ELISA assay used for diagnosis of HIV infection detects antibodies against viruses from the specimen. Although ELISA is fairly specific, it can detect antibodies from a few weeks to months after infection (Farzin et al., 2020). $CD4^+$ cells, viral RNA, HIV-1 tat protein, core protein p24, HIV enzymes are the ligand/markers used for a nanotechnology based platform for rapid detection of HIV infection. Gold nanodots (10–20 nm) were fabricated on indium tin oxide (ITO) coated glass electrode and the surface was modified with fragments of gp120 monoclonal antibody which selectively neutralizing HIV-1. The antigen–antibody reaction shifts the wavelength in the UV-Vis spectrum. The LOD was from 200 fg/ml to 125 pg/ml (Lee et al., 2013).

9.3 APPLICATIONS OF NANOMATERIALS (NMS) IN FOOD SCIENCE

The Global human population is estimated to increase by 2.3 billion from 2009 to 2050 and because of this, the worldwide food demand will rise up to 70% by 2050. Several other factors, such as urbanization, climatic changes, deteriorating soil fertility, insect resistance is continuously decreasing food productivity and quality (Chen & Yada, 2011; Fakhrul & Karim, 2019). Another important issue is food wastage, around 1.3 billion metric tons of food is wasted annually due to poor post-harvesting techniques, transport, and storage. Due to prolonged, inappropriate storage of food, microorganisms contaminated the food products leading to spoilage and deterioration of nutritional value and at the same time increasing the chances of foodborne diseases (Sperber et al., 2009). Nanotechnology dramatically improving different aspects of food science and technology such as enhancing food bioavailability, texture, taste, novel food safety techniques, smart nanopackaging, water solubility, thermal stability and others (Ezhilarasi et al., 2013; Nile et al., 2020). The Market value of nanotechnology products in the food industry is around 1 billion US dollars and expected to reach up to 20 billion US dollars by 2020 (Chau et al., 2007).

9.3.1 NUTRITIONAL VALUE

Food nutrients such as vitamins, proteins, peptides, and flavonoids are susceptible to high temperature used during product manufacturing as well as they can be destroyed in the harsh acidic environment of the stomach and enzymatic degradation in the duodenum (Huang., 2010). Sometimes nutrients are sparsely soluble and hence rapidly eliminated from the body, therefore the body is devoid of its dietary and pharmacological benefits. Nanoencapsulation makes use of organic compounds such as emulsion, liposomes, micelles, and vesicles, which are incorporated with food nutrients like vitamins to protect them from degradation leading to sustained release (Singh et al., 2017). The concept of functional foods to offer health benefits that extend beyond their nutritional value is possible with nanotechnology-based products. Curcumin has a variety of pharmacological actions such as antioxidant, anti-inflammatory, anti-tumor, and antiarthritic but it is very unstable and has a low bioavailability. Nanoemulsion of curcumin was prepared in medium-chain triglyceride oil by ultrasonification. The resulting nanoemulsion was resistant to pepsin digestion, pancreatin releases curcumin from nanoemulsion (Sari et al., 2015). Tan et al. (2014) encapsulates different carotenoids (lutein, β-carotene, lycopene, canthaxanthin) in a liposome delivery system. The antioxidant activity was analyzed using 2,2-diphenyl-1-picrylhydrazyl (DPPH) scavenging, ferric reducing antioxidant power (FRAP) assay, and lipid peroxidation inhibition capacity. It was concluded that the antioxidant activity depends on the incorporation efficacy of liposomes. Liposomes improve stability and solubility of nutrient, and improve bioavailability. Nanoemulsions are capable of delivery of lipid-soluble bioactives and can increase water solubility and bioavailability (Huang et al., 2010). Nanoemulsion was produced using natural surfactant molecules (lecithin and quillaja saponin) for delivery of vitamin E by homogenization technique (Ozturk et al., 2014). The emulsion was stable at temperature from 30 to 90°C but high salt concentration (400 nM NaCl) destabilizes the nanoemulsion. Droplet size of nanoemulsion plays a central role in the stability of emulsion-based delivery systems and also in the bioavailability of encapsulated compounds, decrease in droplet size increases bioavailability.

9.3.2 FOOD PROCESSING

The ideal agent in food processing should possess the following characteristics: (i) ability to deliver additives to product without affecting basic properties of food; (ii) maintain level of bioactive compounds during storage; and (iii) release in desired concentration at a specific rate (Chen et al., 2017; Singh et al., 2017). NMs are used during food processing for improving nutritional value, flavor, color, and food additives as well as increase stability because of their small size (Nile et al., 2020). In some studies, it was found that nanoparticles are more efficiently absorbed as compared to bulk particles with big size (Surendran et al., 2009; Borase et al., 2019). Inorganic NPs were reported to apply in food processing, TiO_2NPs are permitted to add (less than 1% weight of food) in food as color additives. NPs of copper, zinc, iron, aluminum, and silicon dioxide were used as preservative, anticaking agent, dietary supplement and some of them are already permitted by the FDA and EC (He et al., 2019).

9.3.3 FOOD PRESERVATION AND INCREASE IN SHELF-LIFE

Fruits and vegetables are a rich source of antioxidant, vitamins, and polyphenols. They are useful in boosting immunity and age-related chronic diseases. During the storage period, a steady drop in moisture, taste, contents was observed in fruits and vegetables (Kerch, 2015). Edible nanocoatings is an exciting and eco-friendly technique to maintain the nutritional content of the food during prolonged storage. Nanocoating act as a barrier to moisture, flavors, enzymes, antioxidants, and anti-browning agent (Singh et al., 2017). There is range of edible polymers that are available in the market, such as starch alginate, carrageenan, chitosan (CS), Polylactic acid, carboxymethylcellulose, pullulan, zein (Shit & Shah, 2014). Strawberry fruit has a very short postharvest shelf-life and prone to fungal contamination. Fakhouri et al. (2014) coated strawberry with gelatin coatings with cellulose nanocrystals. The nanocoated strawberry show weight loss of 36% as compared with uncoated having 65% weight loss. Moreover, the nanocoatings preserve the ascorbic acid of the fruit and exhibit antimicrobial effect. CS coating was successfully applied on cherries, guava, pineapple, sponge gourd, and others with a general conclusion that CS has food protective and antimicrobial activity, but these properties can be further improved by developing films and coating of CS nanocomposites with several other synergistically helpful ingredients such as rosemary, olive, and carp oils (Kerch, 2015).

9.3.4 FOOD PACKAGING

The type of packaging material used for storage of food products play an important role in maintaining nutritional status as well as prevention from microbial spoilage. Since many decades plastics such as high- and low-density polyethylene, polyethylene terephthalate, polypropylene, and polystyrene are being used although many of them are recyclable but are non-biodegradable. Composite mixed with organic compounds such as essential oils, bacteriocins were used for antimicrobial properties, but these organic compounds are unable to withstand high temperature and pressure used during food processing and manufacturing (Singh et al., 2017). Nanotechnology-based smart food packaging ,have improved mechanical, barrier, and antimicrobial properties. Particularly inorganic metal NPs already known for potent biocidal properties such as silver, copper, and titanium are effective in very small concentration and stable to harsh processing parameters. Metal NPs can be easily incorporated into polymers leading to the formation of functional antimicrobial packaging material that is in contact with food to retard pathogens existing on food surfaces (Duncan et al., 2011). Simple dipping of carrots and pears into an aqueous alginate solution with added biologically synthesized AgNP resulted in low water loss with more consumer acceptability on the basis of sensory analysis as compared to uncoated and alginate coated carrots and pears (Mohammad-Fayaz et al., 2009). In a similar experiment, asparagus spears were stored in bags made by AgNP/polyvinylpyrrolidone (PVP) nanocomposite film. The shelf life was increased up to 25 days with more green color, intact texture and less microbial growth (yeast, mold, and psychotrophic bacteria). In another application, cellulose pads containing AgNPs reduce microbial load from exudates from beef meat (Duncan et al., 2011). Nanocoating delay rate of respiration, slow down biochemical processes and inhibition of pathogenic microbes, leading to an increase in shelf life.

9.3.5 FOOD SAFETY DETECTION SENSORS

Food adulteration as well as food spoilage due to microbial and abiotic factors are the major reasons for foodborne diseases. According to NCEZID, foodborne infections lead to around 48 million cases of illnesses and 1,28,000 hospitalized in the US. Unpalatable food has typical color, odor, and other sensory properties that can be easily identified by consumers to avoid its consumption. But in the case of packaged food products, consumers have to depend on the expiry date. Expiry date is determined by the manufacturer on the basis of ideal storage and transport condition and deviation in those parameters can spoil the product very early like in the case of packaged milk. NMs with unique opto-electronic properties lead to the development of sensitive, selective, low-cost, rapid, and point-of-care (without laboratory) sensors of adulteration, spoilage, gases, and the aroma of food products (Duncan et al., 2011; Borase et al., 2014). NMs-based sensors are useful in quality control of food products during manufacturing, and consumers can purchase the products with best qualities. Melamine is adulterated to increase the protein content of the milk, and consumption of this milk was reported to cause kidney stone in thousands of infants in China (Zhang et al., 2009). AuNPs functionalized with cyanuric acid selectively bind to melamine. AuNPs changes its color from red to blue with LOD 2.5 ppb, biosynthesized AuNPs without any ligand was also reported for melamine detection (Zhang & Zhang, 2008; Borase et al., 2014). Numerous sensors can find in literature utilizing NMs such as detection of gliadin (celiac disease), gluten (immunological response), cyanide (poison), botulinum toxin, aflatoxins (Duncan et al., 2011). Apart from adulterants, and pathogens detection, identification using NMS is also an emerging area. A novel technique called immunomagnetic separation consists of magnetic particles attached with selective antibodies to selectively separate pathogens from the food. *L. monocytogenes* specific antibodies functionalized on magnetic iron oxide nanoparticles were efficiently separate from the target bacteria from contaminated milk and detected using real time PCR. Wang et al. (2011), utilized silica-coated magnetic nanoparticles (MNPs) and AuNPs labeled with Raman reporter molecules for surface-enhanced Raman spectroscopy (SERS) identification of *S. enterica* serovar Typhimurium and *S. aureus* in spinach wash and peanut butter emulsion with a detection limit of 10^3 CFUs/mL. *Mycobacterium avium* spp. *Paratuberculosis, Brucella, Salmonella, Escherichia, S. typhimurium* were among different food pathogens successfully detected using a combination of NMs with other techniques (Duncan et al., 2011). Selective colorimetric detection using AgNPs is represented in Figure 9.2.

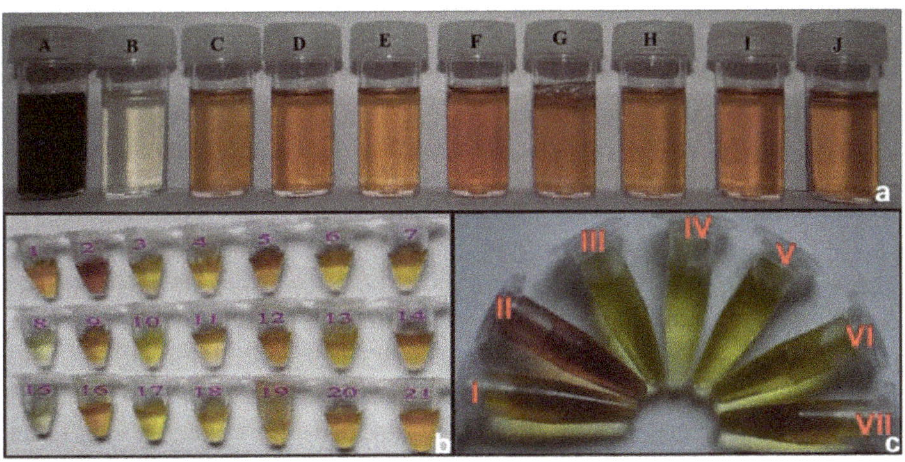

FIGURE 9.2 Selective colorimetric detection using AgNPs. (a) Mercury: only mercury is colorless while other elements give yellowish-brown color after mixing with AgNPs (b) and (c). AgNPs selectively interact with cysteine and melamine to produce orange color and not with other amino acids and normal milk components.

9.4 APPLICATIONS OF NANOMATERIALS (NMS) IN AGRICULTURE

Agriculture and agriculture-associated businesses such as poultry, cattle farming, and dairy are the source of income for more than half of the world's human population (FAO, 2013). But the agriculture sector is facing a lot of challenges, *viz.*, low yield, insect-pest resistance, decrease in soil fertility, climatic variations, resource scarcity, low plant uptake of nutrients, and environmental pollution due to the use of agrochemicals. Nanotechnological inventions such as nanofertilizers, nanopesticides, controlled release of nutrients, sensors for pathogens and agrochemicals detection are trustworthy candidates to above-mentioned agriculture challenges (Acharya & Pal, 2020). Despite of above promising applications of NMs in agriculture, there are very few commercially available nanotechnology-based agro products in the market and this area needs better attention (Acharya & Pal, 2020).

9.4.1 NANOMATERIALS (NMS) AS ELICITOR OF PLANT-SPECIALIZED METABOLITES

Various reports are available highlighting effects of NMs on plant growth and physiology, but there use to enhance production of plant secondary metabolites (SM) is still not much explored (Disfani et al., 2017; Manjaiah et al., 2019). Plant specialized/SM such as alkaloids, sulfur-containing compounds, polyphenols, and terpenoids play a vital role in ecological adaptation of plants to unfavorable environmental conditions, but at the same time, many of them are very useful in therapeutics, cosmetics, and foods (Borase et al., 2014; He et al., 2019). Quercitin, quinine, artemisinin, morphine, resveratrol, and vinblastine have proven biological actions. Plants synthesized these SM in a very less amount which is an important barrier for their commercial uses at low cost (Anjum et al., 2019). NMs such as metallic NPs and organic NMs entered into the plant system via leaf by foliar spray, through soil root uptake, by tissue and callus growing on artificial nutrient medium and act as elicitors of SM. *Capsicum frutescens* cell suspension culture grown with AgNPs (3 mg/Lit) was reported to produce twofold more capsaicin (Bhat & Bhat, 2016). A nearly 4% increase in production of artemisinin was found after treatment of AgNPs (900 mg/Lit) in hairy root culture of *Artemisia annua* (Zhang et al., 2013). In another study, the production of quercetin and glycyrrhizin was improved in seedlings of *Glycyrrhiza glabra* after the addition of AgNPs into the growth media (Tahoori et al., 2019). *Momordica charantia* (bitter melon) showed increase in accumulation (20 to 90%) of anticancer (cucurbitacin-lycopene) and antidiabetic (charantin, insulin) compounds after exposure of fullerenes at a concentration of 10.8 Mm (Kole et al., 2013). Increase in ROS stress, calcium ion and flux movement, upregulatin of mitogen-activated protein kinase (MAPK) cascades are some of the mechanisms by which NMs enhances production of specialized metabolites (Anjum et al., 2019).

9.4.2 NANOMATERIALS (NMS) AS PLANT GROWTH PROMOTER

In this area, an extensive number of studies have shown that the application of NMs has positive effects on germination as well as on plant growth and development. The multiwalled carbon nanotubes (MWCNTs) positively influences seed germination of different crop species, including tomato, corn, soybean, barley, wheat, maize, peanut, and garlic (Lahiani et al., 2013; Srivastava & Rao, 2014; Khodakovshaya & Biris, 2010; Joshi et al., 2018). Nano-SiO_2, TiO_2 and Zeolite positively stimulate seed germination in crop plants (Manjaiah et al., 2019; Lu et al., 2019). Moreover, Fe/SiO_2 nanoalloy have shown significant potential to improve seed germination in barley and maize (Disfani et al., 2017). Despite a significant body of research on NMs-induced positive effects on germination, the underlying mechanisms of how NMs could stimulate germination still remain unclear. Limited studies have demonstrated that NMs have the potential to penetrate the seed coat and enhance the ability of absorption and utilization of water, which stimulates enzymatic system and ultimately improves germination and seedling growth (Lu et al., 2002; Banerjee & Kole, 2016).

9.4.3 NANOPESTICIDES

The use of agrochemicals in controlling plant pathogens, insects, and pest is one of the reasons behind massive environmental pollution. The use of ecofriendly fabricated NMs is verified for their potency to kill pests as well as for environmental safety. AuNPs synthesized from the latex of Jatropha *curcas* inhibit the activity of the enzyme trypsin (responsible for digestion, insecticide resistance) under *in vitro* conditions. Complex formation between trypsin and AuNPs fill the active sites of the enzyme, so that the substrate is no more able to bind enzyme. The positive typsin inhibitory activity of AuNPs was also observed *in vivo* when several insects and pest viz. *Helicoverpa armigera* (a cotton pest), *Callosobruchus maculatus* and *Callosobruchus chinensis* (pest of stored legumes), *Maconellicoccus hirsutus* ingested a definite amount of AuNPs (Patil et al., 2016).

Bean yellow mosaic virus affects legume crops and responsible for lower yield. *Bacillus licheniformis* synthesized AgNPs reported to inhibit this fastidious virus (Elbeshehy et al., 2015). In precision farming concept, maximum crop yield is targeted using minimal inputs of fertilizers, pesticides. Precision farming also aimed to decrease production of agriculture waste and adopt real time monitoring of plant health and weather forecast (Vecchio et al., 2020). Nanoscale carriers such as nanocapsules (poly(epsilon-caprolactone) nanocapsules loaded with atrazine)) and nanocarriers (silica NPs, polymeric NPs) were modified to deliver pesticides in a controlled manner so that lower dosage can be used with the same herbicidal action which benefit environment a lot (Singh et al., 2017; Nile et al., 2020). Several chemical insecticides and biopesticides such as cypermethrin, avermectin, imidacloprid, cyfluthrin, garlic essential oil, neem essential oil was delivered by different types of NMs ranging from nanocomposites, nanocapsules, gum particles (Acharya & Pal, 2020). Apart from these, NMs are also utilized for transfer of genetic materials to produce insect resistant varieties (He et al., 2019).

9.4.4 NANOFERTILIZERS

Despite of using in very high quantities, traditional mineral fertilizers have low plant uptake. Nanofertilizers are found to enhance fertilizers incorporation into plant and soil microflora with the sustained release manner therefore ultimately limiting loss (Askary et al., 2017). Nanofertilizers are expected to transform fertilizer production industries (Chaudhry et al., 2018; He et al., 2019). In an interesting study, when tomato seeds were sowed in a soil containing CNTs, an enhanced plant growth was observed because of increased water uptake due to CNT penetration in the seeds. CNT can also be used as a vehicle to administer the desired molecules into the seed at the time of germination for the protection of the plant from diseases in addition to the plant growth-promoting effect (Khodakovskaya et al., 2009).

Foliar application of nanophosphorus particles significantly increases the yield of cluster bean and pearl millet under arid environment, while ZnONPs promotes synthesis of polysaccharides and phosphatases at a higher rate as compared to bulk ZnO (Tarafdar, 2012). Urea modified hydroxyapatite (HA)

NPs exhibit slow and sustained release of nitrogen for 60 days while other nitrogen fertilizers are failing to have such long persistence (Kottegoda et al., 2011). Treatment of iron oxide nanoparticles (0, 5, 10, 20, 30, and 40 mM) on *Catharanthus roseus* for 70 days resulted in stimulation in protein content and photosynthetic pigments (Askary et al., 2017). Polymeric NPs coated with biofertilizers and nanofertilizers found to have strong desiccation resistance and leaching (Acharya & Pal, 2020).

9.5 APPLICATIONS OF NANOMATERIALS (NMS) IN COSMETICS

The cosmetics industry is the pioneer industry to use NMs in a different cosmetics product. Since last decade renowned cosmetic companies (L'Oreal, Unilever, P&G and Johnson and Johnson) are slowly incorporating NMs in diverse cosmetic formulations (Santos et al., 2019; Fytianos et al., 2020). Sometimes, the term "nano" is often added in the product name to attract consumers. NMs are used in cosmetics for stability enhancement of makeup products, increasing the sun protection factor (SPF), transparency of sunscreens and improved delivery of coloring agents of nail polish and lipsticks, antimicrobial agents in anti-aging creams (Pulit-Prociak et al., 2019; Fytiano et al., 2020). The global cosmetic market is predicted to jump from 435 billion USD to 648 billion USD from 2018 to 2026 (Global Cosmetics Market, 2019). Consumers are now asking for cosmetic products with best properties, use of natural (less side effects, toxicity) ingredients and attractive packaging (Global Cosmetics Market, 2019). UV radiations of the sun are harmful to skin and can cause skin cancer, skin burning, inflammation, and other adverse effects. Food and Drugs Administration (FDA) permitted use of NMs in cosmetics products in 1999. Zinc and titanium oxide nanoparticles are used in sunscreens are now getting replaced by nanosize titanium dioxide (TiO_2) and zinc oxide, but they are having some side effects such as complexion with biomolecules, immunological reactions, and mutation (Holmes et al., 2020).

Nanoparticles of gold and silver synthesized by green chemistry protocols are safer and more effective than ZnONPs and TiO_2NPs (Newman et al., 2009). It was found that the SPF value of commercial sunscreens increased up to 24% of its original value after incorporation of AuNPs (50 nm) biosynthesized using latex of *Jatropha gossypifolia* (Borase et al., 2014). According to the authors, agglomerates of AuNPs in commercial sunscreen formation is the main factor behind significant improvement in SPF values.

The stable nano-cosmetic formulation was prepared using silver and AuNPs with acceptable smell, appearance, consistency with potent inhibitory action against *Aspergillus niger* and *Saccharomyces cerevisiae* (Pulit-Prociak et al., 2019). Apart from silver, gold, ZnO, and TiO_2 NPs several other NMs are also used such as silica NPs, carbon black, HAp for improvement in cosmetics products (Pastrana et al., 2018).

Some organic NMs such as chitin nanofibrils in emulsion form decreased water evaporation and hence help to keep skin hydrated (Morgant & Morganti, 2008). Cellulose nanofibrils obtained from bacterial sources possess a high degree of hydration and used as an effective peeling agent in scrubs, carrier of another active agent in skin pores (Pacheco et al., 2018). As compared with inorganic NMs, the organic biodegradable polymers are safe on skin, non-allergenic, and also harmless of normal skin micro-flora. Despite of the fact that the use of NMs in cosmetics is very normal practice and many nano-cosmetic products are already in consumer market, but we cannot deny the likely toxic consequences due to NMs that are not experimented. Hence, risk assessments of NMs need to perform by product to product basis. Selective colorimetric detection using AgNPs is represented in Figure 9.3.

9.6 CHARACTERISTICS OF NANOMATERIALS (NMS) THAT ARE VITAL FOR APPLYING IN BIOLOGICAL SCIENCE

NMs with a particular set of properties are only useful in specific area of biology. Hence it is necessary to synthesize NMs with specific properties. Numerous studies were done to investigate shape, size, charge, solubility, chemical makeup, stability, aggregation, surface energy, charge, solubility, density, pore size and impurities of NMs and exploring applications on the basis of NMs properties (Gatoo et al., 2014).

9.6.1 SIZE AND SURFACE MODIFICATION

Size is a key factor in determining the utility of NMs in diverse applications. With the intention of using NMs in medicine, the size of the NMs can either increase or decrease not only therapeutic potential, but also the toxicity (Bondarenko et al., 2013; Borase et al., 2019). Being very small in size, NMs have greater surface to volume ratio than macro-particles. Therefore, relatively more molecules with medicinal importance can be conjugated on the nano-surface for targeted drug delivery (Wang et al., 2014).

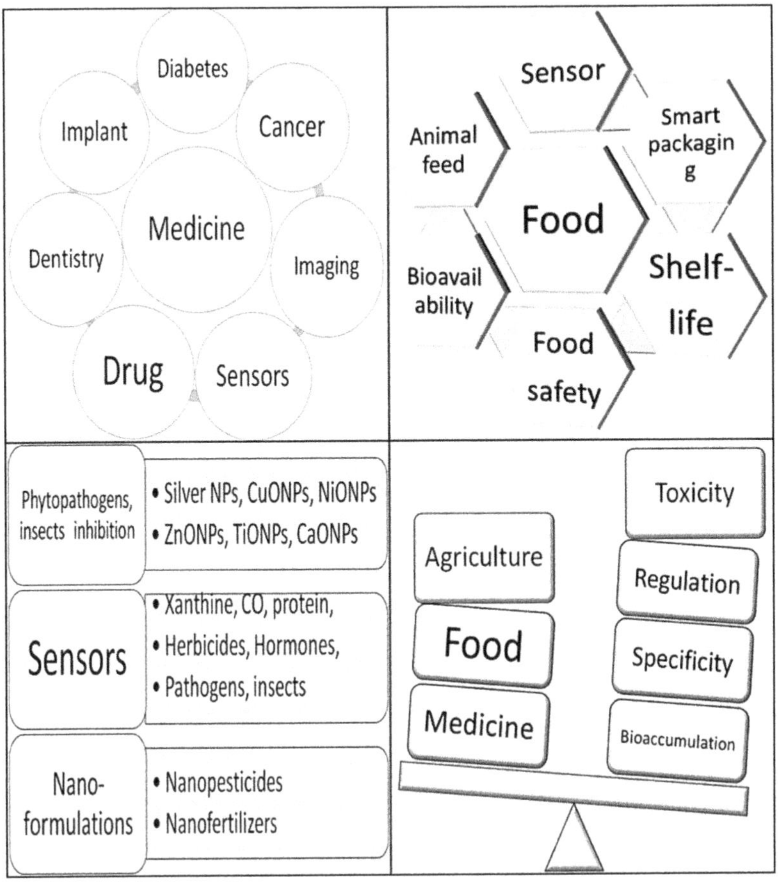

FIGURE 9.3 Applications of NMs and regulatory requirement for use of NMs in biology.

A 15 nm silica nanoparticles were conjugated with penicillin-G and this supramolecular complex have the inhibitory action on methicillin-resistant *Staphylococcus aureus* (MRSA) that is resistant to regularly prescribed antibiotics (Wang et a., 2016). While AuNPs functionalized with three components (paclitaxel derivatives, tumor necrosis factor alpha and thiolated PEG) found to be effective in tumor targeted drug delivery (Paciotti et al., 2017). The transfection efficiency of small interfering RNA (siRNA) was more with 50 nm sphere and 40 nm star shape AuNPs than that of 13 nm spheres. AuNPs, with size of 2 to 15 nm were coated with tiopronin (anticancer drug), it was found that small size NPs were more localized and accumulated in breast cancer cells then large size NPs (Navya et al., 2019).

But, because of this small size NMs can also able to cross cell membranes and enter (interfere?) with biological systems inside the body such as respiration, digestion, and circulatory system. This may be one of the reasons that nowadays researchers are more interested in exploring the toxic effect of NMs on human and non-target animals (Bondarenko et al., 2013; Borase et al., 2019; Holmes et al., 2020).

9.6.2 CHEMICAL COMPOSITION

The toxicity of NMs is largely depends on its synthesis methods. Chemical and physical methods utilized hazardous reducing and stabilizing agents such as sodium borohydrate, hydrazine (Gatoo et al., 2014; Rasmussen et al., 2018). Therefore, use of eco-friendly nanosynthesis methods (use of bacteria, fungi, and plant extracts) can be promising alternative as these methods use safe chemicals such as proteins, phenols, flavonoids for NMs synthesis (Chen et al., 2011; Borase et al., 2014).

9.6.3 SHAPE

Nanoparticles can be synthesized with different shapes such as spherical, triangular, hexagonal, etc. Although there is little definitive evidence, the health effects of nanoparticles are likely to depend also on their shape (Huang et al., 2017). Despite of the confirmed relationship between shape and toxicity, the exact mechanism remains to be elucidated. Rod shaped cerium oxide NPs were more toxic on RAW 264.7 cells as compared to octahedron or cubic shape particles (Forest et al., 2017). Rod shape iron oxide NPs are more toxic than spherical NPs, they exhibit high cytotoxicity by higher levels of LDH leakage, inflammatory response, ROS production, and necrosis (Lee et al., 2014). Based on several finding nowadays "safe by design" approach is popular regarding nanosynthesis.

9.7 REGULATIONS REGARDING USE OF NANOMATERIALS (NMS) IN BIOLOGY

In a broad sense, nanotechnology is making its impact on three aspects of human life, 1. Social (health, education, environment), 2. Ethical (misuse of technology), and 3. Legal (patents, regulations) (Kumar, 2014). Nanotechnology is dramatically improving human life. But, keeping in view of future safety issue that may arise because of global use of nanoparoducts; it is indispensable to develop strong regulatory and monitoring system for NMs (Mishra et al., 2019). The flow of information to formulate regulations over the use of nanotechnology arrived by three different stakeholders, (1) results published by research institutes, government organization, (2) legislation made to government by the heads of the research bodies, and (30 approval by the government and subsequent release of laws in the public domain (He et al., 2019). International and national regulatory agencies such as World Health Organization (WHO), Organization for Economic Co-operation and Development (OECD), European Commission (EC), Nanotechnology Research Coordination Group (NRCG) as well as non-government organizations and politicians urge to produce evidences regarding safety of NMs, permissible limits of NMs in consumer products as well as its environmental risk (Bondarenko et al., 2013; Pastrana et al., 2018). REACH (Registration, Evaluation, Authorization, and Restriction of Chemicals) was coming into action in 2007, it is the regulatory system adopted by European Union to analyze the risk to human health and the environment posed by any chemical used in industrial process as well as in daily life (ECHA, 2020). NMs also covered under REACH framework. Every industry has to demonstrate how the NMs in their product are safe to users. If the risk is more and unmanageable, the agency can restrict production and use of particular NMS and advice to find safe alternatives. These regulatory agencies are also emphasized on establishing sophisticated facilities for synthesis, characterization, and risk assessment of emerging nanoproducts. The general opinion of these regulatory agencies is that apart from the cost of NMs, the manufacturer should give equal attention to safety and alternative *in vitro* toxicity methods, thus minimizing the use of experimental animals. The OECD is improving some of its test guidelines to incorporate NMs under its umbrella and some NMs specific testing methods are formulated by EU observatory for NMs (EUON). EC with the help of the Scientific Committee on Consumer Safety (SCCS) is preparing a list of NMs for the risk assessment (Fytianos et al., 2020). According to EC, it is compulsory to declare the names of the chemical followed by the word (nano) in consumer products. In the United States of America (USA), National Nanotechnology Initiative (NNI) is prime organization working on sustainable development of nanotechnology while Food and Drug Administration (FDA) assigned task of regulating nano-medical products to different research organization such as center for drug evaluation and research (CDER), center for devices and radiological health (CDRH), center for biological evaluation and research (CBER) and several others

(Miller, 2002). FDA of Taiwan considers NMs as novel food contact agent and implements food packaging materials consisting of NMs to go through safety assessment and approval before commercial use (He et al., 2019).

India is the second most populated country and facing rapid depletion of biodiversity. India has an agriculture based economy and open consumer market for the world. Therefore, it is need of time to set up norms for nano-regulation owing to the rapid arrival of nanoproducts in Indian market from national and international manufacturer (Mishra et al., 2019). In India, it seems that we are quite late in development as well as regulatory aspects of nanotechnology as compared to US and European Union. Since the year 2000, the Government of India catalyzes pioneering work to nurture and promote nanotechnology (Kumar, 2014). In 2001, Nano-Science and Technology Initiative (NSTI) were established, followed by Nano-Mission in 2007. The Nano-Mission Council is top advisory and policymaking organization for nanotechnology in India (Srivastava & Chowdhury, 2006). In 2010, discussion was made to establish a Nanotechnology Regulatory Board, but little progress was made in this direction. Central Drugs Standard Control Organization (CDSCO), Food Safety and Standards Authority (FSSA), Central Pollution Control Board (CPCB), and National Environmental Protection Agency (NEPA) are working on different aspects of nanotechnology regulation.

9.8 CONCLUSIONS

The present chapter summarizes numerous applications of nanotechnology in medicine, food science, agriculture, cosmetics, and other fields of biological sciences. However, several future potentials of NMs in gene therapy, water purification, waste management, respirocytes, microbivores, and nanorobots needs further attention. Synergistic use of NMs with antibiotics, biopesticides, and essential oils can be highly beneficial to obtain the desired biological effect at lower doses. Apart from this, for sustainable development of nanobiotechnology, considerable research is required to answer the present and future toxicity issues on human health and the environment raised due to use of nanoproducts. Details study is also must to elucidate the mechanism of nano-toxicity, which will be useful to prepare safe NMs. Nanoproducts manufacturing industries must work out on the ecofriendly route of NMs synthesis with precision over size, shape, and stability. Regulatory agencies need to frame guideline by considering the risk-benefit assessment of NMs.

ACKNOWLEDGMENT

The authors are thankful to Uka Tarsadia University for giving them the necessary facilities. The authors are also like to thank the editors of the book for giving them an opportunity to write this book chapter.

DECLARATION OF COMPETING INTEREST

All authors of the present manuscript declare no conflicts of interest.

KEYWORDS

- **drug resistance**
- **food adulteration**
- **nanofertilizers**
- **nanoparticles**
- **nanosensors**
- **targeted delivery**

REFERENCES

Acharya, S., Dilnawaz, F., & Sahoo, S. K., (2009). Targeted epidermal growth factor receptor nanoparticle bioconjugates for breast cancer therapy. *Biomaterials, 30*, 5737–5750.

Allahverdiyev, A. M., Kon, K. V., Abamor, E. S., Bagirova, M., & Rafailovich, M., (2011). Coping with antibiotic resistance: Combining nanoparticles with antibiotics and other antimicrobial agents. *Expert Rev. Anti-Infe., 9*, 1035–1052.

Anjum, S., Anjum, I., Hano, C., & Kousar, S., (2019). Advances in nanomaterials as novel elicitors of pharmacologically active plant specialized metabolites: Current status and future outlooks. *RSC Advances, 9*, 40404–40423.

Askary, M., Amirjani, M. R., & Saberi, T., (2017). Comparison of the effects of nano-iron fertilizer with iron-chelate on growth parameters and some biochemical properties of *Catharanthus roseus*. *J. Plant Nutr., 40*, 974–982.

Bhat, P., & Bhat, A., (2016). Silver nanoparticles for enhancement of accumulation of capsaicin in suspension culture of *Capsicum* sp. *J. Expt. Sci., 29*, 1–6.

Bondarenko, O., Juganson, K., Ivask, A., Kasemets, K., Mortimer, M., & Kahru, A., (2013). Toxicity of Ag, CuO and ZnO nanoparticles to selected environmentally relevant test organisms and mammalian cells *in vitro*: A critical review. *Arch. Toxicol., 87*, 1181–1200.

Borase, H. P., Patil, C. D., Salunkhe, R. B., Suryawanshi, R. K., Salunke, B. K., & Patil, S. V., (2014). Phytolatex synthesized gold nanoparticles as a novel agent to enhance sun protection factor of commercial sunscreens. *Int. J. Cosmet. Sci., 36*, 571–578.

Borase, H. P., Patil, S. V., & Singhal, R. S., (2019). *Moina macrocopa* as a non-target aquatic organism for assessment of ecotoxicity of silver nanoparticles: Effect of size. *Chemosphere, 219*, 713–723.

Borase, H. P., Salunke, B. K., Salunkhe, R. B., Patil, C. D., Hallsworth, J. E., Kim, B. S., & Patil, S. V., (2014). Plant extract: A promising biomatrix for ecofriendly, controlled synthesis of silver nanoparticles. *Appl. Biochem. Biotechnol., 173*, 1–29.

Chau, C. F., Wu, S. H., & Yen, G. C., (2007). The development of regulations for food nanotechnology. *Trends Food Sci. Tech., 18*, 269–280.

Chen, D., (2020). Nanocontainers for the encapsulation and delivery of antioxidants/nutrients to food. *Smart Nanocontainers*, 119–136.

Chen, H., & Yada, R., (2011). Nanotechnologies in agriculture: new tools for sustainable development. *Trends Food Sci Technol., 22,* 585–594.

Cottet, V., Pariente, A., Nalet, B., Lafon, J., Milan, C., Olschwang, S., Faivre, J., et al., (2006). Low compliance with colonoscopic screening in first-degree relatives of patients with large adenomas. *Aliment Pharmacol Ther., 24,* 101–109.

Despres, L., David, J. P., & Gallet, C., (2007). The evolutionary ecology of insect resistance to plant chemicals. *Trends Ecol. Evol., 22*, 298–307.

Fakhouri, F. M., Casari, A. C. A., Mariano, M., Yamashita, F., Mei, L. I., & Soldi, V., (2014). "Effect of a gelatin-based edible coating containing cellulose nanocrystals (CNC) on the quality and nutrient retention of fresh strawberries during storage. In: *Proceedings of the IOP Conference Series: Materials Science and Engineering, Conference 1, 2nd International Conference on Structural Nano Composites (NANOSTRUC 2014)* (Vol. 64). doi: 10.1088/1757-899X/64/ 1/012024.

FAO, (2013). *Statistical Yearbook.* World Food and Agriculture, FAO, Rome.

Farzin, L., Shamsipur, M., Samandari, L., & Sheibani, S., (2020). HIV biosensors for early diagnosis of infection: The intertwine of nanotechnology with sensing strategies. *Talanta, 206*, 120201.

Forest, V., Leclerc, L., Hochepied, J. F., Trouvé, A., Sarry, G., & Pourchez, J., (2017). Impact of cerium oxide nanoparticles shape on their *in vitro* cellular toxicity. *Toxicol. in Vitro, 38*, 136–141.

Fytianos, G., Rahdar, A., & Kyzas, G. Z., (2020). Nanomaterials in cosmetics: Recent updates. *Nanomaterials, 10*, 1–16.

Gatoo, M. A., Naseem, S., Arfat, M. Y., Dar, A. M., Qasim, K., & Zubair, S., (2014). Physicochemical properties of nanomaterials: Implication in associated toxic manifestations. *Biomed. Res. Int.*, 498420.

Giese, B., Klaessig, F., Park, B., Kaegi, R., Steinfeldt, M., Wigger, H., Von, G. A., & Gottschalk, F., (2018). Risks, release and concentrations of engineered nanomaterial in the environment. *Sci. Rep., 8*, 1–18.

Global Cosmetics Market – Industry Trends and Forecast to 2026. https://www.databridgemarketresearch.com/reports/global-cosmetics-market# (accessed on 13 February 2022).

Hasan, M., Ullah, I., Zulfiqar, H., Naeem, K., Iqbal, A., Gul, H., Ashfaq, M., & Mahmood, N., (2018). Biological entities as chemical reactors for synthesis of nanomaterials: Progress, challenges and future perspective. *Mater. Today, Chem., 1*, 13–28.

He, X., Deng, H., & Hwang, H. M., (2019). The current application of nanotechnology in food and agriculture. *J. Food Drug Anal., 27*, 1–21.

Holmes, A. M., Mackenzie, L., & Roberts, M. S., (2020). Disposition and measured toxicity of zinc oxide nanoparticles and zinc ions against keratinocytes in cell culture and viable human epidermis. *Nanotoxicol., 14*, 263–274.

Hong, H., Shi, J., Yang, Y., Zhang, Y., Engle, J. W., Nickles, R. J., Wang, X., & Cai, W., (2011). Cancer-targeted optical imaging with fluorescent zinc oxide nanowires. *Nano Lett., 11*, 3744–3750.

https://echa.europa.eu/regulations/reach/understanding-reach (accessed on 13 February 2022).

https://www.statista.com/statistics/1073886/global-market-value-nanotechnology (accessed on 13 February 2022).

https://www.who.int/health-topics/cardiovascular-diseases#tab=tab_1 (accessed on 28 August 2023).

https://www.who.int/news-room/fact-sheets/detail/cancer (accessed on 13 February 2022).

Huang, Q. R., Yu, H. L., & Ru, Q. M., (2010). Bioavailability and delivery of nutraceuticals using nanotechnology. *J. Food Sci., 75*, R50–R57.

Huang, Y. W., Cambre, M., & Lee, H. J., (2017). The toxicity of nanoparticles depends on multiple molecular and physicochemical mechanisms. *Int. J. Mol. Sci., 18*, 2702.

Immordino, M. L., Dosio, F., & Cattel, L., (2006). Stealth liposomes: Review of the basic science, rationale, and clinical applications, existing and potential. *Int. J. Nanomed., 1*, 297–315.

Islam, S. M., & Karim, Z., (2019). World's demand for food and water: The consequences of climate change. In: *Desalination-Challenges and Opportunities*. IntechOpen. doi: http://dx.doi.org/10.5772/intechopen.85919.

Jere, D., Jiang, H. L., Kim, Y. K., Arote, R., Choi, Y. J., Yun, C. H., Cho, M. H., & Cho, C. S., (2009). Chitosan-graft-polyethylenimine for Akt1 siRNA delivery to lung cancer cells. *Int. J. Pharm., 378*, 194–200.

Kerch, G., (2015). Chitosan films and coatings prevent losses of fresh fruit nutritional quality: A review. *Trends Food Sci. Tech., 46*, 159–166.

Khodakovskaya, M., Dervishi, E., Mahmood, M., Xu, Y., Li, Z., Watanabe, F., Biris, A. S., & Biris, A. S., (2009). Carbon nanotubes are able to penetrate plant seed coat and dramatically affect seed germination and plant growth. *ACS Nano, 3*, 3221–3227.

Kole, C., Kole, P., Randunu, K. M., Choudhary, P., Podila, R., Ke, P. C., Rao, A. M., & Marcus, R. K., (2013). Nanobiotechnology can boost crop production and quality: First evidence from increased plant biomass, fruit yield and phytomedicine content in bitter melon (*Momordica charantia*). *BMC Biotechnol., 13*, 3–7.

Kottegoda, N., Munaweera, I., Madusanka, N., & Karunaratne, V., (2011). A green slow-release fertilizer composition based on urea-modified hydroxyapatite nanoparticles encapsulated wood. *Current Sci., 101*, 73–78.

Kumar, A., (2014). *Nanotechnology Development in India: An Overview*. Research and Information System for Developing Countries.

Lee, J. H., Ju, J. E., Kim, B. I., Pak, P. J., Choi, E. K., Lee, H. S., & Chung, N., (2014). Rod-shaped iron oxide nanoparticles are more toxic than sphere-shaped nanoparticles to murine macrophage cells. *Environ. Toxicol. Chem., 33*, 2759–2766.

Lee, J. H., Kim, B. C., Oh, B. K., & Choi, J. W., (2013). Highly sensitive localized surface plasmon resonance immunosensor for label-free detection of HIV-1. Highly sensitive

localized surface plasmon resonance immunosensor for label-free detection of HIV-1. *Nanomedicine: NBM., 9*, 1018–1026.
Miller, J., (2002). Beyond biotechnology: FDA regulation of nanomedicine. *Colum. Sci. Tech. L. Rev., 4*, 1–10.
Mishra, M., Dashora, K., Srivastava, A., Fasake, V. D., & Nag, R. H., (2019). Prospects, challenges and need for regulation of nanotechnology with special reference to India. *Ecotoxicol. Environ. Saf., 171*, 677–682.
Mohammed, F. A., Balaji, K., Girilal, M., Kalaichelvan, P. T., & Venkatesan, R., (2009). Mycobased synthesis of silver nanoparticles and their incorporation into sodium alginate films for vegetable and fruit preservation. *J. Agric. Food Chem., 57*, 6246–6252.
Morganti, P., & Morganti, G., (2008). Chitin nanofibrils for advanced cosmeceuticals. *Clin. Dermatol., 26*, 334–340.
Nagamune, T., (2017). Biomolecular engineering for nanobio/bionanotechnology. *Nano Converg., 4*, 9–17.
National Center for Emerging and Zoonotic Infectious Diseases (NCEZID). *Division of Foodborne, Waterborne, and Environmental Diseases (DFWED).*
Nayak, A. P., Tiyaboonchai, W., Patankar, S., Madhusudhan, B., & Souto, E. B., (2010). Curcuminoids-loaded lipid nanoparticles: Novel approach towards malaria treatment. *Colloid Surface B, 81*, 263–273.
Nile, S. H., Baskar, V., Selvaraj, D., Nile, A., Xiao, J., & Kai, G., (2020). Nanotechnologies in food science: Applications, recent trends, and future perspectives. *Nano-Micro Lett., 12*, 1–34.
Ozturk, B., Argin, S., Ozilgen, M., & McClements, D. J., (2014). Formation and stabilization of nanoemulsion-based vitamin E delivery systems using natural surfactants: Quillaja saponin and lecithin. *J Food Eng., 142*, 57–63.
Pacheco, G., De Mello, C. V., Chiari-Andréo, B. G., Isaac, V. L., Ribeiro, S. J., Pecoraro, É., & Trovatti, E., (2018). Bacterial cellulose skin masks—Properties and sensory tests. *J. Cosmet. Dermatol., 5*, 840–847.
Paciotti, G. F., Zhao, J., Cao, S., Brodie, P. J., Tamarkin, L., Huhta, M., Myer, L. D., Friedman, J., & Kingston, D. G., (2016). Synthesis and evaluation of paclitaxel-loaded gold nanoparticles for tumor-targeted drug delivery. *Bioconjugate Chem., 27*, 2646–2657.
Pastrana, H., Avila, A., & Tsai, C. S. J., (2018). Nanomaterials in cosmetic products: The challenges with regard to current legal frameworks and consumer exposure. *Nano Ethics, 12*, 123–137.
Patil, C. D., Borase, H. P., Suryawanshi, R. K., & Patil, S. V., (2016). Trypsin inactivation by latex fabricated gold nanoparticles: A new strategy towards insect control. *Enzy. Microbial Technol., 92*, 18–25.
Patil, S. V., Borase, H. P., Patil, C. D., & Salunke, B. K., (2012). Biosynthesis of silver nanoparticles using latex from few *Euphorbian* plants and their antimicrobial potential. *Appl. Biochem. Biotech., 167*, 776–790.
Piccinno, F., Gottschalk, F., Seeger, S., & Nowack, B., (2012). Industrial production quantities and uses of ten engineered nanomaterials in Europe and the world. *J. Nanoparticle Res., 14*, 1109–1123.
Ranzoni, A., Sabatte, G., Van, I. L. J., & Prins, M. W., (2012). One-step homogeneous magnetic nanoparticle immunoassay for biomarker detection directly in blood plasma. *ACS Nano, 6*, 3134–3141.

Rasmussen, K., Rauscher, H., Mech, A., Sintes, J. R., Gilliland, D., González, M., Kearns, P., et al., (2018). Physico-chemical properties of manufactured nanomaterials – Characterization and relevant methods. An outlook based on the OECD testing program. *Regul. Toxicol. Pharmacol., 92*, 8–28.
Salunkhe, R. B., Patil, S. V., Salunke, B. K., Patil, C. D., & Sonawane, A. M., (2011). Studies on silver accumulation and nanoparticle synthesis by *Cochliobolus lunatus*. *Appl. Biochem. Biotechnol., 165*, 221–234.
Santos, A. C., Panchal, A., Rahman, N., Pereira-Silva, M., Pereira, I., Veiga, F., & Lvov, Y., (2019). Evolution of hair treatment and care: Prospects of nanotube-based formulations. *Nanomaterials, 9,* 903–915.
Sari, T. P., Mann, B., Kumar, R., Singh, R. R., Sharma, R., Bhardwaj, M., & Athira, S., (2015). Preparation and characterization of nanoemulsion encapsulating curcumin. *Food Hydrocoll., 43*, 540–546.
Shit, S. C., & Shah, P. M., (2014). Edible polymers: Challenges and opportunities. *J. Polymers, 2014*.
Singh, S. G., & Tripathy, S., (2017). Cardiac biomarker detection based on nanostructured biosensors. *J. Med. Sci. Res., 5*, 66–71.
Singh, T., Shukla, S., Kumar, P., Wahla, V., Bajpai, V. K., & Rather, I. A., (2017). Application of nanotechnology in food science: Perception and overview. *Front. Microbiol., 8*, 1–7.
Sperber, W. H., & Doyle, M. P., (2009). Introduction to the microbiological spoilage of foods and beverages. In: *Compendium of the Microbiological Spoilage of Foods and Beverages* (pp. 1–40). Springer International Publishing, New York.
Srivastava, N., & Chowdhury, N., (2011). Nanotechnology and state regulation (India). Maclurcan, D., & Radywyl, N., (eds.), *Nanotechnology and Global Sustainability, 12*, 1–12. CRC Press.
Surendiran, A., Sandhiya, S., Pradhan, S. C., & Adithan, C., (2009). Novel applications of nanotechnology in medicine. *Indian J Med Res., 130*, 689–701.
Tahoori, F., Ahmad, M. A., Nejadsattari, T., Ofoghi, H., & Iranbakhsh, A., (2019). Qualitative and quantitative study of quercetin and glycyrrhizin in *in vitro* culture of liquorice (*Glycyrrhiza glabra* L.) and elicitation with $AgNO_3$. *Not. Bot. Horti. Agrobot. Cluj-Napoca, 47*, 143–151.
Tan, C., Xue, J., Abbas, S., Feng, B., Zhang, X., & Xia, S., (2014). Liposome as a delivery system for carotenoids: Comparative antioxidant activity of carotenoids as measured by ferric reducing antioxidant power, DPPH assay and lipid peroxidation. *J. Agric. Food Chem., 62*, 6726–6735.
Tarafdar, J. C., Agrawal, A., Raliya, R., Kumar, P., Burman, U., & Kaul, R. K., (2012). ZnO nanoparticles induced synthesis of polysaccharides and phosphatases by *Aspergillus fungi*. *Adv. Sci. Eng. Med., 4*, 1–5.
Vecchio, Y., Agnusdei, G. P., Miglietta, P. P., & Capitanio, F., (2020). Adoption of precision farming tools: The case of Italian farmers. *Int. J. Environ. Res. Public Health, 3*, 1–11.
Ventola, C. L., (2017). Progress in nanomedicine: Approved and investigational nanodrugs. *P & T J. Formul. Manag., 42*, 742–755.
Wang, L., Chen, Y. P., Miller, K. P., Cash, B. M., Jones, S., Glenn, S., Benicewicz, B. C., & Decho, A. W., (2014). Functionalized nanoparticles complexed with antibiotic efficiently kill MRSA and other bacteria. *ChemComm., 50*, 12030–12033.

Wang, Y., Ravindranath, S., & Irudayaraj, J., (2011). Separation and detection of multiple pathogens in a food matrix by magnetic SERS nanoprobes. *Anal. Bioanal. Chem., 399*, 1271–1278.

Wei, Q., Ji, J., Fu, J., & Shen, J., (2007). Norvancomycin-capped silver nanoparticles: Synthesis and antibacterial activities against *E. coli. Sci. China Ser. B., 50*, 418–424.

Wooley, K. L., (2000). Shell crosslinked polymer assemblies: Nanoscale constructs inspired from biological systems. *J. Polym. Sci. Pol. Chem., 38*, 1397–1407.

Yang, H., Qu, L., Wimbrow, A. N., Jiang, X., & Sun, Y., (2007). Rapid detection of *Listeria monocytogenes* by nanoparticle-based immunomagnetic separation and real-time PCR. *Int. J. Food Microbiol., 118*, 132–138.

Ye, F., Barrefelt, Å., Asem, H., Abedi-Valugerdi, M., El-Serafi, I., Saghafian, M., Abu-Salah, K., et al., (2014). Biodegradable polymeric vesicles containing magnetic nanoparticles, quantum dots and anticancer drugs for drug delivery and imaging. *Biomaterials, 35*, 3885–3894.

Ye, F., Zhao, Y., El-Sayed, R., Muhammed, M., & Hassan, M., (2018). Advances in nanotechnology for cancer biomarkers. *Nano Today, 18*, 103–123.

Zhang, B., Zheng, L. P., Yi, L. W., & Wen, W. J., (2013). Stimulation of artemisinin production in *Artemisia annua* hairy roots by Ag-SiO_2 core-shell nanoparticles. *Curr. Nanosci., 9*, 363–370.

Zhang, L., Wu, L. L., Wang, Y. P., Liu, A. M., Zou, C. C., & Zhao, Z. Y., (2009). Melamine-contaminated milk products induced urinary tract calculi in children. *World J. Pediatr., 5*, 31–35.

Zhang, W. H., & Zhang, W. D., (2008). Fabrication of SnO_2–ZnO nanocomposite sensor for selective sensing of trimethylamine and the freshness of fishes. *Sens. Actuators B Chem., 134*, 403–408.

CHAPTER 10

Clinical Toxicities of Nanomaterials and Remediation

HOSAM M. SALEH and AMAL I. HASSAN

Radioisotope Department, Nuclear Research Center, Egyptian Atomic Energy Authority, Cairo, Egypt

ABSTRACT

Nanoscience has developed considerably throughout the past decade because it has transitioned from high science to applied technology. There are few studies of the fate of nanoparticles and their path in human tissues or organs. Thus, that there are searches to date have addressed the mechanisms that prevent these particles from buildup in the vital organs, and by these ways the body can benefit from it as well as, we can prevent their negative impacts on human health.

Because nanomaterials (NMs) are part of the growing daily consumer products, especially in producing processes and medical merchandise, it's necessary to safeguard each employee and end-users from inhalation of such substances attributable to their potential toxicity. One of the principles of toxins is that the risk of harm is related to the seriousness and exposure of the substance and also the vulnerability of the individual. The assessment of such criteria is complicated by the large diversity of nanoparticles, as well as the unique properties of nanoparticles because of their small size and large surface area, which would lead to a change in physical properties, as well as added chemicals. Procedures for the analysis and measurement of specific nanoparticles in media such as air, water, soil, tissue, blood, and urine should be developed and validated. It should be taken into account that there are properties of nanoparticles that exceed the concentration and that include size, shape, surface electrical charges, crystalline structure, surface chemistry,

Advanced Nanomaterials for Biological, Nutraceutical, and Medicinal Applications.
Sreeraj Gopi, Augustine Amalraj, and T. R. Sreeraj (Eds.)

and conversion, as well as chemical coating. More importantly, the ratio between its small size and its large surface area means that NMs have unique physical-chemical properties compared to similar materials of larger size. The challenge is to determine how NMs interact with vital organs. The classic question relates to its survival and bioaccumulation in animals, humans, and the environment, as well as, may interfere with the genetic material in the cell or cell proteins, leading to an imbalance in the normal functions of the cell. It is necessary to develop a set of criteria to assess the potential risk of nanoparticles. It is important to recognize biological interactions as well as the biocompatibility between vital components within the body and nanoparticles to give a clear picture of the long-term usage of such techniques. In this chapter, we will present the clinical toxicities of NMs, the mechanisms of these toxicities, and how to overcome them. Furthermore, we also outline and present the therapeutic applications of some of these nanoparticles.

10.1 TOXICITY OF NANOPARTICLES

Nanomedicine is a subset of nanotechnology and a new field of science-based on physical and chemical properties. Nanomedicine is an important application of nanotechnology for the diagnosis, treatment or prevention of diseases. This discovery in the field of medical science has helped many developments in the medical field, for example, the delivery of medicine and genes, the discovery of tumors and imaging as well as tissue engineering (TE) (Yang et al., 2017).

Because nanomaterials (NMs) are regarded as suitable vectors because they possess a large proportion of the surface area, it is possible to control the absorbed and chemically related signals on the surface very effectively through the conveyor materials. So, the tiny size and large surface area are unique properties of NMs that make them perform superb functions in the field of biosciences and medicine. But the tiny size is a double-edged sword where NMs can easily penetrate the human body to the places to be treated, while at the same time they can be inhaled, swallowed, or easily penetrated into the skin, which may cause dysfunction within the tissues or penetrated cells for a long time (Molino & Wang, 2014). The applied nanoscience has expedited the manufacture of advanced structures in drug development. However, to use the potential of this technology in pharmaceutical drug development, attention is required to safely and toxicological studies, as, for a few formulations, analysis is required to stop unsought effects and hazards that are needed to safety (Onoue et al., 2014; Schuhmacher et al., 2016).

The previous manners of penetrating NMs of the biological system and generating undesirable results considered as toxic to this system. There is restricted knowledge of the toxicity and fate of nanoparticles within the body is limited, but the size and shape of these particles emphasizes toxicity (Yang et al., 2017). Multiplied production and deliberated exposure (sunscreens, vector medicines) or unwitting exposures whether through the environment or professional work, increases the likelihood of a negative impact on the health of the exposed. It is vital that NMs used in sustainable medical applications should be safe for this purpose. Not all toxicological studies so far deal with nanoparticles as recently defined (size <100 nanometers) or nanoparticles are characterized by modern knowledge (Kumar et al., 2017). However, this does not necessarily contradict the conclusions reached in these studies.

The indiscriminate use of natural rates of engineered particles with unknown toxic properties may pose a variety of risks to the environment, wildlife, and human health. It is important to investigate their transition to the environment and its impact on environmental health. The harmful effects of nanoparticles are still very limited, and no specific regulations have been developed for the use of NPs (Jamil et al., 2018). Because nanomedicine and nanotoxins are two sides of the same coin, the value of this coin depends on their wise use. There is a potential risk of human and environmental exposure to nanoparticles throughout their life cycle, from manufacturing to disposal (Figure 10.1). The cytotoxic effects on the lung, heart, fertility, and kidneys as well as the skin have been documented (Zhang et al., 2018).

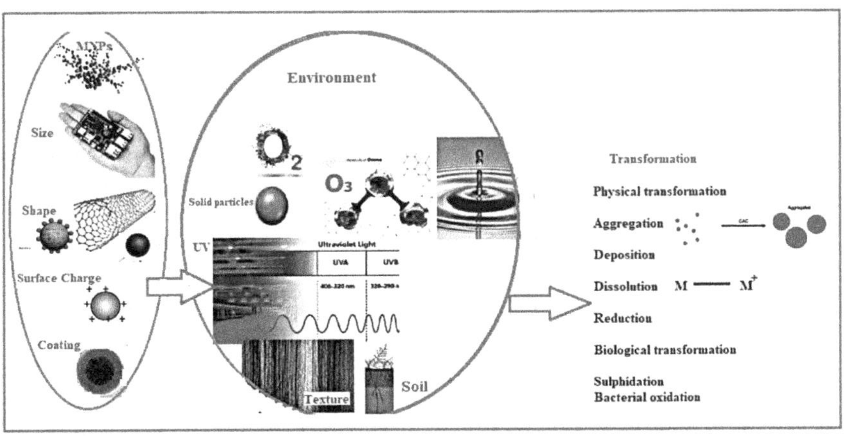

FIGURE 10.1 The potential toxic effects of nanoparticles and mechanisms for the conversion of these effects from the environment to living organisms.

NP toxicity refers to the ability of these molecules to negatively affect the natural physiology of organs and tissues of humans and animals. It is widely known that this toxicity depends on the physical and chemical factors of such particles, such as size, shape, surface charge, chemistry, composition, and subsequent stability of nuclear energy rates (Sohal et al., 2018).

The exact basic mechanism is not yet known; however, recent literature suggests that nanoparticles can infiltrate the human immune system and stimulate cytokines that lead to infections and can also cross the blood-brain barrier (BBB) (Meneses et al., 2019). In addition to other factors associated with cytotoxicity, such as the dose given, the method of administration and the extent of tissue distribution are important factors in determining cytotoxicity. Carbon nanotubes are more harmful than quartz dust that causes silicosis, is a disease that arises from professional activities. This usually occurs in people whose activities are related to the metal, porcelain, engineering, and mining industries (Poland & Duffin, 2019).

Environmentalists emphasize that the dangerous thing is that the uses of this technology have begun to leak into the field of food and agricultural sector, without the knowledge of consumers or even open the door for community discussion about it and conduct adequate tests on its safety. A new experiment from the University of Rochester conducted on mice breathed nanoparticles and was later found to have settled in the brain and lungs, leading to serious health complications. It was found that nanoparticles would be able to spread to other parts of the body. However, knowledge about the harmful effects of this is very limited. The effect of nanoparticles on the airways, factors that ultimately affect them, such as the properties of molecules (composition, size, and solubility), and the effect of bronchial formation on them have been studied and show that unintentional absorption of substances. Nanoparticles can be performed by inhalation, or through the mouth, or through the skin. And through the eye and also other ways we can imagine in the future. Inhalation is currently one of the most important methods of exposure to fine particles, and the first research in the present.

There are two points associated with this issue: (i) Free nanoparticles are released into the air or water during the production process (or production accidents) or as waste from the production process and then accumulated in soil, water, and plant life. (ii) In their fixed state, when they form part of the finished material or finished product, they must then be recycled or disposed of as waste (Westerhoff et al., 2013). It remains to be seen whether nanoparticles will form a new layer of fully biodegradable pollutants. If this is done, it is also unknown how to dispose of these pollutants and remove them from the air or water because conventional filters are not suitable for such tasks as their pores are too large to catch nanoparticles.

It can be said that the impact of NMs on human health is a great danger as nanoparticles are able to enter the human body very easily through the pores and without any resistance and can spread within the body, which is harmful to human health. Imagine that a 300-nanometer particle can easily enter the human body and that a 70-nanometer particle can easily enter the nucleus of the cell. This indicates a great danger to humans. It changes the properties of the cell or its poisoning and death. The Oxford University study showed that nano-titanium dioxide molecules found in sunscreen ointments damaged the skin's DNA.

Despite the theoretically stunning logic, animal, and human studies have highlighted the possibility of linking organ toxicity to a predetermined dose; the extent to which an efficiently administered NP dose to target tissues is determined (Westerhoff et al., 2013). Cytotoxicity-based studies use increased doses of NP in order to monitor the toxicity of dose-associated cells or tissues. These correlations between dose and response are the basis for determining safe limits for particle concentrations in vivo administration.

It was found that these nanoparticles run out of the bloodstream via diploid membranes and enter organelles such as mitochondria and nuclei, which can cause oxidative stress and damage to deoxyribonucleic acid (DNA). Numerous studies using animals or cellular farms have reported oxidative stress, inflammatory responses and rupture of cell membranes due to lipid oxidation following exposure to nanoparticles. In addition to the lungs, the skin forms a surface for potential imaging following skin exposure (as for cosmetics, sunscreens, and nanoparticles). Studies have shown that safe skin effectively protects the body from nanoparticles (TiO_2 in sunscreens). However, particles designed for very specific purposes may actually run out of it, and therefore there is no comprehensive conclusion regarding access through the skin. Access from the dermis to the lymphatic and circulatory system is possible or possible. In the case of inflammatory or traumatic lesions, there is a high probability of access to the skin because this has been proven even for larger particles.

The oral ingestion of NMs has been properly tested to date. According to some scientific studies, once ingested, they are emptied efficiently through the intestine for small particles (<100 nm). Increased metabolism has been observed by the intestinal wall in rats. A number of studies have shown that some NMs are transmitted directly and efficiently from adipose neurons to the central nervous system, exceeding the BBB. Data on translocation among members are based on various methods and therefore cannot be considered as yet confirmed. Various studies report that modified nanoparticles cross the BBB. But there are no reports yet in the public literature on whether the blood-brain or placental barrier can be accessed first, but he thinks this is possible, given the nanometric size of the particles.

10.1.1 TECHNIQUES OF NANOPARTICLE PERNICIOUSNESS

The toxicity of NPs is determined by their physical and chemical properties, such as their size, shape, specific area, surface charge, and chemical activity. The small size of NPS allows it to penetrate through animal tissue and epithelial barriers within the body's liquid and blood material to be transported through the blood and lymphocytes to many organs and tissues, such as the brain, heart, liver, kidney, spleen, bone marrow (Panzarini et al., 2018), and are then transferred to the cells by transcytosis mechanisms or unfold only across the semipermeable membrane (De Matteis, 2017; McCracken et al., 2016; Panzarini et al., 2018). Some nanoparticles will permeate the skin due to their tiny size (Nair et al., 2013), even the biggest microparticles will penetrate the skin although they're bent (Su et al., 2017). Nanoparticles, due to their tiny size, will leak through the liner in inflammatory sites, like animal tissue (e.g., intestinal tract and liver), tumors, or microvascular infiltration (Mauricio et al., 2018). Experiments conducted on the poisonous effects of NPs on the body showed that these particles cause coagulation by promoting platelets buildup (Sharifi et al., 2012). Additionally, disorders related to NPs toxicity, the respiratory system inflammation, degeneration of the nervous system, especially the neurons in the brain, stroke, myocardial infarct, and different disorders (Ramanathan et al., 2018). NPs enter the organs, tissues, and cells, as well as cell organelles, e.g., mitochondria, and nuclei; this could drastically alter cell metabolism and cause DNA lesions, mutations, and necrobiosis that lead to apoptosis (Wang et al., 2018; Chen et al., 2019). To study the toxicity of NPs, it is necessary to conduct analyzes *in vitro* and *in vivo* (Sahu & Hayes, 2017). The laboratory analysis can guide us about many harmful effects at the cellular level. From this knowledge, one will acquire data regarding cell death, organs injury, DNA damage, oxidation, mutations, and down or up-regulation of protein (Sesti et al., 2012). While, *in vivo* assessment should be carried out in an animal model, this technique is much more complex than the cell line (Di Meo et al., 2016). In addition, we can study several other indicators that represent public health, such as behavioral abnormalities, weight loss, mortality, and average life expectancy (Figure 10.2).

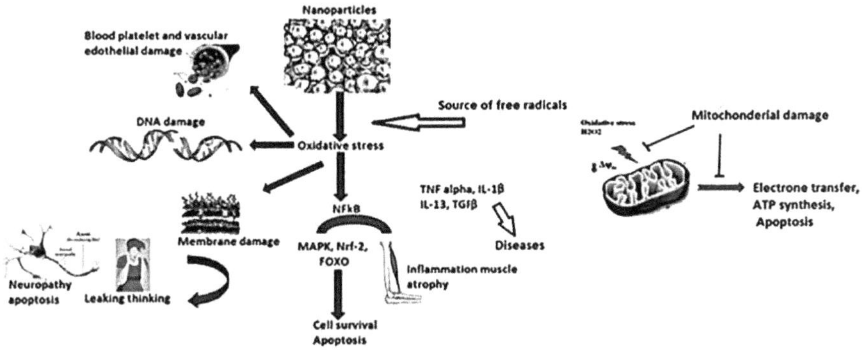

FIGURE 10.2 Toxicity arises by damaging membranes, blood elements, DNA, and mitochondria or by releasing a cell signal (NFkB, MAPK, pro induction, etc.). It should be noted that these abnormalities cause abnormalities in coagulation, apoptosis, neuropathy, mindfulness, etc., leading to diseases.

Figure 10.2 illustrates the development of toxicity caused by the nanoparticles and the cascade of damages inside the cell. Oxidation resulted from the toxicity of NPs affects forthwith platelet damage, vascular endothelial damage, DNA damage, membrane damage, NFkB stimulating signals, pro-cytokine expression, mitochondrial injury, leading to blood clotting abnormalities βof the event, DNA damage, and cell death.

10.1.2 TOXICITY OF NANOPARTICLES IN VITRO

The vast majority of *in vitro* studies on nanoparticle toxicity have used cell lines outside *in vivo*. However, the results of the nanoparticle toxicity test in cell lines or primary cells have shown inconsistencies, highlighting the need to expand the use of primary cells for *in vitro* assays (Joris et al., 2013; Bakand & Hayes, 2016; Bessa et al., 2020).

The use of cell lines in monoculture systems is suggested for the primary stage nanoparticles (NPs) hazard analysis as they permit a far better liableness and equivalence for lay laboratory comparison of either identical NP sort or among totally different NPs (Bahadar et al., 2016; Armand et al.,

2016). Choice of cell lines powerfully depends on the analysis question and therefore the investigated endpoints. For the next tier analysis, the employment of primary human cells and/or co-culture models are most well-liked to raised perceive NM mechanistic behavior in additional advanced systems; but such systems can be troublesome to use in high-throughput assays thanks to cost and reliability (Charles et al., 2018).

Recently, cell lines models for the study of cytotoxic properties of built nanoparticles are valuable tools in cancer analysis. Several techniques and ways are promptly on the market to check the properties of nanoparticles with respect to selective toxicity and/or anti-propagation effects within the cancer treatment field (Concu et al., 2017). However, setting up these techniques requires careful monitoring, coordination of a wide range of available methods is necessary to compare screening and replication. Data from cancer cellular models are useful for further research. Individual or basic laboratory capacities play a role in the selection and availability of technologies (Figure 10.3). As well as a spread of cell lines and a variety of metabolic strategies concerned in cell responses offer acceptable laboratory cell models to check the biological impact of nanoparticles at the cell level and a true approach to more *in vivo* and clinical studies (Table 10.1) (Gioria et al., 2016).

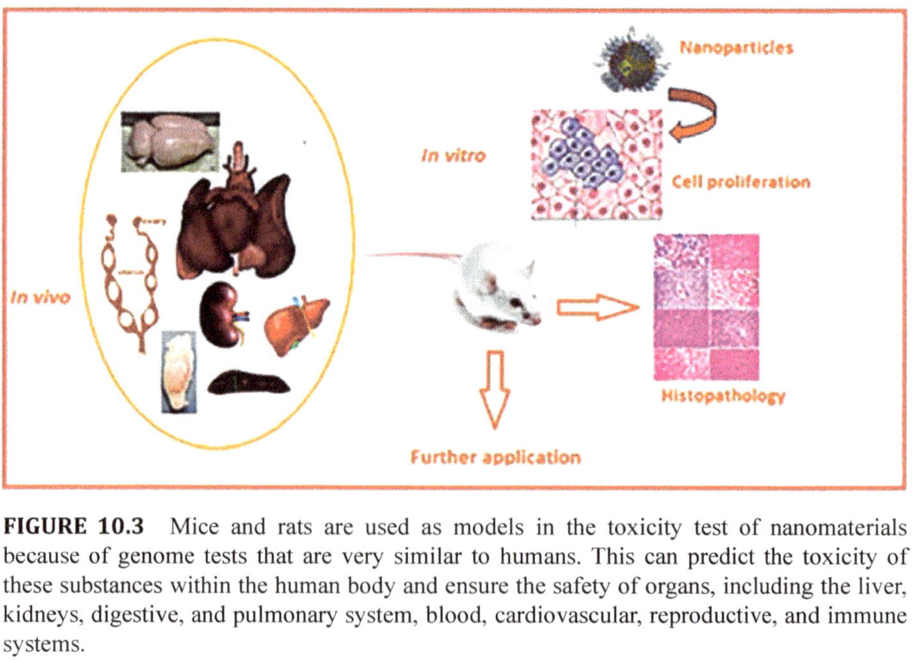

FIGURE 10.3 Mice and rats are used as models in the toxicity test of nanomaterials because of genome tests that are very similar to humans. This can predict the toxicity of these substances within the human body and ensure the safety of organs, including the liver, kidneys, digestive, and pulmonary system, blood, cardiovascular, reproductive, and immune systems.

TABLE 10.1 Some Measurements on Nanotoxicity *In Vitro* Tests, and Cell Types Used for This Purpose

Name of Examination	The Type of Nanoparticles Used	The Cellular System Used	Documentation by References
Microculture Tetrazolium Test (MTT): 3-(4,5-Dimethylthiazol-2-yl)-2,5-diphenyl-tetrazolium bromide.	Quantum dots	Human lung	Grabowska-Jadach et al. (2018)
Test Principle: Which depends on measuring the activity of living cells that in turn dehydrogenate the compound 5655-M by 3-[4,5-dimethylthiazol-2-yl]-2,5-diphenyl tetrazolium bromide to the water-insoluble formazan violet product.	Titanium dioxide (TiO_2)	Human colorectal cancer and umbilical vein endothelial cell lines	Kukia et al. (2018)
Natural red	TiO_2 NPs	Zebrafish embryos	Tang et al. (2019)
Lactate Dehydrogenase (LDH) Testing: to determine the activity of lactate hydrogenase (LDH) from damaged or dead cells in the medium. The enzyme lactate hydrogen is known as the cytoplasm enzyme found in all cells. When the cells are exposed to toxic effects, the integrity of the plasma membrane is disrupted and the LDH enzyme leaks out of the cells and passes to the medium. Thus, the damage is assessed by measuring LDH enzyme activity after exposure.	TiO_2 NPs	Lung epithelial cells	Han et al. (2011)
	CNTs	Human pneumocystis' cells	–
MTS Assay: (3-(4,5-dimethylthiazol-2-yl)-5-(3 carboxymethoxyphenyl)-2-(4-sulfophenyl)-2H-tetrazolium) assay, viable cells will convert tetrazolium salt into a colored soluble formazan product by mitochondrial dehydrogenase enzymes.	Ag NPs	Mouse embryonic fibroblasts	Lee et al. (2014)
	Gold NPs (Au-NPs)	Mammalian cells	Caballero-Díaz & Valcárcel (2014)
Trypan blue dye (Tb)	Au-NPs	Mouse fibroblast	Iswarya et al. (2016)
	TiO_2 NPs	Human lung epithelial cells	Grabowski et al. (2013)

Nanoparticles exhibit many distinct chemical properties because they overlap with other natural properties that make them unique and at the same time make us cautious in dealing with them (Kroll et al., 2009). Such properties include mechanical properties are at the forefront, the reduction of particle size and the presence of large numbers of atoms on their outer surface, resulting in increased surface assimilation capacity. Then completely different optical properties that interfere with visible light or visible radiation absorption detection systems; increased chemical action activity because of enhanced surface energy and magnetic properties that create them to reduce active oxidation thus interfere with strategies that act as catalysts which react strongly with toxic gases; accelerate the oxidation-reduction active (Kroll et al., 2009). Previous studies have shown that the natural defense mechanisms of the human body treat nanoparticles as living organisms, and these particles can bind to each other to form large fibers that cannot be swallowed by white blood cells (macrophages). In addition, nanoparticles with a high content of transition metals with active surfaces can bind to other particles; their ability to damage and damage cells changes.

A study by Sharma et al. (2012) has incontestable that TiO_2 nanoparticles bind to LDH, and consequently, TiO_2 nanoparticle-induced toxicity might be estimated by the LDH activity assay. Titanium dioxide (TiO_2) powders are widely used in the production of dyes, coatings, plastics, paper, and plastics industries, producing large quantities of them globally, and away from the traditional uses. For example, titanium grains with diameters of less than 50 nanometers exhibit superior UV blocking but do not block light from passing through (Tsuzuki & Wang, 2010). It does not cause opacity when used in the purposes of prevention and protection against scorching ultraviolet (UV) rays, and therefore it is now successfully used in the manufacture of skin creams and cosmetics for sun protection, as well as nanoparticles in the manufacture of lacquer paint wood surfaces to maintain, as well as in the manufacture of fibers Textile and toothpaste industry (Lu et al., 2006).

Titanium nanoparticles are one of the most powerful photocatalysts that have oxidizing surfaces, and therefore fight bacterial presence and other organic compounds such as dirt, mildew, and bacteria, if their surfaces are exposed to sunlight or even to a light source such as a fluorescent lamp and due to these unique properties (Tayel et al., 2018).

Cell size, proliferation rate, growth characteristics and expression of the relevant surface receptors are the most important biological factors necessary for the absorption of NPs *in vitro* (Drasler et al., 2017). The absorption of NPs happens via internal pathways, like activity involving bodily process of huge molecules or NP complexes (> 0.25 μm in diameter) and bodily process as well as bodily process of little particles (< 0.15 μm in diameter) (Bastús et al., 2009). The mechanisms of cellular absorption of nanoparticles are numerous: pinocytosis, endocytosis, and receptor-mediated endocytosis (Mahmoudi et al., 2011). The endocytosis approach is the most important pathway for the absorption of nanoparticles in cells, because nanoparticles as they enter *in vivo* are coated with plasma proteins. Phagocytosis occurs when the material entering the cells is greater than 0.5 microns (Baek et al., 2012; Oh & Park, 2014). After absorption of the nanoparticles *in vivo*, they spread in the bloodstream and enter into many types of cells through the plasma membrane. If we look at the plasma membrane, we will find it is a permeable membrane, but at the same time selective transports the necessary materials that maintain the continuity of life of the cell. Thus, the plasma membrane would determine the pathways of nanoparticle molecules according to their chemical and physical properties. It is important to understand cellular absorption and organ distribution of nanoparticles when examining targeting and therapeutic efficacy in drug delivery applications. Nanoparticles administered in the body are eventually cleared by macrophages of the mononuclear phagocyte system in the liver and spleen. However, it is known from previous studies that nanoparticles remain in these organs for a long time, which makes the possibility of nanotoxicity persists (Gustafson et al., 2015). Therefore, it is also important to study the disposal of these inner nanoparticles in many cell types, to assess biosafety. However, studies on the disposal of nanoparticles after they enter the *vivo* are very scarce.

Metallic nanoparticles (MNPs), especially silver nanoparticles (AgNPs), have been known to have powerful toxic effects on a wide range of microorganisms such as Staphylococcus genus isolated from the air at concentrations on top of 30 $\mu g/ml^{-1}$. According to these broad-spectrum antimicrobial properties, silver has been widely used for biomedical applications and other environmental disinfection processes for centuries (Yuan et al., 2017). AgNP have a strong interaction with microorganisms because of their high surface area. Thus, these nanoparticles attack the bacterial surface, by penetrating these organisms and disrupting the vital functions of the organism due to a change in the membrane fluidity, thereby increasing cell permeability (Mpenyana-Monyatsi et al., 2012). Bacterial membrane proteins contain sulfur and phosphorus in its structure. AgNP such as silver ions can interact with these proteins and inhibit DNA functions by interacting with these chemicals. AgNP or silver ions can attack the respiratory tract of bacterial mitochondria and lead to cell death (Le Ouay & Stellacci, 2015). Additionally, the binding of nanoparticles to the bacterial cell wall may be due to negative charges and specific functional groups on the bacterial surface (Figure 10.4). After penetrating into the bacterial cell, AgNPs may disturb the rigidity of the cell wall or lipopolysaccharides, and enzymes, the transport system, and the generation of H_2O_2 that result in bacterial death (Jasuja et al., 2014).

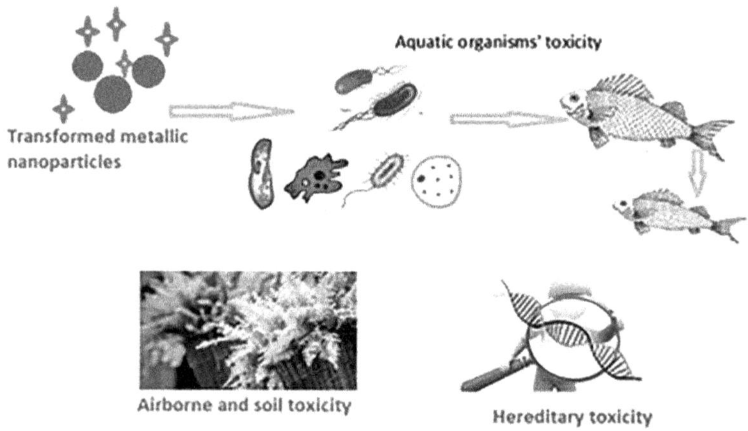

FIGURE 10.4 The cycle of transformed nanomaterials in the environment. The toxicity of nanoparticles on biological systems leads to severe damage at the cellular level either in the terrestrial and aquatic organisms.

MNPs have additionally been ascertained to induce toxicant effects on terrestrial plants, like reduction of seed germination, decrease in plant growth, and alteration of mineral nutrition and chemical action (Zhang et al., 2018). As an example, $nCeO_2$ and nZnO might cut back the grain yield and quality of corn (Zhao et al., 2015). The potential toxicity of MNPs to plants is, to an outsized extent, captivated with the interior properties of the particles. As an example, nFe_3O_4 aroused the basis growth of tomato plants, whereas $nSnO_2$ reduced their root growth (Antisari et al., 2015).

10.1.3 TOXICITY OF NANOPARTICLES IN VIVO

In vivo studies, although several animal models are designed, the quality and therefore the prognostic model remain indeterminate. Besides the foremost common experimental mice and rats, zebrafish, and Caenorhabditis elegance were additionally employed in nanotoxicology studies. The *in vivo* assessments are divided into two substantial classes. One comprises the changes in tissue structure, apoptosis, and infiltration of inflammation within the main organs such as the urinary organ, brain, and heart. The opposite one targets the particular systems whose structural properties are at risk of NPs concentration. As an example, sinus and internal organ cells are the structures of the liver operate in metabolism and detoxification, during which NPs are at risk of deposition, moreover as in urinary organ filtration membrane. Consequently, applicable appraisal models might be established, particularly for the drug delivery of NPs. Biochemical methods are often used to illustrate the acute and chronic toxicity of NMs. However, it is difficult to find mechanisms for free radicals formed *in vivo* and in cells.

The cytotoxicity as a result of NPs also depends on the exposure time of these NPs. The respiratory system, as mentioned on top, a nominated target of the toxicity caused by NPs as they're not solely the entry purpose for indrawn particles however additionally gate to the internal organ outcome (Figure 10.5).

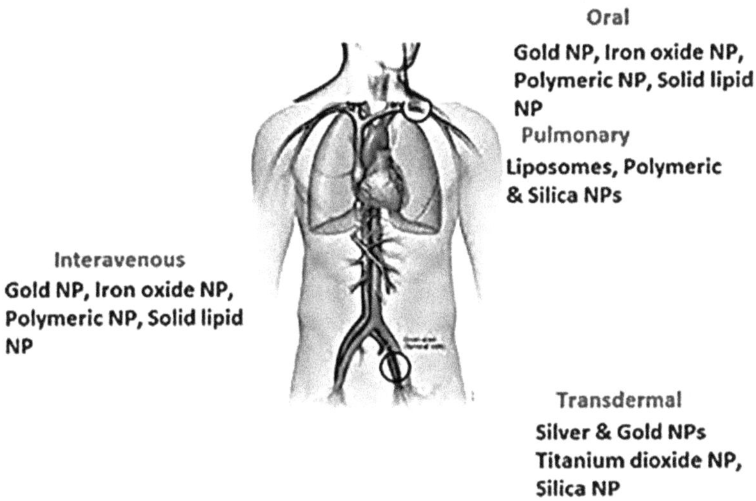

FIGURE 10.5 Pathways taken by nanoparticles in human organs.

The cytotoxicity of nanoparticles also depends on the time of exposure. From the gates of entry to those particles *in vivo* is the respiratory target is the specific toxicity caused by NPs when inhaled (Figures 10.5 and 10.6). Additionally, the enhanced permeability and retention (EPR) effect is a controversial concept by which molecules of certain sizes nanoparticles drugs tend to accumulate in tumor tissue much more than they do in normal tissues (Figure 10.6). For nanoparticles with a diameter of 20 nm: 50% is deposited in the area of the alveoli and 25% in the area of the nasal throat and trachea. In the workplace, the skin may be exposed to NMs during manufacturing, use, or contact with contaminated surfaces. The extent to which nanoparticles are generally able to penetrate the skin and induce extensive adverse effects is still under study. Most studies have been done on individual substances such as ZnO and TiO_2 in their contact with the skin. The effect of skin elasticity on the penetration of NMs to damaged skin should be studied.

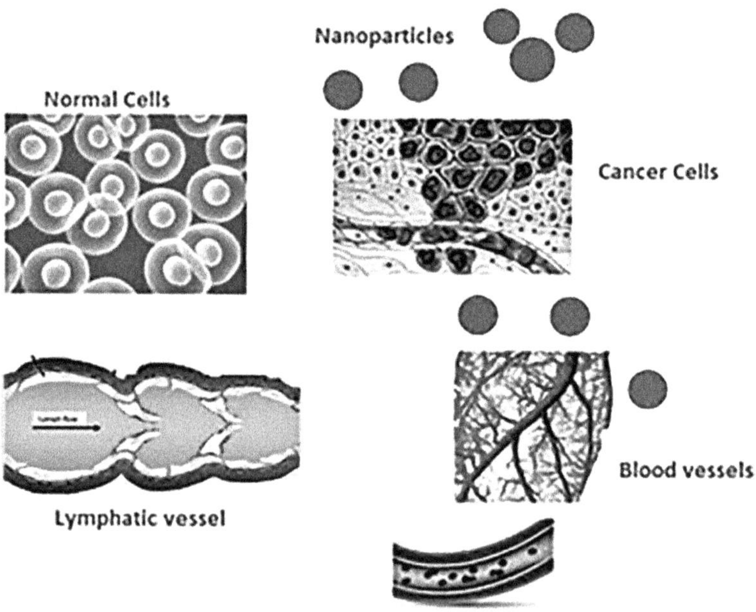

FIGURE 10.6 (a) SEM and (b) TEM images of the prepared silica from rice husk (Al Adham et al., 2018); (c) XRD and EDX (inset) of the prepared silica; and (d) the thermal adsorption isotherm to absorb N_2 nitrogen and pore size distribution of the prepared silica.

10.2 FACTORS AFFECTING BIOLOGICAL RESPONSE

First, the deposition location depends on the aerodynamic particle space, which changes with the degree of agglomeration and aggregation. Second, if the nanoparticles (or clusters) of the nanoparticles are not agglomerated or not fully aggregated in the post-deposition stage, the resulting biological effect is similar to the equivalent exposure to separate nanoparticles. In addition, if the biological response is associated with the surface area of a precipitated aerosol; for a specific volume of material, the response to precipitated clusters and clusters with an open, semi-random composition will be similar to those arising from a dose equivalent to separate particles. However, if nanostructures are agglomerated, they will not be transported to other organs within the body, as do separate nanoparticles, so biological effects are different.

Therefore, given the ease with which specific particles cannot agglomerate, we can determine the appropriate body space and exclude unwanted particles (large particles that are not nanostructured materials) that change with particle properties.

Through the radical free mechanisms of most poisoning caused by NMs, this is a catalyst for oxidative stress mediated by ROS formation. In the first place, it is difficult to clarify the mechanisms of free radicals at the molecular level as a result of highly reactive free radicals, and in most cases, these roots have a very short half-life. These short-lived, free radicals make it difficult to detect them directly within a 109-second period, and therefore difficult to isolate directly for structural identification (Griendling et al., 2016). Therefore, it uses a highly accurate technique to determine the formation of ROS (Griendling et al., 2016) Therefore, it uses a very precise technique to identify the formation of ROS, electron spin resonance spectroscopy (ESR) spin trapping techniques (Scott et al., 2016).

However, it is difficult to determine the original composition of new free radicals due to the reactive reaction of those roots. If we take the formation of the chemical group as a radical example of hydroxyl, it will be created directly: (i) from a photo-excited nanoparticle extracting electrons from water or hydroxide ions; (ii) through the Haber Weiss reaction from the reaction of a redox-type nanoparticle at the start with superoxide, and later on with H_2O_2; and (iii) several alternative pathways. It's even tougher to work out the free-radical mechanism of nanomaterial-induced toxicity in an exceedingly biological system, notably *in vivo*.

10.3 GENOTOXICITY FOR NPS

Genotoxicity describes the potential damage of NMs to DNA. In general, the genetic toxicity of NMs can be classified as incipient or incidental genetic toxicity. The incipient genotoxicity is illustrated as genetic damage caused by NPs itself, not because of the inflammatory response, while the incidental genotoxicity means that the involvement of inflammatory cells, which end up in oxidative damage to DNA by ROS leads to manipulation of genes and chromosomes (Drasler et al., 2017). In theory, incipient genetic toxicity can be evoked through immediate or indirect mechanisms. Immediate genetic toxicity is caused by the physical interfering of the drugs with genomic deoxyribonuclease, which can occur for smaller nanoparticles able to reach the nucleus through the nuclear foramen (Shemetov et al., 2012). Whereas the indirect genetic harm will occur across many pathways; for example, through the genesis of ROS by NM and organic components but not inducement inflammation (Schins & Knaapen, 2007).

10.4 DOSAGE ASSESSMENT (INTERNAL EXPOSURE)

There are three ways to introduce natural or synthetic nanoparticles into the human body: human skin, lungs, and digestive system. Although the skin is mostly an effective barrier to foreign substances, the lungs and digestive system are more susceptible. Injection and implantation are also different ways of exposure and are primarily limited to engineering materials. It is clear that particle size will have a dramatic consequence on the means by which an organism responds when exposed to foreign matter.

There are several aspects of the current problem:

- Size controls where particles are stabilized within the body (exposure). For example, established models have been developed to predict the deposit of respiratory molecules of a certain mechanical diameter (Rzigalinski et al., 2003).
- Size can be considered the body's strength in disinfection or disposal of foreign particles (disinfection mechanisms) without damaging the body (Renwick et al., 2004). The ciliary system within the respiratory system is intended to remove particles from the upper airways (Geiser & Kreyling, 2010). The physical function of macrophages can be a common mechanism in which the innate system of the body attempts to disinfect molecules in a number of microns or less (Powers et al., 2009).

Direct internal exposure reflects health effects. In the workplace, less harmful methods are used, such as collecting hair, urine, and exhaled air (Arnold et al., 2013). However, the dose assessment includes analysis of tissues, body fluids, and exhaled air.

The dose can be determined by measuring the number of nanoparticles involved and/or their biological activity. The term "biomarker" is often used to describe several biological effects resulting from interactions between human biological systems and toxicities. Biomarkers can provide direct evidence of exposure to a particularly toxic substance when there is a single correlation between a specific biological marker and the toxic substance (Mussali-Galante et al., 2013). One advantage of measuring vital signs of exposure is that it gives information about mixed exposure through multiple pathways and includes unpopulated areas. Vital sign measurements are used to survey and monitor workers. Biomarkers for measuring exposure to nanoparticles in the first stage of development are complicated by the wide variation in the physical and chemical properties of nanoparticles, resulting in widespread biological responses (Raj et al., 2012). Exposure to inhalation of low toxic and poorly soluble nanoparticles has been shown to cause an inflammatory response. For example, nitric oxide in exhaled air has been suggested as a biological marker for inflammation.

10.5 NANOMEDICINE

10.5.1 CLINICAL APPLICATION OF NANOMATERIALS (NMS)

NPs have tremendous perspective in the medical field such as drugs and genes conveyance, fluorescent markers, and contrast agents (Lin et al., 2015). So as to qualify as a "real" NP, one among its physical dimensions must have at least a scale of 1 to 100 nm. The medical applications of nanotechnology are the most important of this modern technology, because of its direct link to human life and health, nanotechnology promises a lot of medical applications related to accurate diagnosis and treatment of high efficiency as well as many applications in the field of health care, facing the most-deadly human diseases such as cancer (Nikalje, 2015). Because nanoparticles are very small, the drug can be delivered not only to diseased tissues but to infected cells with great accuracy. As a result, fewer side effects that may occur from the arrival of the drug to other parts that were not intended to be treated (Yang et al., 2017). In cancer treatment where nanoparticles reach and are concentrated in cancer cells and then heated by certain radiofrequency waves, killing cancer cells without damaging normal cells. If this technique proves effective and safe, it may in future dispense with chemo-radiotherapy, which has many side effects. The utilization of NPs as drug-carrying systems, particularly therapeutic medicine, is widespread owing to the power to focus on cancer cells specifically, enhance efficaciousness and scale back general toxicity. NPs Gold (AuNP) offers many useful properties including a non-toxic and biocompatible metal core making it an ideal starting point for NMs (Khan et al., 2019). Moreover, AuNP will operate multi-surface functions that mix different kinds, like medicine and targeting agents, making it an extremely versatile targeting tool (Kong et al., 2017). Other pharmaceutical nanoparticles already used clinically include lipid-based (Yildirimer et al., 2011), polymer-based (Bartlett et al., 2007) and biological NPs (Grenha et al., 2010). Nanotechnology has provided new perspectives and unique additions to early cancer diagnosis through an advanced class of materials known as Nanocrystals, also called quantum dots (Medintz et al., 2008; Gahrouei et al., 2019).

They are prepared in the form of spherical granules with homogeneous dimensions ranging in diameter from 2 to 10 nanometers. Due to the low sizes of these nanocrystals, they exhibit the behavior of a single atom, which enables them to enjoy distinct optical and conductivity properties that no other semiconductor material possesses (Yong et al., 2009). The various fluorescent labels used in medical NPs are one of the few clinically approved NPs such as chemical super magnetic nanoparticles (SPIONs) and are also used in current applications in the medical specialty such as resonance imaging (MRI), drug, and gene delivery coupled with the severe destruction of tumor tissue (Veiseh et al., 2009; Lee et al., 2009). Nanometal oxides are known to be highly toxic and are included in the formulations of topically applicable preparations such as creams and sunscreens such as TiO_2 and zinc oxide NPs (Choksi et al., 2010). It's necessary to work out whether or not NPs will penetrate deep layers of the skin or not, as well as presumably those substances are absorbed into the circulation and accumulate in the tissues. Silver nanoparticles (AgNPs) have become modern because of their (Yong et al., 2009) antimicrobial effect that is successfully exploited in medical applications such as silver saturated wound dressings, contraceptives, surgical instruments, and orthopedic prostheses (Miller et al., 2010). Carbon nanotubes (CNTs) are made up of sheets of graphene wrapped and classified as single-walled or multi-walled carbon nanotubes (MWCNTs) counting on the constituent numbers of graphene layers. Because of its distinctive size and form, an excellent deal of effort has been dedicated to analyzing the medical applications of CNTs (Abdal Dayem et al., 2017) This broad potential needs an intensive assessment of toxicity. In addition to exposure due to the use of the work and manufacturing of such materials, the manufacture and use of nanoparticles inevitably cause extensive activity and serious environmental exposure. Although toxicology studies of nanoparticles are very limited, studies on the toxicity of chemical nanoparticles are of concern, like nano-TiO_2, nano-ZnO, nano-CuO, nano-CuZn, nano-Fe_3O_4, and nano-Fe_2O_3, with nano-TiO_2, nano-ZnO particularly, has been reported by (Fu et al., 2014). For example, studies indicate that the absorption of TiO_2 through healthy skin is very limited. However, studies confirm the release of its molecules through aquatic environments, as can occur during swimming or bathing has been harmful to algae and aquatic organisms. Needless to say, totally different NMs exhibit different harmful efficiency. As an example, Kumar et al. (2020) compared the toxicity of 3 nano-metal oxides, nano-CuO, nano-CdO, and nano-TiO_2. Nano-CuO was resolute to be the foremost potent in toxicity and polymer injury, resulting in 8-hydroxy-20-deoxyguanosine (8-OHdG) formation, whereas nano-TiO_2 was the smallest amount, while not causing a major level of 8-OHdG (Chakraborty et al.). The mechanism by which the inorganic nanoparticles work is shown by the action of toxic action on cancer cells and is summarized in two main mechanisms is the release of toxic metal ions into the cell gradually and thus the formation of toxic free radicals in the cytoplasm. These free radicals are leading to damages in tissues as a result of releasing superoxide anion radicals, OH radicals, singlet oxygen, and H_2O_2 (Gonzalez et al., 2008). In addition, these radicals play an important physiological alterations role in cellular signaling systems and involve mitogen-activated protein kinase (MAPK). Transition metals like Cu and Fe may participate in one-electron redox reactions, resulting in the formation of ROS (Yin et al., 2012). Cells can malfunction to maintain the traditional missions regulated by oxidation and reduction in the instance of excessive ROS, which stimulates oxidative stress (Meng et al., 2009). This damage develops to include oxidative modification of proteins to generate protein roots (Phaniendra et al., 2015), initiation of lipid peroxide activity (Schieber & Chandel, 2014; Hotamisligil & Davis, 2016), breakdown of DNA strips and thus modification of gene expression by activating oxidative-sensitive transcription factors (Jiang & Gao, 2016; Bokhari & Sharma, 2019), modifying inflammatory responses by signaling (Bodamyali et al., 2000) leading to cell death and genetic toxicity (Fu et al., 2012). Recent studies have shown that these toxic-free molecules have the potential to infect humans with some degenerative diseases such as amyotrophic lateral sclerosis, arthritis, cardiovascular disease (CVD), Alzheimer's disease, Parkinson's disease, and diabetes (Burton et al., 2010; Liguori et al., 2018). Besides, 8-OHDG is the predominant induced form of these free radicals which confirms the definitive proof of damage to DNA and mitochondria (Guo et al., 2016; Korkmaz et al., 2018). Damage associated with the release of toxic free molecules is also due to the regression of the body's defense system and the existence of superoxide dioxide (SODs), peroxidase oxide, and catalase are some of the prominent antioxidant enzymes that effectively protect against these harmful biological events (Fu et al., 2014). This means that antioxidants are the only known defense against hydroxyl radicals *in vivo*, and they contribute to oxidation and reduction reactions by giving electrons or hydrogen atoms (Hayyan et al., 2016; Ighodaro & Akinloye, 2018). This procedure allows cells to function normally and avoid the repercussions of oxidation of vital components. A very important point is that chemically engineered NMs have a small size, high surface area, and high surface reactivity, which release higher levels of oxygen molecules than their larger counterparts and thus result in cellular toxicity in the physiological system, such as human erythrocytes and dermal fibroblasts as well as genetic toxicity destroy DNA (Akter et al., 2018). However, size is not the only factor in determining the potential toxicity of these particles, the chemical composition, shape, concentration, surface composition, charge, and solubility also affect toxicity. A large number of variants affecting the toxicity of these particles is also found. Therefore, the health hazards accompanied exposure to NMs, and potential risks should be studied on a case-by-case basis and the outcome verified through long-term studies (Oberdörster et al., 2005; Khanna et al., 2015; Yusupov et al., 2017).

10.5.2 BIOCOMPATIBILITY AND REGULATION THE NANOMATERIALS (NMS)

Biocompatibility is the competence of the material to effectuate with a proper response from the host in specific conditions. The characteristic of compatibility with living tissue or living system is that they are not toxic or harmful and do not cause immune rejection. This is considered a high degree of biocompatibility interacts with the body without producing inadmissible carcinogenic. In addition, biocompatibility is a test to determine the potential toxicity resulting from physical contact with a substance or medical device. Medical compatibility is vital for medical devices (MDs) and substance. The vagueness of the term reflects the continuing evolution of insights into how biomaterials interact with the human body, and how these interactions determine the clinical success of the medical system. By the interaction of nanoparticles with various biological substances such as immune cells, proteins, and fats, the levels of these responses are determined in part (Yoshioka et al., 2017).

There have been some recent experiments to determine the fate of NMs *in vivo*, including one conducted on pregnant mice that were injected intravenously using 70 nm nanosilica, these particles were detected in the maternal's liver, placenta, brain, and fetal liver, while smaller silica particles were not detected (Yamashita et al., 2011). In recent years, very small nanoparticles with a diameter of no more than 10 nm have been examined for possible use as nanoparticles, or carriers of drugs *in vivo*. Dreaden et al. (2012) they performed on the gold nanoparticles (AuNPs), which have different diameters 2, 6, and 15 nm, they have compared their penetration *in vitro* and *in vivo* and studies have shown that smaller AuNPs enter cancer cells more efficiently than large AuNPs in the laboratory *in vitro*.

As well as aggregation is a phenomenon associated with most nanoparticles, however, very few studies performed on the effect of aggregation on cellular absorption and toxicity. One study was conducted by Albanese & Chan (2011) showed that aggregation affects the absorption patterns of various AuNPs. Being able to manipulate the surface chemistry of nanoparticles, such as adding functional groups to minimize surface interaction or enhance stability, will be necessary for the future to develop nanoparticles for use as a treatment for most diseases.

Before accepting new nanomedicine for use by patients, it must be decided, what regulatory system is applied. Additionally, what is the fate of these particles *in vivo*? The medical device performs its function by physical means, such as mechanical or chemical action, while the medical product performs its responsibility exclusively by pharmaceutical, immunological, or metabolic means (Fuschi et al., 2019). The clarification of these issues are crucial not only for patient safety but also for the pharmaceutical industry, which requires further coordination in the current regulatory framework of nanomedicine. In the European Union, the decision to classify as a medical device or medicine is based on the European Union Directive 2001/83/EC on Human Medicines, as amended, and the EU Directive 93/42/EEC on MDs, as amended. Toxic investigations are the basis for assessing the safety and risks of nanomedicines and are therefore necessary for their approval. If we consider them, it becomes comprehensible that the thoroughness and regulation of nanomedicine and nanomedicine is still under development.

10.5.3 NATURAL MATERIALS FOR NANOMEDICINE (GREEN NANOTECHNOLOGY)

Many different natural biological resources have already been used for nanoparticle synthesis, including microorganisms (such as fungi, yeasts, algae, viruses, and bacteria) and plant extracts (Harshiny et al., 2015). Plant extracts can be used to synthesize NMs, an interest-provoking substitutional for acquiring nanoparticles by processes that do not harm the environment, maintaining their desired properties for environmental or other uses (Villaseñor-Basulto et al., 2019). It is necessary to create environmentally friendly procedures for the creation of environmentally friendly, non-toxic nanoparticle processes (NPs) To avoid toxic effects resulting from treatment processes as much as possible (Harshiny et al., 2015).

Over the past few years, the use of intermediate synthetic protocols has been increased that will create nanoparticles that do not harm the environment as they do not use toxic chemicals as well as low-cost (Herlekar et al., 2014).

Recently, we reported in our laboratory that the therapeutic effect of silica particles prepared from rice husk with or without platelet-derived growth factor (PDGF) on liver veno-occlusive disease (VOD) is caused by dactinomycin chemotherapy (DAC) in rats. The protection mechanisms of nano-RHS-PDGF may be attributed to its activity in removing free radicals and their protective effect on antioxidant system damage. Based on the findings obtained, this treatment can be used as a prevention of the undesirable effects of chemical drugs as well as a treatment for a disease such as hepatic veno-occlusive that kills many patients (Al-Adham et al., 2018).

By a 150-watt stainless, the agricultural waste of rice husk was grinded repeatedly, and then refilled with 17% of a hydrochloric acid solution at a rate of 1:10 W/h for one hour. Thereafter, the rice husk was filtered several times with distilled water, for our samples, it was found that the specific surface area is 236 m^2/g. One can observe that the molecules clump in a chip-like structure by SEM image of the prepared silica, while the TEM image reveals that the nano-sized silica molecules range from 5 to 35 nm with a spherical shape (Figure 10.7). This suggests that the procedure used succeeds in producing pure nano-silica molecules from rice husks. Therefore, we can get rid of rice husk in a way that is not harmful to the environment instead of filling the land with waste or incineration, which may produce harmful effects on both the environment and human health, so it is recommended to use this waste in both economic and environmental ethics (Table 10.2) (Totlani et al., 2012).

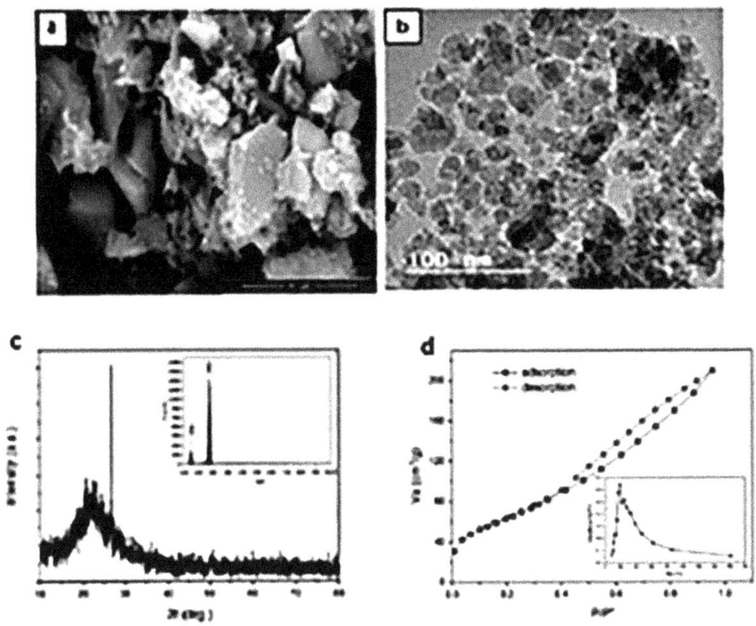

FIGURE 10.7 Graphical illustration of the EPR effect.

TABLE 10.2 The Parameters from N_2 Adsorption–Desorption Analysis of the Prepared Nanosilica from Rice Husks

BET Surface Area (m²/g)	Average Pore Diameter (nm)	External Surface Area (m²/g)	Micropore Area (m²/g)	Total Pore Volume (cc/g)	Micropore Volume (cc/g)
236	5.48	223.9	12.1	0.323	0.005

From the above, it is illustrated that there are experiments being conducted on the synthesis of nanoparticles from natural materials that have no harmful effects on humans and the environment, as well as examined on experimental models to determine their fate *in vivo*. These experiments have also been conducted to find out how effective such substances are in treating many diseases, but this leads us to find out the most vital reason why the overwhelming majority of nanomedicine analysis has been centered on cancer is what has been known as the improved permeation and retention (EPR) impact (Figure 10.7). Angiogenesis is the process of growth cells are able to stimulate the formation of blood vessels to receive enough nutrients and eliminate waste products from metabolism. However, this growth is fast and confusing, unlike the formation of healthy blood vessels, resulting in incomplete blood vessels that are almost disjointed, as well as pores within the walls of capillaries that are larger than those in healthy tissues and organs. The presence of these pores allows the release of nanoparticles into diseased cells, which may not be possible in healthy tissues (Figure 10.6). Nanoparticle accumulation within the growth space by the EPR impact depends on the circulation time of the nanoparticles in the blood. When shot of the nanoparticles, a series of steps should turn up to permit the EPR impact to require place. First, the nanoparticles got to be stable in suspension whereas current within the blood. Furthermore, nanoparticle assembly can lead to clogged blood vessels, putting the patient's life at risk. After that, the nanoparticles should remain inside the blood long enough to emerge the remedial impact (Chen et al., 2019; Rjiba et al., 2021).

Nano-polysaccharides, which have received considerable attention in the past few years, may become more effective as antimicrobials when they are associated with sugars. We have verified the external polysaccharide (EPSST) produced by the isolated Mediterranean streptococcus for the ability to reduce the vigor of mucositis *in vivo*. The isolation of streptococci from the Mediterranean sediments was isolated, identified, and confirmed by molecular 16S rDNA analysis. The characterizations of polysaccharide were discovered, such as functional groups, MW, and chemical assessment via infrared Fourier Transform (FTIR) and high-performance liquid chromatography (HPLC). The isolated strain became recognized as *Streptomyces* sp. based totally on 16S rDNA sequence with accession number SAMN08349905. The chemical appraisals of EPSST were galacturonic, glucose, galactose, mannose, and arabinose with a relative ratio of 2.1: 1: 5.37: 1.62: 1.29 individually, with an average molecular weight (Mw) 9.687×103 g/mol. Also, the EPSST contained uronic acid (16%) and sulfates (12.149%), and no protein was detected. EPSST inhibited the DPPH radical activity. The results of this study indicate that EPSST inhibits mucositis caused by fluorouracil by stopping oxidative stress, apoptosis, decreased inflammatory factors, and activation of antioxidant enzymes, which increases the clinical effectiveness of chemotherapy (Mahmoud et al., 2020).

10.6 CONCLUSIONS

Nanotechnology is a technology that carries a number of great benefits to humans, but it also carries with its negative aspects like any other technology. In spite of the great benefits that mankind expects from this modern technology, it has some negative aspects, and carries with it some risks to human health and its environment, as well as the lack of information available about the potential health and environmental side effects of its applications; Nanotechnology will lead humanity to a long road full of health and environmental problems.

Nanotoxicology is defined as an area that studies the potential health risks of NMs. The nanoparticle size means that NMs have the ability to penetrate the human body from other large particles, and how nanoparticles move and interact within the organism is a big issue. The behavior of nanoparticles is indicative of their function, size, and surface interaction with the abutting tissue. Nanoparticles accumulate within organs as a result of being not decomposing or decomposing slowly and of concern is their interaction. Because of its large surface and once nanoparticles are exposed to tissue and liquids, some of the fine particles they contain are absorbed onto the surfaces of tissues. A large number of variables affecting poisoning means that it is difficult to generalize risks to issues associated with exposure to NMs; each nanomaterial must be evaluated individually. The characteristics of all materials must be taken into account, and both health and environmental issues are combined in the business environment related to the production and use of NMs as well as the laboratory environment.

The world is moving now towards the green nanotechnology, and this technology has two objectives: the first is the production of NMs and their products without harming the environment or human health, and the second is the production of NMs that can provide solutions to environmental problems. This is done by using current principles of green chemistry and green engineering to manufacturing nanomaterial products without toxic components under low temperatures and using as little and renewable energy inputs as possible. Green nanotechnology also means using nanotechnology to make existing manufacturing processes of NMs and materials more environmentally friendly.

KEYWORDS

- **biocompatibility**
- **nanoparticles**
- **toxicity**

REFERENCES

Abdal, D. A., Hossain, M. K., Lee, S. B., Kim, K., Saha, S. K., Yang, G. M., Choi, H. Y., & Cho, S. G., (2017). The role of reactive oxygen species (ROS) in the biological activities of metallic nanoparticles. *Int. J. Mol. Sci., 18*(1), 120.

Akter, M., Sikder, M. T., Rahman, M. M., Ullah, A. K. M. A., Hossain, K. F. B., Banik, S., Hosokawa, T., et al., (2018). A systematic review on silver nanoparticles-induced cytotoxicity: Physicochemical properties and perspectives. *J. Adv. Res., 9*, 1–16.

Al-Adham, E. K., Hassan, A. I., Shebl, A., & Hazem, M. M., (2018). Evaluation of the therapeutic effects of rice husk nanosilica combined with platelet-derived growth factor in hepatic veno-occlusive disease. *Biochem Cell Biol., 96*(5), 682–694.

Albanese, A., & Chan, W. C. W., (2011). Effect of gold nanoparticle aggregation on cell uptake and toxicity. *ACS Nano, 5*(7), 5478–5489.

Antisari, L. V., Carbone, S., Gatti, A., Vianello, G., & Nannipieri, P., (2015). Uptake and translocation of metals and nutrients in tomato grown in soil polluted with metal oxide (CeO_2, Fe_3O_4, SnO_2, TiO_2) or metallic (Ag, Co, Ni) engineered nanoparticles. *Environ Sci. Pollut. Res., 22*(3), 1841–1853.

Armand, L., Tarantini, A., Beal, D., Biola-Clier, M., Bobyk, L., Sorieul, S., Pernet-Gallay, K., et al., (2016). Long-term exposure of A549 cells to titanium dioxide nanoparticles induces DNA damage and sensitizes cells towards genotoxic agents. *Nanotoxicology, 10*(7), 913–923.

Arnold, S. M., Angerer, J., Boogaard, P. J., Hughes, M. F., O'Lone, R. B., Robison, S. H., & Robert, S. A., (2013). The use of biomonitoring data in exposure and human health risk assessment: Benzene case study. *Crit. Rev. Toxicol., 43*(2), 119–153.

Baek, M., Chung, H. E., Yu, J., Lee, J. A., Kim, T. H., Oh, J. M., Lee, W. J., et al., (2012). Pharmacokinetics, tissue distribution, and excretion of zinc oxide nanoparticles. *Int. J. Nanomedicine, 7*, 3081.

Bahadar, H., Maqbool, F., Niaz, K., & Abdollahi, M., (2016). Toxicity of nanoparticles and an overview of current experimental models. *Iran Biomed J., 20*(1), 1.

Bakand, S., & Hayes, A., (2016). Toxicological considerations, toxicity assessment, and risk management of inhaled nanoparticles. *Int. J. Mol. Sci., 17*(6), 929.

Bartlett, D. W., Su, H., Hildebrandt, I. J., Weber, W. A., & Davis, M. E., (2007). Impact of tumor-specific targeting on the biodistribution and efficacy of siRNA nanoparticles measured by multimodality *in vivo* imaging. *Proc. Natl. Acad. Sci., 104*(39), 15549–15554.

Bastús, N. G., Sánchez-Tilló, E., Pujals, S., Farrera, C., Kogan, M. J., Giralt, E., Celada, A., et al., (2009). Peptides conjugated to gold nanoparticles induce macrophage activation. *Mol. Immunol., 46*(4), 743–748.

Bessa, M. J., Brandão, F., Viana, M., Gomes, J. F., Monfort, E., Cassee, F. R., Fraga, S., & Teixeira, J. P., (2020). Nanoparticle exposure and hazard in the ceramic industry: An overview of potential sources, toxicity and health effects. *Environ. Res., 184*, 109297.

Bodamyali, T., Stevens, C. R., Blake, D. R., & Winyard, P. G., (2000). Reactive oxygen/nitrogen species and acute inflammation: A physiological process. In: *Free Radicals Inflamm.* (pp. 11–16). [place unknown]: Springer.

Bokhari, B., & Sharma, S., (2019). Stress marks on the genome: Use or lose? *Int. J. Mol. Sci., 20*(2), 364.

Burton, D. G. A., Matsubara, H., & Ikeda, K., (2010). Pathophysiology of vascular calcification: Pivotal role of cellular senescence in vascular smooth muscle cells. *Exp. Gerontol., 45*(11), 819–824.

Caballero-Díaz, E., & Valcárcel, M., (2014). Toxicity of gold nanoparticles. In: *Compr. Anal. Chem.* (Vol. 66, pp. 207–254). Elsevier.

Chakraborty, A., Ghosh, S., Chakraborty, R., Chatterjee, S. M., & Hopper, W., (2018). Molecular mechanism of nanotoxicity–A critical review. *Int. J. Curr. Biotechnol., 6*(6), 1–12.

Charles, S., Jomini, S., Fessard, V., Bigorgne-Vizade, E., Rousselle, C., & Michel, C., (2018). Assessment of the *in vitro* genotoxicity of TiO_2 nanoparticles in a regulatory context. *Nanotoxicology, 12*(4), 357–374.

Chen, S., Hu, J., Wang, F., & Liu, H., (2019). Preparation and drug release application of pH and light dual-stimuli-responsive nanocarrier based on mesoporous silica nanoparticles. *J. Dispers Sci. Technol., 40*(12), 1725–1735.

Chen, W., Luo, G., & Zhang, X., (2019). Recent advances in subcellular targeted cancer therapy based on functional materials. *Adv. Mater., 31*(3), 1802725.

Choksi, A. N., Poonawalla, T., & Wilkerson, M. G., (2010). Nanoparticles: A closer look at their dermal effects. *J. Drugs Dermatology JDD., 9*(5), 475–481.

Concu, R., Kleandrova, V. V., Speck-Planche, A., & Cordeiro, M. N. D. S., (2017). Probing the toxicity of nanoparticles: A unified in silico machine learning model based on perturbation theory. *Nanotoxicology, 11*(7), 891–906.

De Matteis, V., (2017). Exposure to inorganic nanoparticles: Routes of entry, immune response, biodistribution and *in vitro/in vivo* toxicity evaluation. *Toxics, 5*(4), 29.

Di Meo, S., Reed, T. T., Venditti, P., & Victor, V. M., (2016). Role of ROS and RNS sources in physiological and pathological conditions. *Oxid. Med. Cell Longev., 2016.*

Drasler, B., Sayre, P., Steinhäuser, K. G., Petri-Fink, A., & Rothen-Rutishauser, B., (2017). *In vitro* approaches to assess the hazard of nanomaterials. *Nano Impact., 8*, 99–116.

Dreaden, E. C., Alkilany, A. M., Huang, X., Murphy, C. J., & El-Sayed, M. A., (2012). The golden age: gold nanoparticles for biomedicine. *Chem Soc. Rev., 41*(7), 2740–2779.

Fu, P. P., Xia, Q., Hwang, H. M., Ray, P. C., & Yu, H., (2014). Mechanisms of nanotoxicity: Generation of reactive oxygen species. *J. Food Drug Anal., 22*(1), 64–75.

Fu, P. P., Xia, Q., Sun, X., & Yu, H., (2012). Phototoxicity and environmental transformation of polycyclic aromatic hydrocarbons (PAHs)—Light-induced reactive oxygen species, lipid peroxidation, and DNA damage. *J. Environ. Sci. Heal Part C., 30*(1), 1–41.

Fuschi, P., Maimone, B., Gaetano, C., & Martelli, F., (2019). Noncoding RNAs in the vascular system response to oxidative stress. *Antioxid Redox Signal., 30*(7), 992–1010.

Gahrouei, D. S., Khaniabadi, P. M., Khaniabadi, B. M., & Gahrouei, S. S., (2019). Medical imaging modalities using nanoprobes for cancer diagnosis: A literature review on recent findings. *J. Res. Med. Sci., 24*(16), 2019–2024.

Geiser, M., & Kreyling, W. G., (2010). Deposition and biokinetics of inhaled nanoparticles. *Part Fiber Toxicol., 7*(1), 1–17.

Gioria, S., Lobo, V. J., Barboro, P., La Spina, R., Tomasi, G., Urbán, P., Kinsner-Ovaskainen, A., et al., (2016). A combined proteomics and metabolomics approach to assess the effects of gold nanoparticles *in vitro*. *Nanotoxicology, 10*(6), 736–748.

Gonzalez, L., Lison, D., & Kirsch-Volders, M., (2008). Genotoxicity of engineered nanomaterials: A critical review. *Nanotoxicology, 2*(4), 252–273.

Grabowska-Jadach, I., Zuchowska, A., Olesik, M., Drozd, M., Pietrzak, M., Malinowska, E., & Brzozka, Z., (2018). Cytotoxicity studies of selected cadmium-based quantum dots on 2D vs. 3D cell cultures. *New J. Chem., 42*(15), 12787–12795.

Grabowski, N., Hillaireau, H., Vergnaud, J., Santiago, L. A., Kerdine-Romer, S., Pallardy, M., Tsapis, N., & Fattal, E., (2013). Toxicity of surface modified PLGA nanoparticles toward lung alveolar epithelial cells. *Int. J. Pharm., 454*(2), 686–694.

Grenha, A., Gomes, M. E., Rodrigues, M., Santo, V. E., Mano, J. F., Neves, N. M., & Reis, R. L., (2010). Development of new chitosan/carrageenan nanoparticles for drug delivery applications. *J. Biomed. Mater. Res. Part A An Off. J. Soc. Biomater. Japanese Soc. Biomater Aust. Soc. Biomater. Korean Soc. Biomater., 92*(4), 1265–1272.

Griendling, K. K., Touyz, R. M., Zweier, J. L., Dikalov, S., Chilian, W., Chen, Y. R., Harrison, D. G., & Bhatnagar, A., (2016). Measurement of reactive oxygen species, reactive nitrogen species, and redox-dependent signaling in the cardiovascular system: A scientific statement from the american heart association. *Circ. Res., 119*(5), e39–e75.

Guo, C., Li, X., Wang, R., Yu, J., Ye, M., Mao, L., Zhang, S., & Zheng, S., (2016). Association between oxidative DNA damage and risk of colorectal cancer: Sensitive determination of urinary 8-hydroxy-2′-deoxyguanosine by UPLC-MS/MS analysis. *Sci. Rep., 6*(1), 1–9.

Gustafson, H. H., Holt-Casper, D., Grainger, D. W., & Ghandehari, H., (2015). Nanoparticle uptake: The phagocyte problem. *Nano Today, 10*(4), 487–510.

Han, X., Gelein, R., Corson, N., Wade-Mercer, P., Jiang, J., Biswas, P., Finkelstein, J. N., et al., (2011). Validation of an LDH assay for assessing nanoparticle toxicity. *Toxicology, 287*(1–3), 99–104.

Harshiny, M., Matheswaran, M., Arthanareeswaran, G., Kumaran, S., & Rajasree, S., (2015). Enhancement of antibacterial properties of silver nanoparticles–ceftriaxone conjugate through *Mukia maderaspatana* leaf extract mediated synthesis. *Ecotoxicol. Environ. Saf., 121*, 135–141.

Hayyan, M., Hashim, M. A., & AlNashef, I. M., (2016). Superoxide ion: Generation and chemical implications. *Chem Rev., 116*(5), 3029–3085.

Herlekar, M., Barve, S., & Kumar, R., (2014). Plant-mediated green synthesis of iron nanoparticles. *J Nanoparticles, 2014.*

Hotamisligil, G. S., & Davis, R. J., (2016). Cell signaling and stress responses. *Cold Spring Harb. Perspect. Biol., 8*(10), a006072.

Ighodaro, O. M., & Akinloye, O. A., (2018). First line defense antioxidants-superoxide dismutase (SOD), catalase (CAT) and glutathione peroxidase (GPX): Their fundamental role in the entire antioxidant defense grid. *Alexandria J Med., 54*(4), 287–293.

Iswarya, V., Manivannan, J., De, A., Paul, S., Roy, R., Johnson, J. B., Kundu, R., et al., (2016). Surface capping and size-dependent toxicity of gold nanoparticles on different trophic levels. *Environ. Sci. Pollut. Res., 23*(5), 4844–4858.

Jamil, B., Javed, R., Qazi, A. S., & Syed, M. A., (2018). Nanomaterials: Toxicity, risk management and public perception. In: *Nanomaterials Ecotoxicity, Safety, Public Percept* (pp. 283–304). [place unknown]: Springer.

Jasuja, N. D., Gupta, D. K., Reza, M., & Joshi, S. C., (2014). Green synthesis of AgNPs stabilized with biowaste and their antimicrobial activities. *Brazilian J. Microbiol., 45*(4), 1325–1332.

Jiang, X., & Gao, H., (2016). *Neurotoxicity of Nanomaterials and Nanomedicine*. [place unknown]: Academic Press.

Joris, F., Manshian, B. B., Peynshaert, K., De Smedt, S. C., Braeckmans, K., & Soenen, S. J., (2013). Assessing nanoparticle toxicity in cell-based assays: Influence of cell culture parameters and optimized models for bridging the *in vitro–in vivo* gap. *Chem. Soc. Rev., 42*(21), 8339–8359.

Khan, T., Ullah, N., Khan, M. A., & Nadhman, A., (2019). Plant-based gold nanoparticles; a comprehensive review of the decade-long research on synthesis, mechanistic aspects and diverse applications. *Adv. Colloid Interface Sci., 272*, 102017.

Khanna, P., Ong, C., Bay, B. H., & Baeg, G. H., (2015). Nanotoxicity: An interplay of oxidative stress, inflammation and cell death. *Nanomaterials, 5*(3), 1163–1180.

Kong, F. Y., Zhang, J. W., Li, R. F., Wang, Z. X., Wang, W. J., & Wang, W., (2017). Unique roles of gold nanoparticles in drug delivery, targeting and imaging applications. *Molecules, 22*(9), 1445.

Korkmaz, K. S., Butuner, B. D., & Roggenbuck, D., (2018). Detection of 8-OHdG as a diagnostic biomarker. *J. Lab. Precis Med., 3*, 95.

Kroll, A., Pillukat, M. H., Hahn, D., & Schnekenburger, J., (2009). Current *in vitro* methods in nanoparticle risk assessment: Limitations and challenges. *Eur. J. Pharm. Biopharm., 72*(2), 370–377.

Kukia, N. R., Rasmi, Y., Abbasi, A., Koshoridze, N., Shirpoor, A., Burjanadze, G., & Saboory, E., (2018). Bio-effects of TiO_2 nanoparticles on human colorectal cancer and umbilical vein endothelial cell lines. *Asian Pacific J. Cancer. Prev. APJCP., 19*(10), 2821.

Kumar, G., Nandakumar, K., Mutalik, S., & Rao, C. M., (2020). Biologicals to direct nanotherapeutics towards HER2-positive breast cancers. *Nanomedicine Nanotechnology, Biol Med.*, 102197.

Kumar, L., Kumar, S., & Agarwal, S., (2017). To find the role of DNA damage marker 8-hydroxy 2-deoxy guanosine in patients of prostate cancer, benign prostatic hyperplasia and its association to other prognostic factors of prostate cancer. *J. Integr. Nephrol. Androl., 4*(2), 55.

Le Ouay, B., & Stellacci, F., (2015). Antibacterial activity of silver nanoparticles: A surface science insight. *Nano Today, 10*(3), 339–354.

Lee, J. H., Smith, M. A., Liu, W., Gold, E. M., Lewis, B., Song, H. T., & Frank, J. A., (2009). Enhanced stem cell tracking via electrostatically assembled fluorescent SPION–peptide complexes. *Nanotechnology, 20*(35), 355102.

Lee, Y. H., Cheng, F. Y., Chiu, H. W., Tsai, J. C., Fang, C. Y., Chen, C. W., & Wang, Y. J., (2014). Cytotoxicity, oxidative stress, apoptosis and the autophagic effects of silver nanoparticles in mouse embryonic fibroblasts. *Biomaterials, 35*(16), 4706–4715.

Liguori, I., Russo, G., Curcio, F., Bulli, G., Aran, L., Della-Morte, D., Gargiulo, G., et al., (2018). Oxidative stress, aging, and diseases. *Clin. Interv. Aging., 13*, 757.

Lin, G., Zhang, H., & Huang, L., (2015). Smart polymeric nanoparticles for cancer gene delivery. *Mol. Pharm., 12*(2), 314–321.

Lu, H., Fei, B., Xin, J. H., Wang, R., & Li, L., (2006). Fabrication of UV-blocking nanohybrid coating via miniemulsion polymerization. *J. Colloid Interface Sci., 300*(1), 111–116.

Mahmoud, M. G., Selim, M. S., Mohamed, S. S., Hassan, A. I., & Abdal-Aziz, S. A., (2020). Study of the chemical structure of exopolysaccharide produced from streptomycete and its effect as an attenuate for antineoplastic drug 5-fluorouracil that induced gastrointestinal toxicity in rats. *Anim. Biotechnol., 31*(5).

Mahmoudi, M., Sahraian, M. A., Shokrgozar, M. A., & Laurent, S., (2011). Superparamagnetic iron oxide nanoparticles: Promises for diagnosis and treatment of multiple sclerosis. *ACS Chem. Neurosci., 2*(3), 118–140.

Mauricio, M. D., Guerra-Ojeda, S., Marchio, P., Valles, S. L., Aldasoro, M., Escribano-Lopez, I., Herance, J. R., et al., (2018). Nanoparticles in medicine: A focus on vascular oxidative stress. *Oxid. Med. Cell Longev., 2018.*

McCracken, C., Dutta, P. K., & Waldman, W. J., (2016). Critical assessment of toxicological effects of ingested nanoparticles. *Environ. Sci. Nano, 3*(2), 256–282.

Medintz, I. L., Mattoussi, H., & Clapp, A. R., (2008). Potential clinical applications of quantum dots. *Int. J. Nanomedicine, 3*(2), 151.

Meneses, G., Cárdenas, G., Espinosa, A., Rassy, D., Pérez-Osorio, I. N., Bárcena, B., Fleury, A., et al., (2019). Sepsis: Developing new alternatives to reduce neuroinflammation and attenuate brain injury. *Ann. NY Acad. Sci., 1437*(1), 43–56.

Meng, H., Xia, T., George, S., & Nel, A. E., (2009). A predictive toxicological paradigm for the safety assessment of nanomaterials. *ACS Nano, 3*(7), 1620–1627.

Miller, C. N., Newall, N., Kapp, S. E., Lewin, G., Karimi, L., Carville, K., Gliddon, T., & Santamaria, N. M., (2010). A randomized-controlled trial comparing cadexomer iodine and nanocrystalline silver on the healing of leg ulcers. *Wound Repair Regen., 18*(4), 359–367.

Molino, N. M., & Wang, S. W., (2014). Caged protein nanoparticles for drug delivery. *Curr. Opin. Biotechnol., 28,* 75–82.

Mpenyana-Monyatsi, L., Mthombeni, N. H., Onyango, M. S., & Momba, M. N. B., (2012). Cost-effective filter materials coated with silver nanoparticles for the removal of pathogenic bacteria in groundwater. *Int. J. Environ. Res. Public Health, 9*(1), 244–271.

Mussali-Galante, P., Tovar-Sánchez, E., Valverde, M., & Rojas, D. C. E., (2013). Biomarkers of exposure for assessing environmental metal pollution: From molecules to ecosystems. *Rev. Int. Contam. Ambie., 29*(1), 117–140.

Nair, A., Jacob, S., Al-Dhubiab, B., Attimarad, M., & Harsha, S., (2013). Basic considerations in the dermatokinetics of topical formulations. *Brazilian J. Pharm. Sci., 49*(3), 423–434.

Nikalje, A. P., (2015). Nanotechnology and its applications in medicine. *Med. Chem., 5*(2), 81–89.

Oberdörster, G., Oberdörster, E., & Oberdörster, J., (2005). Nanotoxicology: An emerging discipline evolving from studies of ultrafine particles. *Environ Health Perspect, 113,* 823–839.

Oh, N., & Park, J. H., (2014). Endocytosis and exocytosis of nanoparticles in mammalian cells. *Int. J. Nanomedicine, 9*(Suppl 1), 51.

Onoue, S., Yamada, S., & Chan, H. K., (2014). Nanodrugs: Pharmacokinetics and safety. *Int. J. Nanomedicine, 9,* 1025.

Panzarini, E., Mariano, S., Carata, E., Mura, F., Rossi, M., & Dini, L., (2018). Intracellular transport of silver and gold nanoparticles and biological responses: An update. *Int. J. Mol. Sci., 19*(5), 1305.

Phaniendra, A., Jestadi, D. B., & Periyasamy, L., (2015). Free radicals: Properties, sources, targets, and their implication in various diseases. *Indian J. Clin. Biochem., 30*(1), 11–26.

Poland, C. A., & Duffin, R., (2019). The toxicology of chrysotile-containing brake debris: Implications for mesothelioma. *Crit. Rev. Toxicol., 49*(1), 11–35.

Powers, K. W., Palazuelos, M., Brown, S. C., & Roberts, S. M., (2009). Characterization of nanomaterials for toxicological evaluation. In: Sahu, S., & Casciano, D., (eds.), *Nanotoxicology from Vivo Vitro Model to Heal RISKS* (pp. 1–27).

Raj, S., Jose, S., Sumod, U. S., & Sabitha, M., (2012). Nanotechnology in cosmetics: Opportunities and challenges. *J. Pharm. Bioallied Sci., 4*(3), 186.
Ramanathan, S., Archunan, G., Sivakumar, M., Selvan, S. T., Fred, A. L., Kumar, S., Gulyás, B., & Padmanabhan, P., (2018). Theranostic applications of nanoparticles in neurodegenerative disorders. *Int. J. Nanomedicine, 13*, 5561.
Renwick, L. C., Brown, D., Clouter, A., & Donaldson, K., (2004). Increased inflammation and altered macrophage chemotactic responses caused by two ultrafine particle types. *Occup. Environ. Med., 61*(5), 442–447.
Rjiba, A., Khoder, H., Jelassi, J., Bouguet-Bonnet, S., Gardiennet, C., Bendeif, E. E., Gansmüller, A., & Dorbez-Sridi, R., (2021). Differential scanning calorimetry and NMR study of water confined in a mesoporous bioactive glass. *Microporous Mesoporous Mater., 316*, 110922.
Rzigalinski, B. A., Bailey, D., Chow, L., Kuiry, S. C., Patil, S., Merchant, S., & Seal, S., (2003). In: Faseb, J., (ed.), *Cerium Oxide Nanoparticles Increase the Lifespan of Cultured Brain Cells and Protect Against Free Radical and Mechanical Trauma* (Vol. 17, p. A606). [place unknown]: Federation Amer Soc Exp Biol 9650 Rockville Pike, Bethesda, MD 20814-3998 USA.
Sahu, S. C., & Hayes, A. W., (2017). Toxicity of nanomaterials found in human environment: A literature review. *Toxicol. Res. Appl., 1*, 2397847317726352.
Schieber, M., & Chandel, N. S., (2014). ROS function in redox signaling and oxidative stress. *Curr Biol., 24*(10), R453–R462.
Schins, R. P. F., & Knaapen, A. M., (2007). Genotoxicity of poorly soluble particles. *Inhal. Toxicol., 19*(sup1), 189–198.
Schuhmacher, A., Gassmann, O., & Hinder, M., (2016). Changing R&D models in research-based pharmaceutical companies. *J. Transl. Med., 14*(1), 1–11.
Scott, M. J., Billiar, T. R., & Stoyanovsky, D. A., (2016). N-tert-butylmethanimine N-oxide is an efficient spin-trapping probe for EPR analysis of glutathione thiyl radical. *Sci. Rep., 6*(1), 1–11.
Sesti, F., Tsitsilonis, O. E., Kotsinas, A., & Trougakos, I. P., (2012). Oxidative stress-mediated biomolecular damage and inflammation in tumorigenesis. *In Vivo (Brooklyn), 26*(3), 395–402.
Sharifi, S., Behzadi, S., Laurent, S., Forrest, M. L., Stroeve, P., & Mahmoudi, M., (2012). Toxicity of nanomaterials. *Chem. Soc. Rev., 41*(6), 2323–2343.
Sharma, V., Kumar, A., & Dhawan, A., (2012). Nanomaterials: Exposure, effects and toxicity assessment. *Proc. Natl. Acad. Sci. India Sect. B Biol. Sci., 82*(1), 3–11.
Shemetov, A. A., Nabiev, I., & Sukhanova, A., (2012). Molecular interaction of proteins and peptides with nanoparticles. *ACS Nano., 6*(6), 4585–4602.
Sohal, I. S., O'Fallon, K. S., Gaines, P., Demokritou, P., & Bello, D., (2018). Ingested engineered nanomaterials: State of science in nanotoxicity testing and future research needs. *Part Fiber Toxicol., 15*(1), 1–31.
Su, R., Fan, W., Yu, Q., Dong, X., Qi, J., Zhu, Q., Zhao, W., et al., (2017). Size-dependent penetration of nanoemulsions into epidermis and hair follicles: Implications for transdermal delivery and immunization. *Oncotarget., 8*(24), 38214.
Tang, T., Zhang, Z., & Zhu, X., (2019). Toxic effects of TiO_2 NPs on zebrafish. *Int. J. Environ. Res. Public Health., 16*(4), 523.

Tayel, A., Ramadan, A. R., & El Seoud, O. A., (2018). Titanium dioxide/graphene and titanium dioxide/graphene oxide nanocomposites: Synthesis, characterization and photocatalytic applications for water decontamination. *Catalysts, 8*(11), 491.

Totlani, K., Mehta, R., & Mandavgane, S. A., (2012). Comparative study of adsorption of Ni (II) on RHA and carbon embedded silica obtained from RHA. *Chem. Eng. J., 181*, 376–386.

Tsuzuki, T., & Wang, X., (2010). Nanoparticle coatings for UV protective textiles. *Res. J. Text Appar.*

Veiseh, O., Gunn, J. W., Kievit, F. M., Sun, C., Fang, C., Lee, J. S. H., & Zhang, M., (2009). Inhibition of tumor-cell invasion with chlorotoxin-bound superparamagnetic nanoparticles. *Small, 5*(2), 256–264.

Villaseñor-Basulto, D. L., Pedavoah, M. M., & Bandala, E. R., (2019). Plant materials for the synthesis of nanomaterials: greener sources. *J. Handb. Ecomater.*, 105–121.

Wang, R., Song, B., Wu, J., Zhang, Y., Chen, A., & Shao, L., (2018). Potential adverse effects of nanoparticles on the reproductive system. *Int. J. Nanomedicine, 13*, 8487.

Westerhoff, P. K., Kiser, M. A., & Hristovski, K., (2013). Nanomaterial removal and transformation during biological wastewater treatment. *Environ. Eng. Sci., 30*(3), 109–117.

Yamashita, K., Yoshioka, Y., Higashisaka, K., Mimura, K., Morishita, Y., Nozaki, M., Yoshida, T., et al., (2011). Silica and titanium dioxide nanoparticles cause pregnancy complications in mice. *Nat. Nanotechnol., 6*(5), 321–328.

Yang, Y., Qin, Z., Zeng, W., Yang, T., Cao, Y., Mei, C., & Kuang, Y., (2017). Toxicity assessment of nanoparticles in various systems and organs. *Nanotechnol. Rev., 6*(3), 279–289.

Yildirimer, L., Thanh, N. T. K., Loizidou, M., & Seifalian, A. M., (2011). Toxicology and clinical potential of nanoparticles. *Nano Today, 6*(6), 585–607.

Yin, J. J., Liu, J., Ehrenshaft, M., Roberts, J. E., Fu, P. P., Mason, R. P., & Zhao, B., (2012). Phototoxicity of nano titanium dioxides in HaCaT keratinocytes—Generation of reactive oxygen species and cell damage. *Toxicol. Appl. Pharmacol., 263*(1), 81–88.

Yong, K. T., Ding, H., Roy, I., Law, W. C., Bergey, E. J., Maitra, A., & Prasad, P. N., (2009). Imaging pancreatic cancer using bioconjugated InP quantum dots. *ACS Nano, 3*(3), 502–510.

Yoshioka, Y., Kuroda, E., Hirai, T., Tsutsumi, Y., & Ishii, K. J., (2017). Allergic responses induced by the immunomodulatory effects of nanomaterials upon skin exposure. *Front Immunol., 8*, 169.

Yuan, Y. G., Peng, Q. L., & Gurunathan, S., (2017). Effects of silver nanoparticles on multiple drug-resistant strains of *Staphylococcus aureus* and *Pseudomonas aeruginosa* from mastitis-infected goats: An alternative approach for antimicrobial therapy. *Int. J. Mol. Sci., 18*(3), 569.

Yusupov, M., Wende, K., Kupsch, S., Neyts, E. C., Reuter, S., & Bogaerts, A., (2017). Effect of head group and lipid tail oxidation in the cell membrane revealed through integrated simulations and experiments. *Sci. Rep., 7*(1), 1–14.

Zhang, J., Guo, W., Li, Q., Wang, Z., & Liu, S., (2018). The effects and the potential mechanism of environmental transformation of metal nanoparticles on their toxicity in organisms. *Environ Sci. Nano, 5*(11), 2482–2499.

Zhao, L., Sun, Y., Hernandez-Viezcas, J. A., Hong, J., Majumdar, S., Niu, G., Duarte-Gardea, M., et al., (2015). Monitoring the environmental effects of CeO_2 and ZnO nanoparticles through the life cycle of corn (*Zea mays*) plants and in situ μ-XRF mapping of nutrients in kernels. *Environ. Sci. Technol., 49*(5), 2921–2928.

CHAPTER 11

Art and Future Perspective in the Application of Nanomaterials in Nutraceutical, Biological, and Medicinal Fields

A. RAJESWARI,[1] E. JACKCINA STOBEL CHRISTY,[1] SHINTU JUDE,[2] AUGUSTINE AMALRAJ,[2] and ANITHA PIUS[1]

[1]*Department of Chemistry, The Gandhigram Rural Institute – DU, Gandhigram, Dindigul, Tamil Nadu, India*

[2]*R&D Center, Aurea Biolabs (P) Ltd., Kolenchery, Cochin, Ernakulam, Kerala, India*

ABSTRACT

Below 100 nanometer scale are nanoparticles (NPs). Recently, much progress has been made in the field of nanotechnology towards the various developments and applications in nanomaterials (NMs). Nanoparticles have attracted enormous research attention due to their unique properties in various fields like biological, modern medicine, etc. In this chapter, we discussed about an introduction to nanotechnologies and applications of NMs in various fields such as nutraceuticals, biological, and medicinal fields.

11.1 INTRODUCTION

Nanotechnology is a recently developed field of science that emphasizes the rational synthesis and study of nanoscale materials. Functional nanomaterials (NMs) are especially an attractive topic because they enable to

Advanced Nanomaterials for Biological, Nutraceutical, and Medicinal Applications.
Sreeraj Gopi, Augustine Amalraj, and T. R. Sreeraj (Eds.)

create improved property materials with the combination of exploiting synergistic effects and multiple constituents, such as electronic, optical, magnetic, catalytic properties or bioactivity, selective permeation, adsorption, etc., with unique properties. Modified NPs are a type of high value-added materials possessing potential applications in specific fields (Ning Lin et al., 2012). In electronics, medicine, agriculture, biotechnology, and food industries fields nanotech is extensively worked (Huang et al., 2012; Lugli et al., 2010; Subramanian et al., 2015; Tibbals et al., 2010; Xie et al., 2012) because of different chemical and physical properties of NMs in comparison to microscale and bulk materials. These unexpected properties are mainly due to the higher surface atoms fraction and the surface energy of small particles in comparison to bulk materials (Khajeh et al., 2013). High surface atoms fraction is the most interesting properties of NMs. This means more active atoms are available to interact with free molecules, atoms, and ions.

In many applications like material fabrication, pharma, and nutraceuticals, innovative science and technologies have transformed research and development. Nanotechnology is an enable to change the structure, texture, quality, and can help to understand the physicochemical characteristics of nano-sized substances.

Many of the applications in the food sector, medicine, pharmaceutical, cosmetics, and nutraceuticals related to nanotechnology. In recent decades, the development of nanotechnology is wide-ranging in multibillion-dollar global industry. Evidence for this development is technology be reaching one trillion US dollars by 2015, with around 2 million workers (Roco et al., 2001; Chaudhry et al., 2008). Food, nutraceutical, and pharmaceutical industries have impact of applications of nanotechnology, which is clear from various reviews and reports (Chen et al., 2006a).

11.2 NUTRACEUTICALS

In 1989, De Felice was defined the term 'nutraceutical' from 'nutrition' and 'pharmaceutical.' It provides the healthiest medical benefits with the treatment or prevention of a disease (Kalra et al., 2003). In nutraceutical some unique component and nutrient foods are present which also known as dietary supplements, medical foods and nutritional supplements. Nutraceuticals ranges are decided from the presence of products of herbal and foods which are rich in nutrients. Due to their presumed safety as well as the potential nutritional and therapeutic effects they have received considerable interest (Rajasekaran et al., 2008).

One of the significant areas of study with the research having wide ranging inferences for healthcare providers and regulators (Bagchi et al., 2006). For the regular functioning body, foods and nutrients have an important role to maintain the human health and to reduce the infection of diseases. The concept of "nutraceuticals" evolved Worldwide in the form of "nutrition" and "health" (Rama et al., 2006).

11.2.1 CLASSIFICATION OF NUTRACEUTICALS

Natural sources are used in nutraceutical foods, and they can be listed below (Kalia et al., 2005):

- Dietary fiber;
- Probiotics;
- Prebiotics;
- Polyunsaturated fatty acids;
- Antioxidant vitamin;
- Polyphenols;
- Spices.

11.2.2 BENEFITS OF NUTRACEUTICALS

Recent research, nutraceuticals to come out effectively with nutritional supports and medical proofs. Nutraceuticals can be fabricated from plants and animals. It is used for food manufactures which are useful for the future creating novel foods (Gupta et al., 2015). We consume about 64% of dietary supplements and it plays an important role in avoiding the chronic diseases. Food as medicine and it is prevented from diseases. Figure 11.1 shows the nutraceutical and its benefits.

11.2.3 HEALTH BENEFITS OF NUTRACEUTICALS

At present, causes of diseases are consumption of junk foods, which is a result of unsuitable intake of nutrition. Various harmful diseases such as cancer, arthritis, osteoporosis, heart problems are formed in developing countries due to this unhealthy food.

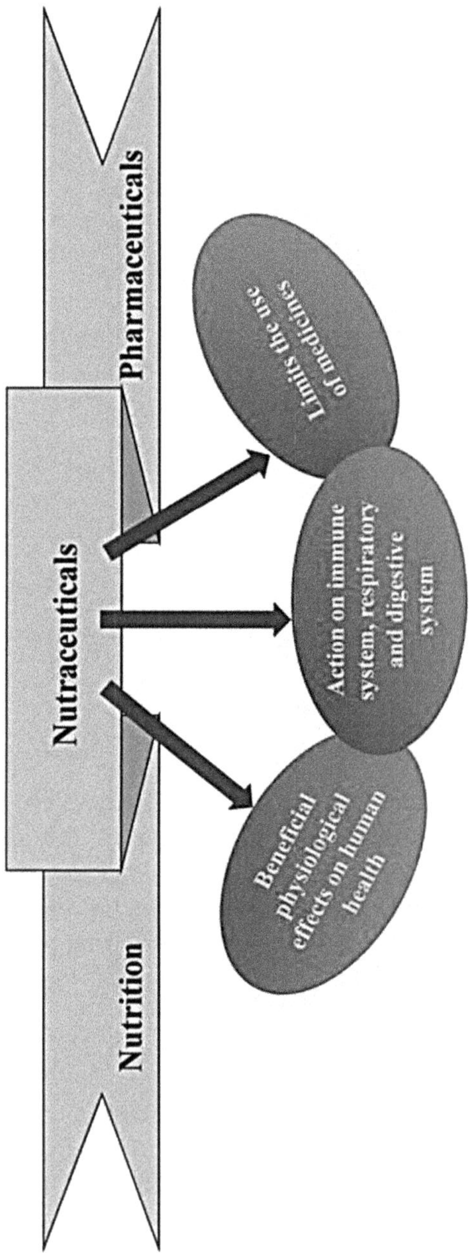

FIGURE 11.1 Nutraceutical and its benefits.

11.2.3.1 CARDIOVASCULAR DISEASES (CVDS)

The nutraceuticals such as antioxidants, dietary fibers, omega-3 fatty acids, vitamins, and minerals are used for the treatment and prevention of cardiovascular disease (CVD). Arterial diseases are prevented and controlled by polyphenols which are highly present in grapes. Some vegetables and fruits contain high content of flavonoids which block the angiotensin-converting enzyme and strengthen oxygen carriers also it is important to all cells (Srivastava Shubhra et al., 2015; German et al., 2000).

11.2.3.2 DIABETES

For diabetic treatment lipoic acid is used as an antioxidant. Diabetic patients glucose and lipid levels are reduced by dietary fibers from psyllium (Coleman et al., 2001; Baljit et al., 2007). Also ethyl esters of n–3 fatty acids are useful in diabetic patients (Sirtori et al., 2002). For the development of neurovisual, docosahexaenoic acid modulates used to resist the level of insulin (Thomas et al., 2006).

11.2.3.3 OBESITY

Natural herbal stimulants are utilized for loss the body weight. Significant dietary containers are a mixture of glucomannan, chitosan (CS), fenugreek, and vitamin C used for weight loss (Woodgate et al., 2003). While the former decreases appetite, became increases the energy disbursement (Bell et al., 2002).

11.2.3.4 CANCER

Flavonoids which block the enzymes that produce estrogen can reduce estrogen induced cancers. Phytoestrogens are recommended to prevent prostate/breast cancer. Isoflavones, curcumin, Lycopene is containing anticancer activities. Saponins possess antitumor and antimutagenic activities. Diferuloylmethane (curcumin) is a polyphenol of turmeric which have anticarcinogenic, antioxidative, and anti-inflammatory properties. Beet roots, cucumber fruits, spinach leaves, and turmeric rhizomes were reported to possess anti-tumor activity (Limer et al., 2004; Mandel et al., 2005; Kucuk et al., 2002; Gulcin et al., 2006; Mshvildadze et al., 2003).

11.2.3.5 ANTI-INFLAMMATORY ACTIVITIES

Curcumin is an important bioactive component isolated from turmeric, which has various biological activities, particularly anti-inflammatory, anticarcinogenic, and anti-oxidative activities. Green leafy vegetables are the source of linoleic acid and used as a medicine for treating problems with autoimmune and inflammation diseases.

11.2.3.6 VISION IMPROVING AGENTS

Lutein and zeaxanthin are xanthophylls which are present in high concentrations in green leafy vegetables such as spinach and kale and also present in egg yolks. Lutein is an essential pigment found in the macula of the retina and possesses a wide range of biological functions, include cataracts, scavenging free radicals, anti-inflammatory, protection of retinal cells from photo-oxidation and anti-angiogenic, anti-mutagenic, and anti-cancer. Zeaxanthin is also used in traditional Chinese medicine and Ayurveda treatment for eye problems (Kumar et al., 2016; Varma et al., 2021).

11.2.3.7 OSTEOARTHRITIS

Glucosamine (GLN) and chondroitin sulfate (CS) is used for treatment of osteoarthritis (Kalioraa et al., 2006). The extracts of 3-O-acetyl-11-keto-boswellic acid from *Boswellia serrata*, -caryophyllene from *Piper nigrum*, aqueous extract of Kaempferia galangal and curcumin from Curcuma longa has also been used to treat osteoarthritis (Amalraj et al., 2019).

11.2.3.8 ALZHEIMER'S DISEASE

Carotene, curcumin, lutein, lycopene, tamarind, etc., have been utilized to manage central nervous disorders with positive properties on specific (Glenville et al., 2006). Bacosides extracted from *Bacopa monnieri* Linn and utilized as a medicine for central nervous related problems. *Rosmarinus officinalis* L. and ginger (*Zingiber officinale*) are also used to improve the neuroprotective action (Amalraj et al., 2021).

11.2.4 NANOTECHNOLOGY IN NUTRACEUTICALS

In previous years, nanotechnology has been used in the food industry. Mayonnaise diluted emulsified form of fat droplets with water in nanosized to get lower fat products, but the process is not disturbing the nature of the product compared with higher fat products without compromising the consumer health. Because of no need of the usage of additional surfactants, water insoluble NPs are used as additives in food products. The flavor, taste, consistency, and texture attributes to obtain through the improvements of nanostructures with constant emulsions. It is the goal of food stuffs nano-texturing (Smolkova et al., 2015). The development of health and well-being is attained through a major application in the health food sector.

In NPs delivery system, nanocarriers are utilized to bioactive compounds encapsulation. It delivers these compounds to improve either bioactivities in body or gastrointestinal (GI) tract (Müller et al., 2000). The focal point is to develop the micronized food materials in Asian countries and medicines for food and medical applications (Table 11.1) (Sk et al., 2012; Peters et al., 2014).

TABLE 11.1 Role of Nanotechnology in Food Industry

Category	Applications	Significance
Food ingredients	• Color enhancers and flavoring agents; • Nanocapsules; • Nanogel; • Detoxification of chemicals and pathogens; • Anticaking agents/thicken pastes; • Packaging materials.	• Dispersion, bioavailability, and absorption of nutrient; • Improve consistency, taste, flavor, and texture attributes selective binding; • Nano-encapsulated; • Flavor enhancers; • Antimicrobials and antifungal to prevent spoilage and oxygen absorption.
Novel food	• Micronized starch; • Cellulose; • Wheat and rice flour, etc.; • Health supplements.	• Improve health and well-being.
Food supplement	• Nano-encapsulation for targeted delivery of nutraceuticals; • Nanosensors/oxygen sensors.	• Stability and protection from degradation in the digestive systems; • Storage enhancement; • Detect chemicals or foodborne; • Pathogens; • Temperature and moisture monitoring.

Bioactive compounds and micronutrients permanence can be preserved by nanotechnology during distribution, processing, and storage with enhanced solubility and improved bioavailability (Chen et al., 2006b). In the food sector, supplements and additives have been achieved the prolonged shelf-life of fresh and processed products based on NPs. Food items storage history is also presented using nanosensors. NPs are leads to reduce of CO_2 leakage from bottles and inhibit the bacterial growth (Neethirajan et al., 2011; Silvestre et al., 2009). Improvement of food flavor and coloring can be attained by the encapsulation of nutritional additives with sustained release nature by using nanotechnology. Traces of NPs found in vegetables due to the utilization of fertilizers and agrochemicals (Bouwmeester et al., 2009). In previous years, NPs are used as metals and MOs (metal oxides) in food production. Generally, NMs are used as nano-encapsulates and MO in food (Peters et al., 2016). Due to the development of foodgrade component NPs are used in various applications in pharma and food industries (Acosta et al., 2009). At the same time, there has been a most important research attempt related to the decrease the toxicity of edible nanoparticles and their understanding (Borel et al., 2014).

11.2.5 NANOMATERIALS (NMS) FOR THE DELIVERY OF NUTRACEUTICALS AND NANOCARRIERS

Nanotechnology is being utilized to devise novel formulations increased bioavailability for oral delivery of nutraceuticals. Recently, nano/micro-encapsulations have been developing applications in pharmaceutical, nutraceutical, and food industries.

For food applications, these nano/micro-encapsulations have interest for the following reasons (Quintanilla-Carvajal et al., 2010):

- Bioactive materials protection from the degradation and undesirable connections with the atmosphere during the processing and storage;
- Mask astringency tastes;
- Make possible improving solubility and disperse ability;
- Control and/or prolong its release;
- Improve its oral bioavailability.

Food grade NPs may be prepared from proteins, carbohydrates, lipids, polyphenols, vitamins, and minerals like building blocks by a number of various preparation techniques like ultrasonication, grinding, antisolvent precipitation, spontaneous emulsification and coacervation) (McClements et al., 2013). Generally utilized nanocarriers for the delivery of nutraceuticals are overviewed in Figure 11.2.

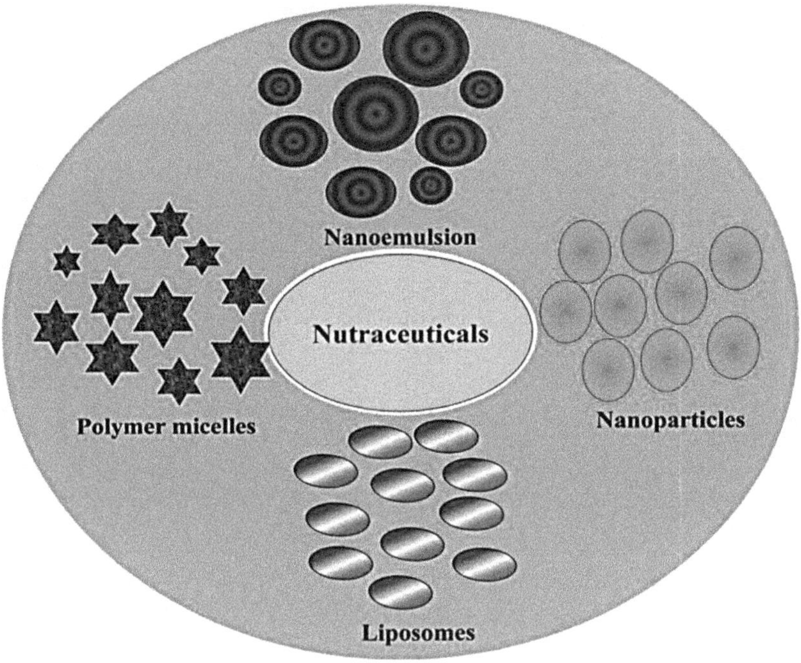

FIGURE 11.2 Diagrammatic representation of some commonly used nanocarriers for the delivery of nutraceuticals.

Bala et al. (2006) developed ellagic acid (EA) loaded poly(D,L-lactide-co-glycolide) (PLGA) NPs for oral administration by emulsion diffusion evaporation method. Polyethylene glycol-400 (PEG) used for drug solubilizing, didodecyldimethylammomium bromide (DMAB) and polyvinyl alcohol (PVA), CS were employed.

From Lin et al. (2012) research reveals that liposome is a high-quality carrier for berberine to inhibit the growth of tumor in HepG2 tumor-bearing mice. A phenolic phytonutrient is ferulic acid (FA) that contains a high anticancer effect and because of reduced bioavailability at the tumor site. Merlin et al. (2012) group research studied the FA-PLGA and application of anticancer treatment. PVA combined particles had low surface charge (Figure 11.3(A)). TEM analysis showed FA-PLGA nanoparticles with smooth surface morphology. No breakage or tear was noticed. Figure 11.3(B) illustrates the TEM image of the FA-PLGA nanoparticles, which shows the spherical NPs with a size of 200 nm. The FT-IR spectrum of FA and FA-PLGA were shown in Figures 11.3(C) and 11.3(D) shows the percentage of FA release from FA-PLGA nanoparticles during different time intervals in 10% FBS/PBS.

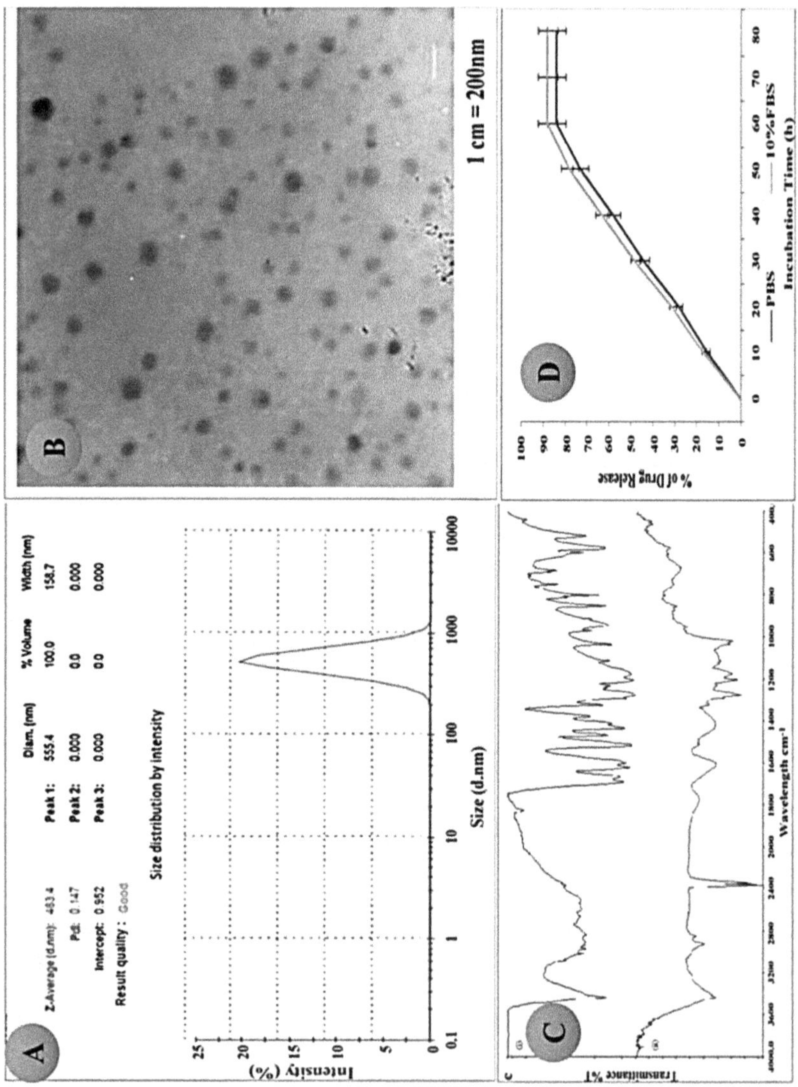

FIGURE 11.3 (A) The size distribution of FA-PLGA nanoparticles by DLS; (B) morphology of FA-PLGA nanoparticles by TEM (spherical in shape and 150 nm in size); (C) FT-IR spectra of FA and FA-PLGA nanoparticles; and (D) the percentage of FA release from FA-PLGA nanoparticles during different time intervals in 10% FBS/PBS.

Source: Merlin et al. (2012). Reproduced with permission from: Copyright © 2012, Elsevier.

11.3 BIOLOGICAL FIELD

Biology and modern biology are the study of life, structure, function, growth, distribution, evolution, and other features of living organisms. General concepts that manage all investigations, which are the cell, all organisms survive by consuming as well as transforming energy, maintain a stable internal environment.

11.3.1 NANOMATERIALS (NMS) IN BIOLOGICAL APPLICATIONS

The smaller size of living organism cells comparison provides an idea of utilizing nanoparticles as very tiny probes that would enter cellular machinery without interference (Taton et al., 2002). Understanding of biological routes in a nanoscale range is a powerful force behind the development of nanotechnology (Whitesides et al., 2003). The biological application of nanoparticles is of growing importance in biotechnology.

11.3.2 DIFFERENT TYPES OF NPS USED IN BIOLOGICAL APPLICATION

NPs are should be inert, stable, freely move, protectively encapsulate chemical entities and contain a surface which is easily conjugated to the desired target antibody. The properties of the NPs have to manage the targeting antibody, where appropriate the encapsulated cargo. Nanocarriers can be generally categorized as organic or inorganic forms and shown in Figure 11.4.

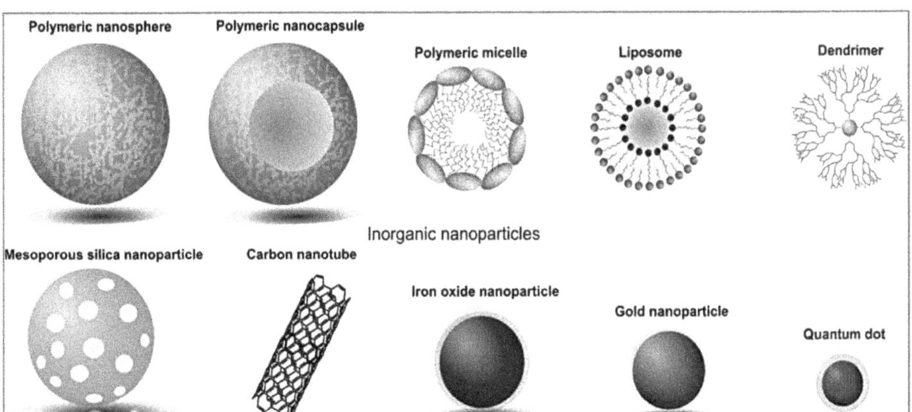

FIGURE 11.4 Pictorial representation of different types of nanoparticles used in biomedical applications.

11.3.2.1 LIPOSOMES

Liposomes were the first NPs platform and first described in 1965 as a model of cellular membranes (Bangham et al., 1993). Then it moves on model in biophysical research, applied for gene and drug delivery. Liposomes are spherical vesicles, single or multiple bilayered structures, self-assemble in nature, diverse range of compositions, abilities to carry and protect many types of biomolecules (Torchilin et al., 2005; Felgner et al., 1989). They are widely studied and utilized as transfection agents of genetic material into cells (Felgner et al., 1987; Bangham et al., 1993). Liposomes have been conjugated with biocompatible polymers like PEG which is enhancing the half-life and stability.

11.3.2.2 POLYMERIC NANOPARTICLES

Polymeric NPs are formulated through block-copolymers, biocompatible, and biodegradable polymers (Chan et al., 2010), which are extensively investigated as therapeutic carriers (Gref et al., 1994). These copolymers impulsively assemble into a core-shell micelle arrangement in the environment (Torchilin et al., 2007). They have been encapsulating hydrophilic/hydrophobic small drug molecules, nucleic acid macromolecules and proteins (Wang et al., 2008).

11.3.2.3 POLYMERIC MICELLES

Polymeric micelles consist of hydrophilic polymeric chains surrounded by a core of aggregated hydrophobic polymers. Due to their size and hydrophilic in nature allow them to avoid uptake by the reticuloendothelial system (Lavasanifar et al., 2002) and polymeric micelles to efficiently and safely encapsulate especially hydrophobic drugs for safe transport through the body (Kawano et al., 2006). Liposomal NPs with polymeric micelles demonstrate excellent biocompatibility (Ahmad et al., 2014). They have high sensitivity and poorly controlled release profiles of encapsulated cargo (Yao et al., 2016).

11.3.2.4 DENDRIMERS

Dendrimers are highly controlled and have successive polymerization steps. Medina et al. (2009) reported that these polymer chains are generally synthetic that employ natural polymers. In medical applications, they are highly regulated, have an excellent control (Tomalia et al., 2007), and they also exhibit good solubility and non-immunogenic nature (Shukla et al., 2006).

11.3.2.5 IRON OXIDE NANOPARTICLES

Iron oxide NPs are widely studied as a passive and active marking imaging agent due to their superparamagnetic. The superparamagnetic iron oxide NP (SPION) generally has an iron oxide core with a hydrophilic coating of dextran or other bio-compatible compound to increase their stability (Weissleder, 2006; Daniel et al., 2003) and it is used SPIONs consist of a magnetite (Fe_3O_4) and/or maghemite (Fe_2O_3) core. They showed size oriented super-paramagnetism, which permit them to become magnetized by the application of an external magnetic field (Bulte et al., 2004). Advantages of SPIONs are sensitivity, decreased toxicity and increased imaging specificity. They can also be degraded to iron and iron oxide molecules that are metabolized, saved in cells as ferritin and incorporated into hemoglobin (Wang et al., 2001). Currently, two SPIO agents, ferumoxides (120–180 nm) and ferucarbotran (60 nm) and they are agreed by medically for magnetic resonance imaging (MRI). SPIONs have also been used in molecular imaging applications such as the detection of apoptosis (Zhao et al., 2001) and gene expression (Weissleder et al., 2000). SPIONs can be functionalized with magnetic, optical, radionuclide, and specific targeting ligands for multimodal imaging. They can also efficiently be utilized as non-invasive diagnostic equipment and drug delivery systems (Mahmoudi et al., 2011).

11.3.2.6 QUANTUM DOT

Quantum dots (QDs) are semiconductor particles, size is < 10 nm in diameter, which is first discovered in 1980. Properties of QDs depend on its unique size, electronic, and optical properties (Collier et al., 1998). Studies of QDs consist of a cadmium selenide (CdSe) core and a zinc selenide (ZnS) cap (Jovin et al., 2003). QDs have broad absorption spectra and confined to narrow emission band which can also release light colors, have high efficiencies, long lifetimes and are constant against photo bleaching. QDs can form different bio-chemical specificities, concurrently excited and detected with significant positives in organic fluorophore dyes for optical field. In fluorescence imaging tools applications cell labeling and biomolecules tracking QDs used in biological research also (Chang et al., 2008; Chan et al., 2002; Michalet et al., 2005).

11.3.2.7 GOLD NANOPARTICLES (AUNPS)

Gold nanoparticles (AuNPs) are larger in size and shape with specific physical and chemical properties, bio-compatibility, and simplistic surface alteration. AuNPs can powerfully increase light absorption, scattering, fluorescence, and surface improvement Raman scattering (SERS). They may be due to the unique contact of the free electrons in the NP with light (Huang et al., 2007). Because of these properties, AuNPs are utilized in various implications, particularly in biochemical sensors and detectors, biological imaging, therapeutic, and diagnostics applications. AuNPs are used in sensing techniques by colorimetric arrays, SERS to considerably advance Raman scattering, allowing for spectroscopic detection and prediction of proteins and single molecules at the surface of NPs (Nie et al., 1997). AuNPs are also used to detect heart disease and cancer biomarkers, and transform absorbed light into heat with high potential for infrared phototherapy (Peng et al., 2009; Liu et al., 2008).

11.3.2.8 MESOPOROUS SILICA NANOPARTICLES (MSNS)

Mesoporous silica nanoparticles (MSNs) consist of mesopores (250 nm pores) bounded with a Si (silica) framework. NPs contain a high surface with large loading capacity. MSNs also established better bio-compatibility and bio-degradability and having desirable features in biomedical applications (Igor et al., 2008; Charu Bharti et al., 2015). Though, constancy matters and fast permission rates considerably limit the utilization of MSNs for a few applications (Fu et al., 2013; He et al., 2011; Tarn et al., 2013).

11.3.2.9 CARBON NANOPARTICLES

Carbon NPs contain only one layer of graphite in either a sheet or cylindrical conformation like carbon nanotubes (CNTs). Outstanding packing qualities, distinctive optical and electrical properties, and low fabrication costs make them capable materials for various implications (Serpell et al., 2016). But the drawbacks are less biodegradability (Seabra et al., 2014), pulmonary harm (Chou et al., 2008), and adverse organ growth (Li et al., 2014; Liu et al., 2014) and have hindered the adoption of carbon-based NPs for *in vivo* applications.

11.3.3 NANOMATERIALS (NMS) IN BIOLOGICAL RESEARCH

11.3.3.1 DETECTION OF BIOMOLECULES

NPs are utilized as signal reporters to analyze bio-molecules in DNA assay, cell bio-imaging, and immunoassay. Generally, functional groups nucleic acid-targeted oligonucleotide probes, antibodies, and protein to produce nanoprobes are usually derivatized NPs (Figure 11.5). AuNPs-based probes have been used in the identification of pathogenic bacteria in DNA-microarray technology (Wen-Tso Liu et al., 2006). Results indicated that the melting profiles of target/probe hybrids can be modified considerably to bring a sharper decrease in the slope of melting curves for mismatched DNA hybrids than for perfectly matched duplexes, owing to the specific aggregation property of nanoparticles. At the melting temperature (Tm), where 50% of the DNA hybrid is dissociated, at least a 3-fold higher sensitivity can be obtained to discriminate a perfectly matched duplex from a 1-mismatched hybrid than that obtained using a conventional fluorophore-based labeling system. In addition, the AuNPs detection system provides another advantage over the signal detection scheme, because visible light can be used to observe a color shift from red to blue, instead of fluorescence light excitation and emission, when AuNPs form aggregates. Thus, the cost of the detection system is significantly reduced. Alternatively, nanoparticles can be utilized as fluorophores in fluorescence *in situ* hybridization (FISH). QDs attached to a specific oligonucleotide probe or immunoglobulin G (IgG) have been used to successfully detect the human Y chromosome (Pathak et al., 2001) and to locate cancer markers in cellular imaging in cancer studies (Wu et al., 2002). To further facilitate *in vivo* real-time imaging, biocompatible, and soluble QDs can be prepared by encapsulating them in a n-poly(ethylene glycol) phosphatidylethanolamine (PEG-PE) phospholipid blockcopolymer micelle (Dubertret et al., 2002). To demonstrate *in vivo* imaging, these QDs containing micelles were microinjected into early-stage Xenopus embryos. Results suggested that these QDs containing micelles could be injected and confined within an embryo cell, and were stable and nontoxic to the cells. Fluorescence was visible in different embryonic cell types of Xenopus. A similar application of QD-based tracers was reported for targeting and imaging cancer cells *in vivo* (Gao et al., 2004).

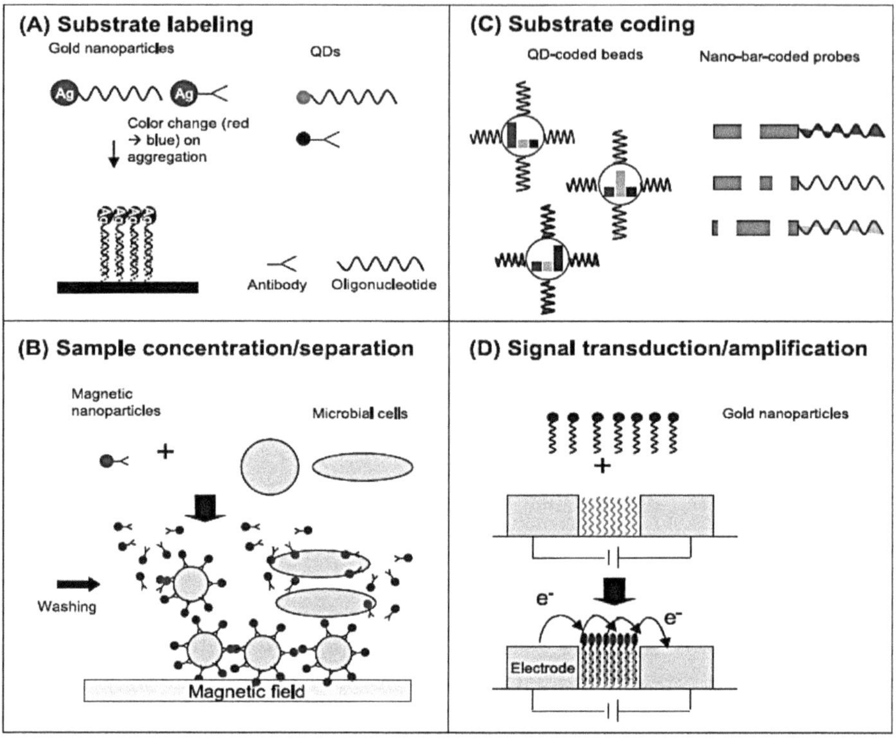

FIGURE 11.5 Applications of nanoparticles in different biological research studies including substrate labeling (A); sample concentration/separation (B); substrate coding (C); and signal transduction and amplification (D).

Source: Wen-Tso Liu et al. (2006); Reproduced with permission from: Copyright © 2006, Elsevier.

11.3.3.2 SAMPLE SEPARATION, PURIFICATION, AND CONCENTRATION

Magnetic nanoparticles (MNPs) are mainly used to assist the separation, purification, and concentration of different biomolecules (Figure 11.5(B)). To do so, different capturing molecules, such as antibodies and oligonucleotide probes can be immobilized on the surface of MNPs. Alternatively, MNPs can be embedded inside a microsphere, such as a polystyrene microsphere, which is further immobilized with capturing probes on the surface. Another approach is to immobilize MNPs and capturing probes together on the surface of the microspheres. After reacting these MNPs/microspheres with sample solutions, targeted molecules can be bound by or captured on the surface of these magnetic microspheres. By further applying a magnetic field, these particles can be attracted and concentrated at a specific location and separated from the bulk solution.

Many studies have demonstrated the use of this type of MNPs or microspheres in biological studies. These include:

- The recovery and tracking (*in vivo*) of progenitor cells;
- The concentration of antigens from bulk solution;
- The use of enzyme-entrapped nanoparticles as nanocatalysts;
- The separation/concentration of microbial cells from water samples;
- The distillation of ultrapure supercoiled plasmid DNA with high yields from bacterial cultures (Yang et al., 2004; Ho et al., 2004; Chiang et al., 2005).

11.3.3.3 SUBSTRATE CODING

Another unique feature of nanoparticles is the ability to code substrates with different colors during multiplexing assays (Figure 11.5(C)). The QbeadTM is a new technique used for high-throughput genotyping of SNPs (single nucleotide polymorphisms) (Xu et al., 2003). Fluorescent QdotTM semiconductor nanocrystals were incorporated in the QbeadTM system, which are used as multiplexed assays. Here, encoding microspheres conjugated to allele-speci®c oligonucleotides genotyping by Qbead system to SNP. Capability of high-throughput automation, the Qbead system is cost-effective and has the energy to be a robust and with various applications.

11.3.3.4 SIGNAL TRANSDUCTION AND AMPLIFICATION

NPs can play as signal amplifiers to reach a fluorescence label-free assay (Figure 11.5(D)). Park et al. (2002) studied about an array of DNA discovery method. Functionalized AuNPs and binding of oligonucleotides are used to leads to changes of conductivity associated with target-probe. An electrode gap binding is localized with AuNPs, silver deposition facilitated by these NPs bridges the gap and conductivity changes leads to readily measurable. Without thermal stringency wash selectivity was achieved by a remarkable salt concentration based on the NPs probes with hybridization behavior. We have identified particular DNA at low concentration (500 femtomolar) and significant point mutation selectivity factor by this method.

11.3.3.5 NANOPARTICLES FOR PATHOGEN DETECTION

Different types of NPs were used in the case of sensors for detection and separation of pathogens. Particularly, in pathogen detection, optical, and magnetic properties of NP platforms are used. Magnetic biosensors are one of the most common methods for the detection of bacteria which directly involve in immunological reactions against surface antigens with antibodies coated magnetic NPs (Varshney et al., 2007; Xia et al., 2006). A novel microfluidic device based on immune-magnetic to attract molecules bound with magnetic NPs from laminar flow path to a local magnetic field gradient. Escherichia coli labeled with biotinylated anti-Escherichia coli antibody using the above device contains red blood cells similar to blood. Antibody coated NPs in a micro-fluidic and core/shell NPs with Fe metal devices are used to detect bacteria which are reported by Weissleder group (Lee et al., 2009). A single-gene mutations detecting device is developed by the same group. The device is assay based on a magnetic barcoding strategy that did not need antibodies (Liong et al., 2013). Polymerase chain reaction (PCR)-amplified mycobacterial genes are modified with complementary DNA captured on polymeric beads which is labeled by SPIONs and identified by NMR (nuclear magnetic resonance) and find mycobacterium tuberculosis (TB), identify drug resistance strains from sputum samples (2.5 h). The researchers also extended rRNA system that uses as a target marker for NP labeling (Chung et al., 2013), which is used a universal and specific nucleic acid probes and that detect 16S rRNA. That is plentiful and regular to many bacterial species. The sensitive device was enough to detect Escherichia coli bacteria in 10 ml of blood, accurately estimate bacterial load.

11.4 NANOMATERIALS (NMS) IN MEDICINAL FIELD

In recent years, NMs have emerged as important role in modern medicine, with applications. NPs have a number of properties, but they are differentiated by their size and other properties like biological mobility, chemical reactivity and energy absorption. In modern medicine, NPs give many benefits. Definitely few cases NPs where make possible analyzes and therapies. However, NPs also bring with them exclusive environmental and social challenges mainly in the view of toxicity. This chapter aims to emphasize the main contributions of NPs to modern medicine and other uses in the environment.

Lots of reports are available to demonstrate the noble metals like silver, gold, and platinum, etc., and their analogs for medicinal uses. Nanotechnology possesses the countless potential as believed to be the future technology and innumerable applications (Hoshika et al., 2010). Metal nanoparticles have been broadly studied due to their large surface to volume ratios which are useful to make them in various applications, in photography, non-linear optics, electronics, photonics, chemical, and cancer therapy, biological sensing, and clinical diagnosis (Gopal et al., 2013). Noble metal nanoparticles have demonstrated many biomedical applications (Yavuz et al., 2009). They have been the center of attraction due to their important role in disease diagnosis, drug delivery, imaging, and in medicine. Metal nanoparticles can be modified using various molecules of desired choice, so that the bio-conjugated metal nanoparticles can be used for specific functions (Figure 11.6).

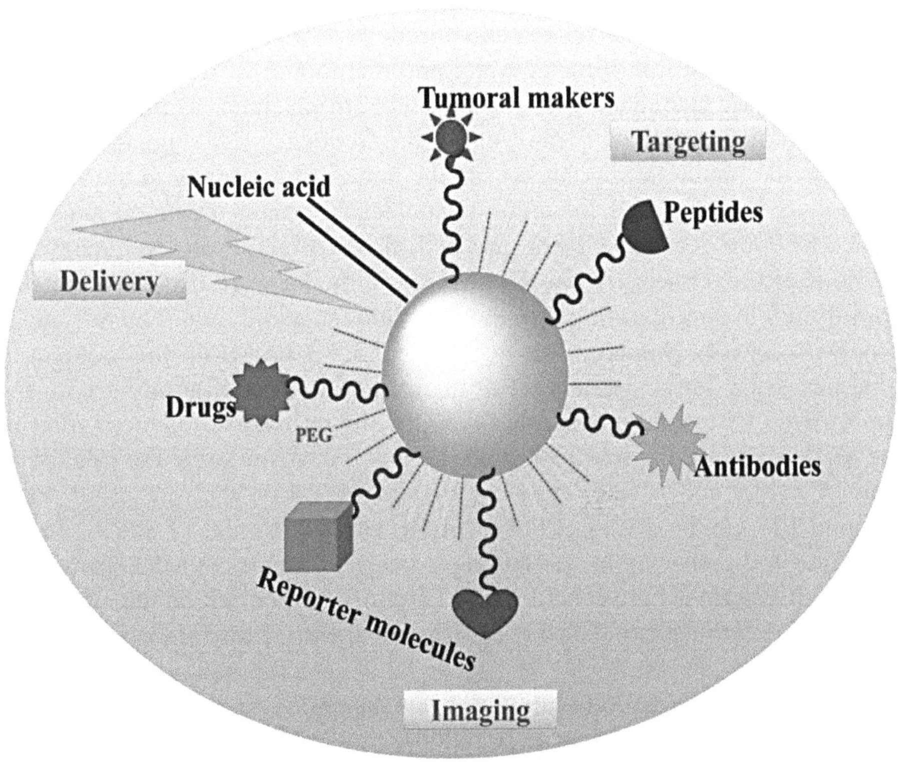

FIGURE 11.6 Bioconjugated metal nanoparticle with different molecules.

Metal nanoparticles, silver, gold, and platinum, are used in biomedical applications like diagnosis and treatment of diseases such as cancer, HIV, TB, and Parkinson's disease. Currently, AgNPs used as anticancerous and antimicrobial agents, AuNPs as an efficient drug carrier, uses of platinum nanoparticles (PtNPs) in bone allograft, dentistry, etc., has been critically reviewed. In addition, the cytotoxic effects of AgNPs, AuNPs, and PtNPs have been elaborately discussed in various studies.

11.4.1 NANOPARTICLES APPLICATIONS IN MEDICAL FIELD

11.4.1.1 MEDICAL IMAGING APPLICATIONS

In MRI technique, three-dimensional (3D) type non-invasive scans of the body which is extensively utilized in modern medicine. It is predominantly used in treating and identifying many infections or syndromes like the musculoskeletal system, spine, and brain. Hydrogen atoms are naturally present in tissue in the form of water and membrane which MRI examines under magnetic resonance spectroscopy. The strong static magnetic field is a sample holding place and a transverse radio frequency (RF) signal. It is used to excite the magnetic dipoles within the nuclei of hydrogen in the sample. Previous to the RF signal, the revolving nuclei are arranged with the magnetic field. The hydrogen nuclei were spun at different direction and frequency due to the additional energy from RF also, these nuclei were back to a state of equilibrium in configuration through the static magnetic field. The relaxation process is usually characterized by T1 and T2. T1 stands for the moment in time necessary for reinstatement of nuclear spins in configuration with the static field; T2 refers to the time taken for the transverse magnetization of the nuclei disappear. Based on both T1 and T2, distinguish the relaxation time of nuclei and various types of tissues. Based on the relaxation times (T1 and T2), can be distinguish the various types of tissue. T1 and T2-based weighing and gathering of spatial-based images includes in MRI scans.

The distinctions between T1 and T2 are tiny particularly in the medicinal field requiring the use of contrast or different agents. These are paramagnetic molecules and can modify T1 as well as T2 of selected regions. An example of contrast agents is gadolinium which has the ability to resolve problems in the brain and kidney (Mornet et al., 2004).

Currently, SPIO (Superparamagnetic iron oxide) NPs act as contrast agents instead using of gadolinium-based agents for T2-weighting. It is significantly employ imaging born marrow, nodes, liver, and lymph. Compared to gadolinium-based agents, superparamagnetic NPs have much higher relaxation times of like iron oxide.

11.4.1.2 THERAPEUTIC APPLICATIONS

11.4.1.2.1 Sustained Gene Delivery

An important application of nanoparticles is drug delivery. Recently, polymer and liposome-based drug delivery system are very successful in NPs used clinical applications. Based on polymer drug delivery systems, can be classified as polymer-protein conjugates, polymeric drugs, polymer drug conjugates and polymeric micelles (Duncan et al., 2003). NPs encapsulated polymers can also be used in drug delivery applications. Biopolymers or natural polymers are anticancer and antiviral properties in drug delivery application. Water soluble polyethylene glycol (PEG) is widely used in drug delivery because of its water soluble and biocompatibility properties.

Chemical composition differentiates the NPs employed in these two techniques. Merging the advantages of medical imaging modalities get through of molecules and molecular imaging, cell biology can imagine, exemplify, and measure biological methods at a molecular and cellular level in a non-invasive manner (Rudin et al., 2003), allowing timely, accurate diagnosis and significant treatment of devastating diseases (McDonald et al., 2003; Weissleder et al., 2008). MRI operation is based on the mechanism of the relaxation of hydrogen proton spins and NMR in an applied magnetic field. Longitudinal relaxation (T1-recovery) and transverse relaxation (T2-decay) can be acquired to generate an MRI. The differences in MRI, the signal variations between adjacent regions, arises from relaxation time among protons associated with the atmosphere in the tissue shorten the longitudinal and transverse relaxation time of their surrounding protons, therefore amplifying the signal variation in between lesions and healthy tissues.

Iron NPs used as contrast agents for MRI and cancer treatment which influence scrutinizes the *in vivo* applications for resolving current and future opportunities. In the physical viewpoint an especial emphasis is made on magnetic NPs. Enhancing the half-life period in the blood compartment and targeting tumor cells due to the bio-distribution and the solutions envisaged. Mornet et al. (2006) improved the fabrication methods on covalent platforms capable of being tailor-derivative by surface molecular chemistry.

11.4.1.3 THERAPEUTIC APPLICATIONS

11.4.1.3.1 Cancer

In the treatment of various types of cancer, NPs have a great impact, particularly in drug delivery applications. Duncan et al. (2003) reviewed and described different examples of various liposomes and polymer-based drugs as some therapeutic agents. An anticancer agent paclitaxel is employed to care for lung, skin, and ovarian cancers (Kikuchi et al., 2005; Koziara et al., 2006). For the most part of administration of this drug which is water insoluble as dissolution in ethanol (Taxol®). The administered blended with a solvent, polyoxyethylated castor oil (Cremophor® EL). A most important drawback of this process is the side effects associated with Cremophor®, addition of hypersensitivity reactions, necessitating the administration of steroids and antihistamines as pre-medications (Micha et al., 2006; Zhang et al., 2005). Before the year 2005, in the medical field a paclitaxel known as Abraxane® was accepted. This paclitaxel is blended with NPs of a natural polymer (albumin) with a high-pressure emulsification method. The NPs loaded paclitaxel has been reduce the side effect also provides some additional benefits coupled with Cremophor® (Ibrahim et al., 2002). Though, paclitaxel loaded NPs did not address multidrug resistance, an ordinary difficulty in tumor therapy that takes place when cancer cells get used to stimuli via efflux transporters or additional proteins on the surface.

Koziara & colleagues (2006) have examined to overcome the problem by loading paclitaxel onto emulsifying wax NPs. Commercially existing product wax (Tween 80®) is alternatively known as polyoxyethylene 20-sorbitan monooleate. The NPs were synthesized under thermal condition with a mixture of the drug, wax, and surfactant. Drug-loaded NPs was assessed in a murine xenograft model (HCT-15) in which tumor cells express p-glycoprotein, an efflux transporter. With the help of a control experiment using Taxol®, the resulting cessation of tumor growth was judged to be due to a combination of overcoming resistance (by nonspecific cytoskeletal disturbance) and the antiangiogenic effect of paclitaxel.

Ali et al. (2014) investigated the anticancer activity of rifampicin-conjugated AuNPs in multidrug-resistant (MDR) cancer cells. Rifampicin has improved the deposition of anticancer drugs in cancer cells. They reported that rifampicin-conjugated AuNPs significantly enhanced the rate and increased their accumulation inside the cancer cells. The cell viability analysis is noted a remarkable improvement in the photothermal therapeutic effect of AuNPs in the presence of rifampicin. Hence, it will help to reduce the unavailability on the overall amount of AuNPs needed for treating cancer and thus decreasing its toxicity. Therefore, through all such applications AuNPs have shown to be target specific and thereby reducing the exposure of potentially harmful drugs to healthy tissues.

11.4.1.3.2 HIV/AIDS

De Jaeghere and colleagues (De Jaeghere et al., 2000) examined the delivery of an HIV-1 protease inhibitor, CGP 70726, by pH-sensitive NPs. This is prepared by a copolymer of methacrylic acid and ethyl acrylate. This copolymer is commercially available under the name Eudragit® L100-55 is commercially available which has chosen because of its pH-dependent solubility. CGP 70726 and other similar antiviral agents are known to disturb the imitation series of HIV-1 (Robins et al., 1993). A main challenge in delivering agents such as CGP 70726 is poorly water soluble. De Jaeghere and colleagues fabricated NPs by emulsifying a mixture of copolymers such as CGP 70726 and benzyl alcohol. The NPs were administered orally to dogs and successful drug release was observed by analysis of blood samples.

Recently, the HIV-1 Tat protein has been promoted as an efficient therapeutic vaccine against HIV-1/AIDS (Caputo et al., 2004). Solid lipid nanoparticles (SLNs) were manufactured from cationic lipid with the addition of surfactant. DNA compacted tat protein consistS SLN based system (Rudolph et al., 2004) and also SLNs incorporated with DNA were shown to transfect mammalian cells *in vitro* (Tabatt et al., 2004). Tat peptide and DNA were adsorbed on the surface of NPs using electrostatic forces. When analyzed to the mice lungs by aerosol application. It confirms the successful transfection of the SLNs. However, several DNA degradations were noted, whereas there is moderately information of *in-vivo* NP-based drug delivery in HIV/AIDS areaS than cancer (Sawant et al., 2006).

11.5 CONCLUSIONS

Nanoparticles (NPs) are being used in several technological applications and have attracted huge interest from the scientific community. As mentioned above, NPs enable novel techniques in imaging, sensing, drug delivery, and understanding of basic biological processes. Since the interest in the biology application for NPs is very recent, because of the exciting and innovative applications of NPs in biology, there is also a lesser amount of technical barrier, and mature methods. In biological research, NP technology will require chemists and nanotechnologists to standardize and simplify the preparation techniques. For a healthy diet of human, nutraceuticals provide all the essential substances. Also, it is widely used in the food and pharmaceutical industries. In order to improve solubility, stability, and bioavailability of active ingredients, nutraceuticals play an important role. In clinical medicine, medical imaging, and drug delivery NPs are having a major role.

ACKNOWLEDGMENTS

The authors AR, EJSC, and AP would like to thank the authorities of Gandhigram Rural Institute – DU for their encouragement.

KEYWORDS

- **biological**
- **medicinal**
- **nanomaterials**
- **nanoparticles**
- **nutraceuticals**

REFERENCES

Acosta, E., (2009). Bioavailability of nanoparticles in nutrient and nutraceutical delivery. *Curr. Opin. Colloid & Interface Sci., 14,* 3–15.

Aggarwal, B. B., Kumar, A., & Bharti, A. C., (2003). Anticancer potential of curcumin: Preclinical and clinical studies. *Anticancer Res., 23,* 363–398.

Ahmad, Z., Shah, A., Siddiq, M., & Kraatz, H. B., (2014). Polymeric micelles as drug delivery vehicles. *RSC Adv., 4,* 17028–17038.

Ali, M. K. R., Panikkanvalappil, S. R., & El-Sayed, M. A., (2014). Enhancing the efficiency of gold nanoparticles treatment of cancer by increasing their rate of endocytosis and cell accumulation using rifampicin. *J. Am. Chem. Soc., 136,* 4464–4467.

Amalraj, A., Jacob, J., Varma, K., Kunnumakkara, A. B., Divya, C., & Gopi, S., (2019). Acujoint™, a highly efficient formulation with natural bioactive compounds, exerts potent anti-arthritis effects in human osteoarthritis – A pilot randomized double blind clinical study compared to combination of glucosamine and chondroitin, *J. Herb. Med., 17, 18*, 100276.

Amalraj, A., Sukumaran, N. P., Nair, A., & Gopi, S., (2021). Preparation of a unique bioavailable bacoside formulation (Cognique®) using polar-nonpolar-sandwich (PNS) technology and its characterization, *in vitro* release study, and proposed mechanism of action. *Regen. Eng. Transl. Med., 7,* 379–392.

Bagchi, D., (2006). Nutraceuticals and functional foods regulations in the United States and around the world. *Toxicology, 221,* 1–3.

Bala, I., Bhardwaj, V., Hariharan, S., Kharade, S. V., Roy, N., & Ravi, K. M. N., (2006). Sustained release nanoparticulate formulation containing antioxidant-ellagic acid as potential prophylaxis system for oral administration. *J. Drug Target, 14,* 27–34.

Baljit, S., (2007). Psyllium as therapeutic and drug delivery agent. *Int. J. Pharm., 334,* 1–14.

Bangham, A. D., (1993). Liposomes: The Babraham connection. *Chem. Phys. Lipids, 64,* 275–285.

Bell, S. J., & Good, R. G. K., (2002). A functional food product for the management of weight. *Crit. Rev. Food Sci. Nutr., 42,* 163–178.

Bharti, C., Nagaich, U., Pal, A. K., & Gulati, N., (2015). Mesoporous silica nanoparticles in target drug delivery system: A review. *Int. J. Pharm. Invest., 5,* 124–133.

Borel, T., & Sabliov, C. M., (2014). Nano delivery of bioactive components for food applications: Types of delivery systems, properties, and their effect on ADME profiles and toxicity of nanoparticles. *Annu. Rev. Food Sci. Technol., 5,* 197–213.

Bouwmeester, H., Dekkers, S., Noordam, M. Y., Hagens, W. I., Bulder, A. S., De Heer, C., Ten, V. S. E., et al., (2009). Review of health safety aspects of nanotechnologies in food production. *Regul. Toxicol. Pharmacol., 53,* 52–62.

Bulte, J. W. M., & Kraitchman, D. L., (2004). Iron oxide MR contrast agents for molecular and cellular imaging. *NMR in Biomedicine, 17,* 484–499.

Caputo, A., Gavioli, R., & Ensoli, B., (2004). Recent advances in the development of HIV-1 Tat-based vaccines. *Curr. HIV Res., 2,* 357–376.

Chan, J., Valencia, P., Zhang, L., Langer, R., Farokhzad, O., Grobmyer, S. R., & Moudgil, B. M., (2010). *Cancer Nanotechnology* (Vol. 624, pp. 163–175). Humana Press.

Chan, W. C. W., Maxwell, D. J., Gao, X. H., Bailey, R. E., Han, M. Y., & Nie, S. M., (2002). Luminescent quantum dots for multiplexed biological detection and imaging. *Curr. Opin. Biotechnol., 13,* 40–46.

Chang, Y. P., Pinaud, F., Antelman, J., & Weiss, S., (2008). Tracking biomolecules in live cells using quantum dots. *J. Biophotonics, 1,* 287–298.

Chaudhry, Q., Scotter, M., Blackburn, J., Ross, B., Boxall, A., Castle, L., Aitken, R., & Watkins, R., (2008). Applications and implications of nanotechnologies for the food sector. *Food Addit. Contam. B., 3,* 241–258.

Chen, H., Weiss, J., & Shahidi, F., (2006a). Nanotechnology in nutraceuticals and functional foods. *Food Tech., 603,* 30–36.

Chen, L., Remondetto, G. E., & Subirade, M., (2006b). Food protein-based materials as nutraceutical delivery systems. *Trends Food Sci. Technol., 17,* 272–283.

Chiang, C. L., Sung, C. S., Wu, T. F., Chen, C. Y., & Hsu, C. Y., (2005). Application of superparamagnetic nanoparticles in purification of plasmid DNA from bacterial cells. *J. Chromatogr. B, 822,* 54–60.

Chou, C. C., Hsiao, H. Y., Hong, Q. S., Chen, C. H., Peng, Y. W., Chen, H. W., & Yang, P. C., (2008). Single-walled carbon nanotubes can induce pulmonary injury in mouse model. *Nano Lett., 8,* 437–445.

Chung, H. J., Castro, C. M., Im, H., Lee, H., & Weissleder, R., (2013). A magneto-DNA nanoparticle system for rapid detection and phenotyping of bacteria. *Nat. Nano, 1,* 12–21.

Coleman, M. D., Eason, R. C., & Bailey, C. J., (2001). The therapeutic use of lipoic acid in diabetes: A current perspective. *Environ. Toxicol. Pharmacol., 10,* 167–172.

Collier, C. P., Vossmeyer, T., & Heath, J. R., (1998). Nanocrystal super lattices. *Annu. Rev. Phys. Chem., 49,* 371–404.

Daniel, M. C., & Astruc, D., (2003). Gold nanoparticles: Assembly, supramolecular chemistry, quantum size-related properties, and applications toward biology, catalysis, and nanotechnology. *Chem. Rev., 104,* 293–346.

De Jaeghere, F., Allémann, E., Kubel, F., Galli, B., Cozens, R., Doelker, E., & Gurny, R., (2000). Oral bioavailability of a poorly water soluble HIV-1 protease inhibitor incorporated into pH-sensitive particles: Effect of the particle size and nutritional state. *J. Control Release, 68,* 291–298.

Dubertret, B., Skourides, P., Norris, D. J., Noireaux, V., Brivanlou, A. H., & Libchaber, A., (2002). *In vivo* imaging of quantum dots encapsulated in phospholipid micelles. *Science, 298,* 1759–1762.

Duncan, R., (2003). The dawning era of polymer therapeutics. *Nat. Rev. Drug Discov., 2,* 347–360.

Felgner, P. L., & Ringold, G. M., (1989). Cationic liposome-mediated transfection. *Nature, 337,* 387–388.

Felgner, P. L., Gadek, T. R., Holm, M., Roman, R., Chan, H. W., Wenz, M., Northrop, J. P., et al., (1987). Lipofection: A highly efficient, lipid-mediated DNA-transfection procedure. *Proc. Natl. Acad. Sci., 84,* 7413–7417.

Fu, C., Liu, T., Li, L., Liu, H., Chen, D., & Tang, F., (2013). The absorption, distribution, excretion and toxicity of mesoporous silica nanoparticles in mice following different exposure routes. *Biomaterials, 34,* 2565–2575.

Gao, X., Cui, Y., Levenson, R. M., Chung, L. W. K., & Nie, S., (2004). *In vivo* cancer targeting and imaging with semiconductor quantum dots. *Nat. Biotechnol., 22,* 969–976.

German, J. B., & Walzem, R. L., (2000). The health benefits of wine. *Annu. Rev. Nutr., 20,* 561–593.

Glenville, M., (2006). Nutritional supplements in pregnancy: Commercial push or evidence based. *Curr. Opin. Obstet. Gynecol., 18,* 642–647.

Gopal, J., Hasan, N., Manikandan, M., & Wu, H., (2013). Bacterial toxicity/compatibility of platinum nanospheres, nanocuboids and nanoflowers. *Sci Rep., 3,* 1–7.

Gref, R., Minamitake, Y., Peracchia, M., Trubetskoy, V., Torchilin, V., & Langer, R., (1994). Biodegradable long-circulating polymeric nanospheres. *Science, 263,* 1600–1603.

Gulcin, M. V., Gepdiremen, A., & Elias, R., (2006). The antioxidant activity of a triterpenoid glycoside isolated from the berries of *Hedera colchica*: 3-O-(bdglucopyranosyl) – hederagenin. *Phytother Res., 20,* 130–134.

Gupta, S. K., Yadav, S. K., & Patil, S. M. M., (2013). Nutraceutical – a bright scope and opportunity of Indian healthcare market. *Inter. J. Res. Develop. Pharm. Life Sci., 2,* 478–481.

He, Q., Zhang, Z., Gao, F., Li, Y., & Shi, J., (2011). *In vivo* biodistribution and urinary excretion of mesoporous silica nanoparticles: Effects of particle size and PEGylation. *Small, 7,* 271–280.

Ho, K. C., Tsai, P. J., Lin, Y. S., & Chen, Y. C., (2004). Using biofunctionalized nanoparticles to probe pathogenic bacteria. *Anal. Chem., 76,* 7162–7168.

Hoshika, S., Nagano, F., Tanaka, T., Ikeda, T., Wada, T., Asakura, K., Koshiro, K., et al., (2010). Effect of application time of colloidal PtNPs on the microtensile bond strength to dentin. *Dental Mater. J., 29,* 682–689.

Huang, Q., (2012). *Nanotechnology in the Food, Beverage and Nutraceutical Industries.* Woodhead Publishing, Oxford.

Huang, X., El Sayed, I. H., Qian, W., & El-Sayed, M. A., (2006). Cancer cell imaging and photothermal therapy in the near-infrared region by using gold nanorods. *J. Am. Chem. Soc., 128,* 2115–2120.

Huang, X., Jain, P. K., El-Sayed, I. H., & El-Sayed, M. A., (2007). Gold nanoparticles: Interesting optical properties and recent applications in cancer diagnostics and therapy. *Nanomed., 2,* 681–693.

Ibrahim, N. K., Desai, N., et al., (2002). Phase I and pharmacokinetic study of ABI-007, a cremophor-free, protein-stabilized, nanoparticle formulation of paclitaxel. *Clin. Cancer Res., 8,* 1038–1044.

Igor, I. S., Vivero-Escoto, J. L., Chia-Wen, W., & Victor, S. Y. L., (2008). Mesoporous silica nanoparticles as controlled release drug delivery and gene transfection carriers. *Adv. Drug Delivery Rev., 60,* 1278–1288.

Jovin, T. M., (2003). Quantum dots finally come of age. *Nat. Biotech., 21,* 32, 33.

Kalia, A. N., (2005). *Textbook of Industrial Pharmacognosy.* CBS Publisher and Distributor, New Delhi.

Kalioraa, A. C., Dedoussisa, G. V. Z., & Schmidtb, H., (2006). Dietary antioxidants in preventing atherogenesis. *Atherosclerosis, 187,* 1–17.

Kalra, E. K., (2003). Nutraceutical – definition and introduction. *AAPS Pharmsci., 3,* 25–27.

Kawano, K., Watanabe, M., Yamamoto, T., Yokoyama, M., Opanasopit, P., Okano, T., & Maitani, Y., (2006). Enhanced antitumor effect of camptothecin loaded in long-circulating polymeric micelles. *J. Controlled Release, 112,* 329–332.

Khajeh, M., Laurent, S., & Dastafkan, K., (2013). Nanoadsorbents: Classification, preparation, and applications (with emphasis on aqueous media). *Chem. Rev., 113,* 7728–7768.

Kikuchi, A., Sakamoto, H., & Yamamoto, T., (2005). Weekly carboplatin and paclitaxel are safe, active, and well tolerated in recurrent ovarian cancer cases of Japanese women previously treated with cisplatin containing multidrug chemotherapy. *Int. J. Gynecol. Cancer, 15,* 45–49.

Koziara, J. M., Whisman, T. R., Tseng, M. T., et al., (2006). *In-vivo* efficacy of novel paclitaxel nanoparticles in paclitaxel-resistant human colorectal tumors. *J. Control Release, 112,* 312–319.

Kucuk, O., Sarkar, F. H., Sakr, W., Khachik, F., Djuric, Z., Banerjee, M., Michael, N. P., et al., (2002). Lycopene in the treatment of prostate cancer. *Pur. Appl. Chem., 74,* 1443–1450.

Kumar, P., Kumar, N., & Omer, T., (2016). A review on nutraceuticals critical supplement for building a healthy world. *World J. Pharm. Pharm. Sci., 3,* 579–594.

Lavasanifar, A., Samuel, J., & Kwon, G. S., (2002). Poly (ethylene oxide)-block-poly (L-amino acid) micelles for drug delivery. *Adv. Drug Delivery Rev., 54,* 169–190.

Lee, H., Yoon, T. J., & Weissleder, R., (2009). Ultrasensitive detection of bacteria using core-shell nanoparticles and an NMR-filter system. *Angew. Chem. Int., 48,* 5657–5660.

Li, R., Wu, R., Zhao, L., Qin, H., Wu, J., Zhang, J., Bao, R., & Zou, H., (2014). *In vivo* detection of magnetic labeled oxidized multi-walled carbon nanotubes by magnetic resonance imaging. *Nanotechnology, 25,* 1–8.

Limer, J. L., & Speirs, V., (2004). Phyto-oestrogens and breast cancer chemoprevention. *Breast Cancer Res., 6,* 119–127.

Lin, N., Huang, J., & Dufresne, A., (2012). Preparation, properties and applications of polysaccharide nanocrystals in advanced functional nanomaterials: A review. *Nanoscale, 4*, 3274–3294.

Lin, Y. C., Kuo, C. Y., Hsu, C. C., Tsai, W. C., Li, W. C., Yu, M. C., & Wen, H. W., (2012). Optimizing manufacture of liposomal berberine with evaluation of its antihepatoma effects in a murine xenograft model. *Int. J. Pharm., 441,* 381–388.

Liong, M., Hoang, A. N., Chung, J., Gural, N., Ford, C. B., Min, C., Shah, R. R., et al., (2013). Magnetic barcode assay for genetic detection of pathogens. *Nat. Commun., 4,* 1752.

Liu, J. H., Yang, S. T., Wang, X., Wang, H., Liu, Y., Luo, P. G., Liu, Y., & Sun, Y. P., (2014). Carbon nanoparticles trapped *in vivo* similar to carbon nanotubes in time-dependent biodistribution. *ACS Appl. Mater. Interfaces, 6,* 14672–14678.

Liu, X., Dai, Q., Austin, L., Coutts, J., Knowles, G., Zou, J., Chen, H., & Huo, Q., (2008). A one-step homogeneous immunoassay for cancer biomarker detection using gold nanoparticle probes coupled with dynamic light scattering. *J. Am. Chem. Soc., 130,* 2780–2782.

Lugli, P., Locci, S., Erlen, C., & Csaba, G. (2010). Molecular Electronics: Challenges and Perspectives. In: Korkin, A., Krstić, P., Wells, J. (eds). *Nanotechnology for Electronics, Photonics, and Renewable Energy.* Nanostructure Science and Technology. Springer, New York, NY. pp. 1–40. https://doi.org/10.1007/978-1-4419-7454-9_1.

Mahmoudi, M., Sant, S., Wang, B., & Laurent, S. T., (2011). Superparamagnetic iron oxide nanoparticles (SPIONs): Development, surface modification and applications in chemotherapy. *Drug Deliv. Rev., 63*, 24–46.

Mandel, S., Packer, L., Youdim, M. B. H., & Weinreb, O., (2005). Proceedings from the third int. conf. mechanism of action of nutraceuticals. *J. Nutritional Biochem., 16,* 513–520.

Mc Clements, D. J., (2013). *Nanoparticle and Microparticle-based Delivery Systems: Encapsulation, Protection and Release of Active Compounds.* CRC Press, Boca Raton, FL.

Mc Donald, D. M., & Choyke, P. L., (2003). Imaging of angiogenesis: From microscope to clinic. *Nat. Med., 9,* 713–725.

Medina, S. H., & El-Sayed, M. E. H., (2009). Dendrimers as carriers for delivery of chemotherapeutic agents. *Chem. Rev., 109,* 3141–3157.

Merlin, J. P. J., Prasad, N. R., Shibli, S. M. A., & Sebeela, M., (2012). Ferulic acid loaded Poly-$_{d,l-}$lactide-co-glycolide nanoparticles: Systematic study of particle size, drug encapsulation efficiency and anticancer effect in non-small cell lung carcinoma cell line *in vitro. Biomedicine & Preventive Nutrition, 2,* 69–76.

Micha, J. P., Goldstein, B. H., Birk, C. L., Rettenmaier, M. A., & Brown, III. J. V., (2006). Abraxane in the treatment of ovarian cancer: The absence of hypersensitivity reactions. *Gynecol. Oncol., 100,* 437, 438.

Michalet, X., Pinaud, F. F., Bentolila, L. A., Tsay, J. M., Doose, S., Li, J. J., Sundaresan, G., et al., (2005). Quantum dots for live cells, *in vivo* imaging, and diagnostics. *Science, 307,* 538–544.

Mornet, S., Vasseur, & Grasset, F., (2006). Magnetic nanoparticle design for medical applications. *Prog. Solid. State Ch., 34,* 237–247.

Mornet, S., Vasseur, S., Grasset, F., & Duguet, E., (2004). Magnetic nanoparticle design for medical diagnosis and therapy. *J. Mater Chem., 14,* 2161–2175.

Müller, R. H., Mader, K., & Gohla, S., (2000). Solid lipid nanoparticles (SLN) for controlled drug delivery e a review of the state of the art. *Eur. J. Pharm. Biopharm., 1*, 161–177.

Neethirajan, S., & Jayas, D. S., (2011). Nanotechnology for the food and bioprocessing industries. *Food Bioproc. Tech., 4*, 39–47.
Nie, S., & Emory, S. R., (1997). Probing single molecules and single nanoparticles by surface-enhanced Raman scattering. *Science, 275,* 1102–1106.
Park, S. J., Taton, T. A., & Mirkin, C. A., (2002). Array-based electrical detection of DNA with nanoparticle probes. *Science, 295,* 1503–1506.
Pathak, S., Choi, S. K., Arnheim, N., & Thompson, M. E., (2001). Hydroxylated quantum dots as luminescent probes for in situ hybridization. *J. Am. Chem. Soc., 123,* 4103–4104.
Peng, G., Tisch, U., Adams, O., Hakim, M., Shehada, N., Broza, Y. Y., Billan, S., et al., (2009). Diagnosing lung cancer in exhaled breath using gold nanoparticles. *Nat. Nano, 4,* 669–673.
Peters, R. J. B., Bouwmeester, H., Gottardo, S., Amenta, V., Arena, M., Brandhoff, P., Marvin, H. J. P., et al., (2016). Nanomaterials for products and application in agriculture, feed and food. *Trends Food Sci. Tech., 54,* 155–164.
Peters, R., Brandhoff, P., Weigel, S., Marvin, H., Bouwmeester, H., & Aschberger, K., (2014). Inventory of nanotechnology applications in the agricultural, feed and food sector. *Sci. Rep., 621,* 1–125.
Quintanilla-Carvajal, M. X., Camacho-Díaz, B. H., Meraz-Torres, L. S., Chanona-Pérez, J. J., Alamilla-Beltrán, L., Jimenéz-Aparicio, A., & Gutiérrez-López, G. F., (2010). Nanoencapsulation: A new trend in food engineering processing. *Food Eng. Rev., 1,* 39–50.
Rajasekaran, A., Sivagnanam, G., & Xavier, R., (2008). Nutraceuticals as therapeutic agents: A review. *Res. J. Pharm. Sci. Technol., 4,* 328–340.
Rama, C. S., Shirode, A. R., Mundala, A. S., & Kadam, V. J., (2006). Nutraceuticals an emerging era in the treatment and prevention of cardiovascular diseases. *Curr. Pharm. Biotechnol., 1,* 15–23.
Robins, T., & Plattner, J., (1993). HIV protease inhibitors – their anti-HIV activity and potential role in treatment. *J. Acquir. Immune Defic. Syndr. Hum. Retrovirol., 6,* 162–170.
Roco, M. C., & Bainbridge, W. S., (2001). *Societal Implications of Nanoscience Nanotechnology*. Kluwer Academic Publishers: Boston.
Rudin, M., & Weissleder, R., (2003). Mesoporous silica nanoparticle nanocarriers: Biofunctionality and biocompatibility. *Nat. Rev. Drug. Discov., 2,* 123–131.
Rudolph, C., Schillinger, U., Ortiz, A., Tabatt, K., Plank, C., Müller, R. H., & Rosenecker, J., (2004). Application of novel solid lipid nanoparticle (SLN)-gene vector formulations based on a dimeric HIV-1 TAT-peptide *in vitro* and *in vivo*. *Pharm. Res., 21,* 1662–1669.
Sawant, R. M., Hurley, J. P., Salmaso, S., Kale, A., Tolcheva, E., Levchenko, T. S., & Torchilin, V. P., (2006). "SMART" drug delivery systems: Double-targeted pH-responsive pharmaceutical nano carriers. *Bioconjugate Chem., 17,* 943–949.
Seabra, A. B., Paula, A. J., De Lima, R., Alves, O. L., & Durán, N., (2014). Nanotoxicity of graphene and graphene oxide. *Chem. Res. Toxicol., 27,* 159–168.
Serpell, C. J., Kostarelos, K., & Davis, B. G., (2016). Can carbon nanotubes deliver on their promise in biology? Harnessing unique properties for unparalleled applications. *ACS Cent. Sci., 2*, 190–200.
Shubhra, S., Sharma, P. K., & Guru, K., (2015). Nutraceuticals: A review. *J. Chronotherapy and Drug Deliv., 6,* 1–10.
Shukla, R., Thomas, T. P., Peters, J. L., Desai, A. M., Kukowska-Latallo, J., Patri, A. K., Kotlyar, A., & Baker, J. R., (2006). HER2 specific tumor targeting with dendrimer conjugated anti-HER2 mAb. *Bioconjugate Chem., 17*, 1109–1115.

Silvestre, C., Duraccio, D., & Cimmino, S., (2011). Food packaging based on polymer nanomaterials. *Prog. Polym. Sci., 36,* 1766–1782.
Sirtori, C. R., & Galli, C., (2002). Fatty acids and the omega 3. *Biomed. Pharmacother., 56,* 397–406.
Sk, M. P., Jaiswal, A., Paul, A., Ghosh, S. S., & Chattopadhyay, A., (2012). Presence of amorphous carbon nanoparticles in food caramels. *Sci. Rep., 2,* 383–387.
Smolkova, B., El Yamani, N., Collins, A. R., Gutleb, A. C., & Dusinska, M., (2015). Nanoparticles in food. Epigenetic changes induced by nanomaterials and possible impact on health. *Food Chem. Toxicol., 77,* 64–73.
Subramanian, K. S., (2015). *Nanotechnology in Agriculture*. New India Publishing Agency. New Delhi, India.
Tabatt, K., Sameti, M., Olbrich, C., Müller, R. H., & Lehr, C. M., (2004). Effect of cationic lipid and matrix lipid composition on solid lipid nanoparticle-mediated gene transfer. *Eur. J. Pharm. Biopharm., 57,* 155–162.
Tarn, D., Ashley, C. E., Xue, M. I. N., Carnes, E. C., Zink, J. I., & Brinker, C. J., (2013). Mesoporous silica nanoparticle nanocarriers: Biofunctionality and biocompatibility. *Acc. Chem. Res., 46,* 792–801.
Taton, T. A., (2002). Nanostructures as tailored biological probes. *Trends Biotechnol., 20,* 277–279.
Thomas, B., Ghebremeskel, K., Lowy, C., Crawford, M., & Bridget Affley-Shore, R. N., (2006). Nutrient intake of women with and without gestational diabetes with a Specific focus on fatty acids. *Nutrition, 22,* 230–236.
Tibbals, H. F., (2010). *Medical Nanotechnology and Nanomedicine*. CRC Press.
Tomalia, D. A., Reyna, L. A., & Svenson, S., (2007). Dendrimers as multi-purpose nanodevices for oncology drug delivery and diagnostic imaging. *Biochem. Soc. Trans., 35,* 61–67.
Torchilin, V. P., (2005). Recent advances with liposomes as pharmaceutical carriers. *Nat. Rev. Drug Discov., 4,* 145–160.
Torchilin, V. P., (2007). Micellar nanocarriers: Pharmaceutical perspectives. *Pharm. Res., 24,* 1–16.
Varma, K., Jude, S., Nair, R. V. R., Varghese, B. A., Jacob, J., Amalraj, A., & Kuttappan, K., (2021). Novel formulation of liposomal lutein using nanofiber weaving (NFW) technology: Antioxidant property and *in vitro* release studies. *Food Hydrocolloids Health, 1,* 100025.
Varshney, M., & Li, Y., (2007). Interdigitated array microelectrode-based impedance biosensor coupled with magnetic nanoparticle–antibody conjugates for detection of *Escherichia coli* O157:H7 in food samples. *Biosens. Bioelectron., 22,* 2408–2414.
Wang, A. Z., Gu, F., Zhang, L., Chan, J. M., Radovic-Moreno, A., Shaikh, M. R., & Farokhzad, O. C., (2008). Biofunctionalized targeted nanoparticles for therapeutic applications. *Expert. Opin. Biol. Ther., 8,* 1063–1070.
Wang, Y. X., Hussain, S., & Krestin, G., (2001). Superparamagnetic iron oxide contrast agents: Physicochemical characteristics and applications in MR imaging. *Eur. Radiol., 11,* 2319–2331.
Weissleder, R., & Pittet, M. J., (2008). Imaging in the era of molecular oncology. *Nature, 452,* 580–589.
Weissleder, R., (2006). Molecular imaging in cancer. *Science, 312,* 1168–1171.
Weissleder, R., Moore, A., Mahmood, U., Bhorade, R., Benveniste, H., Chiocca, E. A., & Basilion, J. P., (2000). *In vivo* magnetic resonance imaging of transgene expression. *Nat. Med., 6,* 351–354.

Wen-Tso, L., (2006). Nanoparticles and their biological and environmental applications. *J. Biosci. Bioengin., 1*, 1–7.

Whitesides, G. M., (2003). The 'right' size in nanobiotechnology. *Nature Biotechnol., 21*, 161–1165.

Woodgate, D. E., & Conquer, J. A., (2003). Prevalence of self-treatment with complementary products and therapies for weight loss: A randomized, cross-sectional study in overweight obese patients in Colombia. *Curr. Ther. Res. Clin. E., 64,* 248–262.

Wu, X., Liu, H., Liu, J., Haley, K. N., Treadway, J. A., Larson, J. P., Ge, N., et al., (2002). Immunofluorescent labeling of cancer marker Her2 and other cellular targets with semiconductor quantum dots. *Nat. Biotechnol., 21,* 41–46.

Xia, N., Hunt, T., Mayers, B., Alsberg, E., Whitesides, G., Westervelt, R., & Ingber, D., (2006). Combined microfluidic-micromagnetic separation of living cells in continuous flow. *Biomed. Microdevices, 8,* 299–308.

Xie, Y., (2012). The *Nanobiotechnology Handbook.* Handbook Boca Raton, Fla. CRC.

Xu, H., Sha, M. Y., Wong, et al., (2003). Multiplexed SNP genotyping using the Qbead™ system: A quantum dot-encoded microsphere-based assay. *Nucleic Acids Res., 31,* 40–43.

Yang, S. Q., Zhang, X. L., Chen, Z. X., Zhuang, Xu, J. G., & Wang, X. R., (2004). Magnetite-containing spherical silica nanoparticles for biocatalysis and bioseparations. *Anal. Chem., 76,* 1316–1321.

Yao, V. J., D'Angelo, S., Butler, K. S., Theron, C., Smith, T. L., Marchiò, S., Gelovani, J. G., et al., (2016). Ligand-targeted theranostic nanomedicines against cancer. *J. Controlled Release, 240,* 267–286.

Yavuz, M. S., Cheng, Y., Chen, J., et al., (2009). Gold nanocages covered by smart polymers for controlled release with near-infrared light. *Nat. Mater., 8,* 935–939.

Zhang, J. A., Anyarambhatla, G., Ma, L., Ugwu, S., Xuan, T., Sardone, T., & Ahmad, I., (2005). Development and characterization of a novel cremophor (R) EL free liposome-based paclitaxel (LEP-ETU) formulation. *Eur. J. Pharm. Biopharm., 59,* 177–187.

Zhao, M., Beauregard, D. A., Loizou, L., Davletov, B., & Brindle, K. M., (2001). Non-invasive detection of apoptosis using magnetic resonance imaging and a targeted contrast agent. *Nat. Med., 7*, 1241–1244.

Index

C

D

E

N

Q

R

T

U

V

W

For Product Safety Concerns and Information please contact our EU
representative GPSR@taylorandfrancis.com
Taylor & Francis Verlag GmbH, Kaufingerstraße 24, 80331 München, Germany

www.ingramcontent.com/pod-product-compliance
Ingram Content Group UK Ltd.
Pitfield, Milton Keynes, MK11 3LW, UK
UKHW022209190926
475702UK00003B/278

* 9 7 8 1 7 7 4 9 1 3 4 6 8 *